A SHEARWATER BOOK

A PLAGUE OF RATS AND RUBBERVINES

A PLAGUE OF RATS AND RUBBERVINES

The Growing Threat of Species Invasions

YVONNE BASKIN

A SCOPE–GISP Project

Island Press / SHEARWATER BOOKS
Washington • Covelo • London

A Shearwater Book
Published by Island Press

Library of Congress
Cataloging-in-Publication Data
Baskin, Yvonne.
A plague of rats and rubbervines : the growing threat
of species invasions / Yvonne Baskin.
p. cm.
"A project of SCOPE, the Scientific Committee on Problems of the Environment, in
collaboration with the World Conservation Union (IUCN) and CAB International, on
behalf of the Global Invasive Species Programme (GISP).
"Shearwater books."
Includes bibliographical references.
ISBN 1-55963-051-5 (paper : alk. paper)
1. Biological invasions. 2. Nature conservation. I. International Council of Scientific
Unions. Scientific Committee on Problems of the Environment. II. International
Union for Conservation of Nature and Natural Resources. III. C.A.B. International.
IV. Title.
QH353 .B28 2002
577'.18—dc21 2002004029

British Cataloguing-in-Publication Data available.

Printed on recycled, acid-free paper

Manufactured in the United States of America
09 08 07 06 05 04 03 8 7 6 5 4 3 2

CONTENTS

A PLAGUE OF RATS AND RUBBERVINES

ONE | Introduction: Confronting a Shrinking World

"We must make no mistake: we are seeing one of the great historical convulsions in the world's fauna and flora. We might say, with Professor Challenger, standing on Conan Doyle's 'Lost World,' with his black beard jutting out: 'We have been privileged to be present at one of the typical decisive battles of history—the battles which have determined the fate of the world.' But how will it be decisive? Will it be a Lost World?"

—Charles Elton, *The Ecology of Invasions by Animals and Plants,* 1958

"In contrast with the aftermath of prehistoric mass extinctions, human-dominated landscapes will encourage the generalist species to proliferate—all the more so as natural controls (predators, parasites) are preferentially eliminated. The upshot could well be a 'pest and weed' ecology, with all that implies for evolutionary history."

—Norman Myers, in the journal *Science,* 1997

Just a twenty-minute drive from downtown Auckland, on a steep slope behind Mick Clout's home, a lush remnant of primeval New Zealand forest remains. Spared from fire, ax, and plow because of its rugged aspect, the site still shelters two hectares of towering tree ferns, nikau palms, and the native evergreen trees the Maori, New Zealand's Polynesian settlers, call kahikatea. One fall day, as a light rain dripped soundlessly through the dense canopy and onto the rust-colored duff of the forest floor, several of us ventured into those woods hoping to see a pair of New

Zealand pigeons that had taken up residence. The legendary chorus of native birdsong that greeted the first European colonists has all but vanished, and what's left of New Zealand's forests are now disquietingly silent. As we listened for the cooing of pigeons, we suddenly heard instead a high-pitched call, eerily familiar to me yet startlingly out of place on this Southern Hemisphere island: the bugling of a North American elk.

Fortunately, the elk stag was confined on a neighboring game farm, along with the European red deer hinds imported to breed with him. Only a few times each year does a hind jump the fence and invade the forest to strip the bark from Clout's palms or damage the understory. Elsewhere in New Zealand, however, invading alien deer and elk cause severe damage to forests, rivaling the destructiveness of the invasive brushtail possums brought here long ago from Australia, which strip some 18,000 metric tons of leaves each night from forests like these. Clout, a professor of ecology at the University of Auckland and chair of the Invasive Species Specialist Group of the World Conservation Union (IUCN), maintains poison bait stations throughout this forest patch to kill the possums, which threaten not only the trees but also nesting pigeons and other native birds. It is possums, European ferrets, rats, and other furry alien invaders that have helped to silence the birdsong in New Zealand's forests. Along a creek at the foot of the slope, Clout pointed out other, more benign-looking invaders that nevertheless menace his remnant forest: recent garden fugitives such as wandering jew, willow, pampas grass, and privet now advancing along the streambed or upslope into the forest, threatening to choke out the native plant life that provides shelter and sustenance for the pigeons and other surviving birds.

As we walked back out of the woods and toward the house, we could see North American mallards dabbling about in rain puddles on the road below. The lush hills beyond were forested with California Monterey pines and Australian eucalyptus. It could have been a scene in San Diego but for the elk. An American or European visitor can easily feel at home amid the biota of Auckland and, indeed, much of the rest of New Zealand. That's because half the plant species and all the mammals (except for two native bat species) came from somewhere else. And New Zealand is not the only place to which many of these same plants and animals have been moved.

You will see many of the same beasts and much of the greenery in Cape Town or Sydney, Kuala Lumpur or Paris, San Francisco or Santiago.

The biological déjá vu of travelers today is the result of a massive game of musical chairs we have played with life on the earth, especially during the past 500 years. The extent and thoroughness of this rearrangement of plants, animals, and microbes is stunning, yet far from finished. We can find American beavers in Tierra del Fuego, African antelopes in New Mexico, Madagascar rubbervines in Queensland, and European pines in South Africa. On our increasingly connected planet, global trade and travel are accelerating the movement of organisms to places they could not have reached without our help. Their arrival is not always a cause for lament. We have transformed the living world in many ways that greatly enrich and sustain us, filling fields the world over with apples and wheat and gardens with geraniums and roses. But much of the transformation has been clumsy and careless at best, and we have created a growing litany of self-inflicted wounds. Among the freshest are the intercontinental movements of tree-killing Asian long-horned beetles, crop-devastating citrus canker, unstoppable zebra mussels, and deadly West Nile encephalitis and foot-and-mouth pathogens. These high-impact newcomers are called invasive alien species. It is the urgent need to reduce the ecological and economic fallout from the ongoing tide of invaders that is the subject of this book.

On the ecological side, the unique natural heritage that each region enjoys is increasingly besieged, not only by direct human activities but also by the overwhelming tide of new life we are introducing, deliberately or accidentally. Most alien creatures that escape or are loosed into the wild either perish or settle into new communities with little disruption. But a significant number—including the possums and privet shrubs, deer and willows, and myriad other species introduced into New Zealand—spread aggressively and invade in their new environments. These invaders dominate, disrupt, outcompete, prey on, hybridize with, or spread disease among native species or alter the terms of life in the community by changing the soil, the available light or water, the frequency of fire, or even the structure of the landscape.

Ecologists now rank biological invasions second only to habitat loss as a threat to native biodiversity in much of the world. (*Biodiversity* is a shorthand term ecologists use for biological diversity, the rich web of life in a community or region.) The threats come from an unlikely array of misplaced creatures, from rust fungus and avian malaria parasites to rubbervines, melaleuca trees, blackberry bushes, goats, snails, and tiny scale insects that can suck the life from trees and shrubs. Few places on the earth remain untouched by such invaders. Even in the Antarctic, seals have been exposed to cattle diseases and penguins to poultry virus. On a tiny scale, the crowd of strangers that threatens to overrun Clout's bit of forest exemplifies the beleaguered status of natural areas worldwide, from Yellowstone National Park to the Everglades, from Hawaii to the Galápagos Islands, from the mountains of the South African Cape provinces to the Italian Apennines.

On the economic and social side, the organisms ecologists call invaders are called weeds, pests, or emerging diseases when they threaten human enterprise and well-being. Invasive alien species create hardships across a spectrum of human activities, altering the character and economic potential of our lands and waters; threatening our health and that of our crops, forests, and livestock; diminishing recreational values and even our sense of place. In the United States alone, ecologist David Pimentel estimates, invasive species cause $137 billion per year in losses, damage, and control expenses.

Most of us have heard *something* about biological invasions. The topic is hitting the headlines and television news reports with increasing frequency. News, by nature, focuses on the striking, the singular, and the menacing new arrivals: West Nile encephalitis virus striking down people and birds in the Northeast, Asian long-horned beetles denuding parks and boulevards of beloved old shade trees, zebra mussels choking off water pipes along the Great Lakes, Formosan termites attacking the historic French Quarter of New Orleans, Africanized bees advancing across the Southwest, and Asian gypsy moths and Mediterranean fruit flies (Medflies) breaching the border.

Even as I listened to the elk bugling through the tree ferns near Auckland, the *New Zealand Herald* was trumpeting an alarm about the third snake in a month to have wriggled out of a shipping container in that snake-free land. Behind the scenes, quarantine inspectors in New Zealand were concerned about an outbreak of foot-and-mouth disease in Japan, the same strain that would strike South Africa six months later and finally burst into world headlines when it hit Great Britain and parts of Europe.

What you will seldom learn from sporadic news accounts, however, is that foot-and-mouth disease and Asian gypsy moths, as well as a host of weeds and pests that will never make headlines, are all manifestations of the same growing problem—the uncontrolled movement of species worldwide, driven by the increase in global trade and travel. Only a small fraction of species that invade new regions spur rapid and dramatic transformations of landscapes, devastating disease outbreaks, crop failures, or other misery. Instead, most invaders manifest themselves in slower, more subtle ways, such as chronic degradation of habitats and landscapes, attrition of native plants and animals, or deterioration of the ecological life-support services that regulate soil fertility, plant growth, and water quality and flow.

Unfortunately, it is difficult to spark a sense of urgency or lasting commitment to action in the face of chronic problems, although they can sometimes be more devastating in the long run than the headline grabbers. Even more troubling is that this kind of gradual ecological degradation can be literally invisible to those of us who cannot easily tell one plant or insect from another, at least beyond our gardens. Nearly half of the world's people now live in urban areas; thus, many of us confront invaders directly only when new termites attack our homes or bugs our boulevard trees or microbes our health. Even those of us who spend a fair amount of time outdoors in parks, forests, or wilderness areas often cannot recognize subtle changes taking place on the land. Because most of us lack a detailed knowledge of the natural plant and animal communities around us, few can spot the strangers or detect the decline of natives as long as landscapes are still green and humming with life. I still cannot name or cite the origin of most of the grasses and wildflowers I see around my Montana home. For years, I never gave much thought to the tall, yellow-flowered plants that grew denser each summer around a foothill trail near Bozeman.

I have since learned that this plant is Eurasian leafy spurge, and now I can see that its invasion has crowded out the lupines, harebells, yarrow, horsemint, and other native plants that once flourished along the trail. What I could not see from the trail is that this invasion is causing more than just a change in the scenery. Spurge now dominates some 728,000 hectares of land in Montana and North Dakota, robbing native deer and elk as well as exotic cattle of palatable forage and depressing the value and productivity of rangelands. What's more, leafy spurge is just one in a lineup of invading weeds—most of them still invisible to me—that continue to degrade the American West both economically and ecologically.

Another stumbling block to recognition and then to possible remedy is that some regions have been so utterly transformed for so long that few people know what lived there before or how the native plant and animal community once functioned ecologically. For those who are familiar with the native flora and fauna of a region, however, invasions are often the most visible element of biological change today, far more apparent than the marginalization or impending elimination of native biodiversity that the invaders may be hastening. Many heavily invaded regions may even host more species than ever before, at least temporarily, a fact that causes some skeptics to ask "What's the problem?" But in many localities, the species count will fall back as natives disappear. What's more, a steady or even increased local species count can mask a global loss of species. Too often, unique, rare, and localized species have been replaced by a cosmopolitan set of species that can be found the world over: eucalyptus and Monterey pines, brown trout and mosquito fish, starlings and bulbuls, Medflies and gypsy moths, black rats and feral goats, lantana and water hyacinth. (A feral animal or plant is one that has escaped from domestication or cultivation and become wild.) While these replacements keep the local numbers high, the earth's total tally of living species declines. What's more, these cosmopolitan replacements homogenize our experience of the world.

The very look and feel of any given place, along with the life in it, is much like any other today, and we are all poorer for it, whatever the species count. The same forces that are rapidly "McDonaldizing" the world's diverse cultures are also driving us toward an era of homogenized, weedy, and uniformly impoverished plant and animal communities that ecologist Gordon Orians has dubbed the Homogocene. It is a play on the naming of geologic time periods that I translate loosely from its Greek roots as "the epoch of sameness or monotony." The invasive species phenomenon poses such a threat to human health and livelihoods that ecologist Michael Soulé wonders why it has not become a "motherhood issue."

Few governments have raised the issue to motherhood priority, although New Zealand and Australia have come close. Nevertheless, the political will to act on the problem of biological invasions (bioinvasions, for short) is growing. Heightened awareness of bioinvasions has developed at the same time that nations have been implementing an unprecedented round of global trade liberalization agreements. These agreements have accelerated the worldwide movement of vessels, cargo, and people—and, as a consequence, the risk of new invasions. As the volume and value of goods traded soared during the 1990s, so did the number of organisms in motion, incidentally or intentionally. Accelerating and costly invasions have caused many governments to begin to rethink their quarantine systems for excluding unwanted organisms and their often lax oversight of deliberate imports of new plants and animals. In the United States, the arrival of the zebra mussel in the Great Lakes in the late 1980s—with its damaging and costly habit of encrusting and fouling everything from boat hulls to industrial water intake pipes—was the first of a number of incursions that brought the issue of bioinvasions into the spotlight. In 1999, at the urging of 500 scientists and land managers, President Bill Clinton issued an executive order creating the National Invasive Species Council. In early 2001, the council released a management plan designed to improve the country's capacity to prevent the introduction of invasive alien species and control their spread.

At the international level, the Convention on Biological Diversity, or Biodiversity Treaty, signed at the United Nations' Earth Summit in Rio de Janeiro in 1992, recognized the threat that invading species pose to biodiversity. One provision, Article 8h, calls on member nations to "prevent the introduction of, control or eradicate those alien species which threaten ecosystems, habitats or species." The treaty took effect in 1993, and some

180 countries have ratified it (the United States, unfortunately, is one of the few that have not). One responsibility of each treaty nation is to prepare a national biodiversity strategy and action plan; a key issue for nations in the early 1990s was how to approach the implementation of Article 8h. Few countries at the time had the awareness or knowledge to address the problem of invasive alien species. In 1996, the Norway/United Nations Conference on Alien Species brought representatives from eighty nations together with scientists and technical experts on bioinvasions. At that meeting, the concept for the Global Invasive Species Programme (GISP) was born. GISP was established in 1997 to gather an international team of biologists, natural resource managers, economists, lawyers, and policy makers who could help bring the issue of bioinvasions to the forefront of the international agenda and support the implementation of Article 8h of the Biodiversity Treaty. This book is part of the GISP effort.

GISP is operated by a consortium (rich in both acronyms and experts) consisting of the Scientific Committee on Problems of the Environment (SCOPE), a nongovernmental scientific organization; CAB International (CABI), an organization long involved in on-the-ground management of invading alien species; and IUCN, the international conservation organization whose specialist group on invaders is headed by Mick Clout. This consortium operates in partnership with the United Nations Environment Programme (UNEP). The goal of the scientists involved in this project has been to develop new tools, evaluate best management practices, and articulate a new global strategy and action plan to help nations come to grips with the problem of bioinvasions. With those goals accomplished, GISP is now involved in helping nations put the tools and the strategy to work in protecting not only biodiversity but also human health and well-being.

Although scientific and technical approaches are indispensable in managing the problem, bioinvasions are fundamentally a human phenomenon, driven by economic activity and by our choices as consumers, travelers, gardeners, pet owners, fishermen, and so on. We are the ones who set species in motion, and all of us, as individuals, families, communities, and nations, must be involved in the solution. For that reason, the GISP partners wanted a book that would reach a broader audience with their findings. Stanford University biologist Harold A. (Hal) Mooney, chairman of the GISP Executive Committee, asked me to write it. Over a two-year period, I shadowed the experts as they gathered to devise early warning systems, analyze the pathways and vectors by which species are moving, consider the problems of risk assessment and economic analysis, examine the status of laws and international instruments, and spell out the best management practices for both preventing new invasions and controlling or eradicating established invaders. In addition, I visited sites around the world, from New Zealand and Australia to South Africa and the Galápagos Islands, where people are already putting such practices to work.

Chapters 2 through 5 of this book detail the extent and consequences of our rearrangement of the earth's species, exploring the often colorful history of past plant and animal movers, from fifteenth-century seafarers to contemporary plant hunters and reptile fanciers. As we'll see throughout the book, the human context and human motivations have changed surprisingly little over the centuries. Besides the flowers, snakes, and exotic game animals that we are still moving deliberately between continents, a vastly greater number of creatures is hitchhiking in ballast tanks and shipping containers and tucked in with the cut flowers, timber, grain, and fruit that is constantly in motion around the world. These first chapters detail both the economic and the social toll of invaders and their adverse effects on native species and natural communities. Trying to capture the economic and ecological effects of bioinvasions throughout the world in a few chapters involved difficult choices about what to include and what to leave out. It would be all too easy to expand the litany of loss and woe, disaster and degradation. Instead, I have chosen to devote more space to the search for solutions.

Chapters 6 through 12 explore the search for solutions, from sophisticated efforts to stop invaders at the border to strategies for preserving what we value in severely invaded lands. They detail the efforts of scientists to identify patterns and clues that could tell us which few species among the vast numbers of those in motion will succeed as invaders. This elusive predictive power has taken on new importance as countries attempt to clamp down on the importation of risky new species. These chapters also examine the trend pioneered in Australia and New Zealand toward a guilty-until-proven-innocent approach to the introduction of new species and the tensions between this type of precautionary approach and the drive for freer trade. We'll visit with inspectors in one of the world's strictest border quarantine systems as they check everything from incoming mail to used car imports to the boot treads of tourists. We'll follow the work of frontline surveillance teams that provide early warning of pest, weed, and disease incursions and allow nations to respond rapidly to nip invasions in the bud. We'll investigate innovative efforts to minimize the damage caused by established invaders and examine a growing shift in management focus away from simply defeating invaders and toward restoring ecosystems. This approach is the key to optimism, for even much-abused and invaded lands harbor species, communities, and processes that people value and want to preserve. One chapter focuses entirely on the Galápagos, where scientists and national park administrators, faced with a rising human presence in the islands, are working on all fronts at once, from quarantine to control, to try to avoid the mistakes that long ago caused massive swamping of native biodiversity in most of the world's other archipelagos. The final chapter recaps the challenges we face as individuals and nations in preserving what we value in an increasingly homogenized world.

Because we'll be traveling far and wide in the pages ahead, the following section provides a road map to some of the underlying themes and values I will be bringing to the issue of bioinvasions.

We are all in this together. The same forces bring invaders to all countries and inflict invasions on all social and economic sectors, from agriculture, industry, and conservation to public health.

There are no remaining forbidden cities or untouched places. Any of us can travel almost anywhere in thirty hours or less, and more and more of us do. The middle of nowhere takes MasterCard and serves McDonald's hamburgers and KFC chicken. Trade and tourism link us together in the need to address the global movement of damaging species. Your weeds or moths can be mine with a crate of grapes, my mites and mildews yours with a garden cutting slipped through the airport in an otherwise law-abiding gardener's carry-on.

Conflicts between various economic sectors remain, of course. Agriculture, forestry, horticulture, the pet trade, and other economic enterprises not only are plagued by exotic weeds, pests, and pathogens but also thrive on exotic plant and animal species. Most of the world's food comes from crops grown and livestock raised beyond their centers of origin. Some of these species, like the farmed elk and deer in New Zealand, escape to invade natural areas and damage conservation, recreational, and aesthetic values. Because agriculture makes little distinction between exotic and native weeds and pests, agricultural solutions to invasions, from herbicides and pesticides to biological controls, or biocontrols, have often been applied with scarcely a nod to biodiversity, public health, and other values of society.

It is vital, however, to acknowledge our common interest in the prevention of new invasions. In most countries, departments of agriculture and, to a lesser extent, of public health have the most experience, funding, and responsibility for whatever quarantine and control infrastructure exists. This experience and organization needs to be enlisted in support of broader social and ecological concerns. At the same time, we cannot let the larger issue of invasive alien species be absorbed and lost within narrowly focused weed and crop pest programs.

My intent is not to condemn all alien or imported species, only the invaders, defined as those that escape control and cause ecological or economic harm.

Exotics feed the world, and often the soul. No one is advocating that we confine wheat to the Levant and potatoes to the Andes or take the cuisine of Italy back to its pre-tomato past. And this book is certainly not about giving up lilacs, peonies, and tomatoes in my Montana garden. Its topic is the thousands of misplaced organisms that are ecological or economic scourges in new settings: zebra mussels, rabbits, goats, mosquito fish, cheatgrass, water hyacinth, lantana, miconia, cottony cushion scales, gypsy moths, Dutch elm disease.

It is also important to avoid treating the depredations of invasive alien species such as the ones just listed as a morality play or simplifying a complex reality into a tale of good and evil. Invaders are not a fixed set of "bad" plants and animals. Their identities change from place to place, although some notorious invaders have wreaked havoc almost everywhere they have been introduced. Nearly all invaders are integral elements of a community somewhere, and some are even becoming rare or endangered at home—ironically, often at the hands of another collection of invaders. The brushtail possum that is so destructive in New Zealand, for instance, is protected in its native Australia.

The language of invasive species discourse is sloppy and value-laden.

Biologists have a vocabulary for newly arrived species that is effectively neutral and does not indicate behavior or consequences: *adventive* or *nonindigenous* species, *archaeophytes* and *archaeozoons, neophytes* and *neozoons.* Even biologists, however, frequently turn to more vigorous and less obscure terms borrowed from the language of war (*invasion, beachhead, battle, kill, eradicate, overrun, explode, run amok*), disease (*infestation, infection, outbreak, epidemic, metastatic, cancerous*), or toxicity (*contaminate, degrade, pollute*) to dramatize the behavior of those "adventive" species that, well, run amok. There are also agricultural terms such as *pest, weed,* and *noxious species* that indicate unacceptable behavior, though not necessarily origin. A weed in that context is any plant, native or exotic, that grows where a farmer or gardener does not want it. The terms *weed* and *pest* also have legal meanings in many jurisdictions, defining which species cannot be grown, which must be controlled, which trigger quarantines. Likewise, the definition of terms such as *alien species* and *invader* are still in flux under the Biodiversity Treaty and in other international contexts in which words underpin legal powers and obligations. (In fact, the word *invader* is the subject of vigorous advocacy by some ecologists who want to preserve its scientific usage as a word that says nothing about behavior.) GISP and a number of other policy- and standard-setting groups have chosen to use the term *invasive alien species* to capture the threat that Article 8h addresses, and I have chosen to follow suit. Invaders, in this book, are high-impact newcomers.

Words from the language of human migration (*alien, immigrant, foreign, colonist, exotic, mongrel, stranger*) get thrown into the semantic mix, too.

Most of us who deal with the topic of bioinvasions are tempted to create linguistic maelstroms of war and pestilence, pollution and upheaval when we try to describe particularly threatening and frustrating invaders. We battle noxious, exotic green cancers that establish beachheads, infest habitat, contaminate communities, and overrun natives. In this book, I have tried to curb such lurid excesses yet use words that commonly appear in news accounts and policy statements.

One caution: the fact that words such as *alien* do multiple duty in our language does not mean that the same social values or context underlie all uses. By reading for metaphor, some social critics dismiss concerns about bioinvasions with the casually repeated but unsupported assertion that these concerns are merely manifestations of xenophobia or anti-immigrant sentiment in society. The limitations of language, however, do not negate the biological phenomenon of invasions. When you read here about problems caused by Eurasian zebra mussels, Australian melaleucas, or Asian gypsy moths, I am speaking of the ecological or economic effects of mussels, trees, and moths, not condemning their origins.

Migration, dispersal, and colonization are natural processes that play key roles in evolution, but the pace and scale of human-caused "migrations," along with our ability to virtually eliminate geographic barriers, make the current wave of species movements quite unprecedented and, in places, enormously damaging to the species, systems, and natural processes we value.

We are not "just speeding up" natural processes. We are swamping ecosystems with an unprecedented scale and variety of newcomers. The rate of introduction of new species is thousands of times faster at our hand. Timing and scale are fundamental in biology: change them and you change a phenomenon profoundly. Everyone knows that poison is in the dose. Seventy-five heartbeats per minute are healthy; 150 are not. A hurricane is not just early delivery on all the month's breezes.

Further, the crowd of new arrivals we are loosing on ecosystems includes not just neighbors that might have wandered in anyway but also plants and animals from other continents that we have helped across previously insurmountable oceanic barriers. No natural force would have put the loblolly pines of the American Southeast in the unlikely embrace of Japanese honeysuckles or rafted a population of pigs to the Galápagos or allowed an Asian tiger mosquito to meet up with an African encephalitis virus in the tunnels of New York.

There are limits to which and how many species can pack together into the world's shrinking supply of natural habitat and limits to how fast ecosystems can adjust to new faces without jettisoning current occupants. Every time a wave of cosmopolitan invaders eliminates a rare, uniquely local species, the diversity of life on the earth is diminished. Packing the maximal number of creatures into one place is not the way to preserve biodiversity. If it were, a crowded zoo would suffice. When it comes to the diversity of life, we can have it all, but we cannot have all of it everywhere.

Genetically modified organisms (GMOs) are potentially a subset of the invasive alien species problem, but this book deals only with the "natural" aliens that are already threatening biodiversity, human health, and economic activities.

The "Aliens with a capital *A*" are getting most of the attention on the international stage. The laws of a number of nations regard GMOs as alien species, although molecular biologists who perform genetic manipulations dispute whether a corn plant carrying a bacterial gene is any more alien than its unmanipulated parent. The more important question is whether some of these species, alien or not, will be invasive. A genetically modified corn or soybean plant, for instance, might become invasive in the sense that the "transgene" it has received makes it more likely than its parent to escape cultivation and become a weed. Molecular biologists are working to build a "better" trout and a "better" Monterey pine, yet both of these species are already invasive in many parts of the world. That means there is every reason for concern about where we release genetically modified trout and pines. A GMO could also contribute to invasions by hybridizing with wild species and passing on a transgene that renders those wild species weedier. In some quarters, a GMO might be considered an invader in another sense of that word, simply because it has been put out into the world by industry amid a public that isn't happy to see it.

GMOs pose an array of potential problems that ought to be dealt with responsibly, and the Biodiversity Treaty nations have concluded a legally binding agreement known as the Cartagena Protocol on Biosafety to do that. I believe that in a more rational world, the nature-made and human-moved invasive species that are already costing countries billions of dollars in damage would be getting the lion's share of the "biosafety" attention. It is a further irony that most of the genetically engineered crops on the market today—the herbicide- and pest-resistant varieties—were created to avert the ravages of the "natural" invasive alien species already plaguing our fields.

A significant proportion of invaders, even some quite noxious ones, benefit or please someone, and this makes regulation politically and socially complicated. Invaders are taking a very real and growing economic and ecological toll, but that toll is not evenly distributed. Those who benefit from an invader are seldom the same ones who bear the cost of its damages.

Many invading plants and animals got where they are because someone wanted them and imported them. As Hal Mooney says, one person's pest is another's livelihood. The Nile perch that have driven many native fish extinct and destroyed village fisheries around Africa's Lake Victoria have brought wealth to owners of trawlers and perch-processing plants. The black wattles and pines that have invaded watersheds and reduced water supplies throughout South Africa sustain a valuable timber industry. The water hyacinth that clogs lakes and rivers the world over got its passport as a valued ornamental. So did lantana shrubs, miconia trees, pampas grass, kahili ginger, and a world of other invasive plants we will hear about in this book. Many of the furbearers, such as beavers and nutrias (coypu), that cause devastation in wetlands from northern Europe to Tierra del Fuego have lost their advocates as the market for wild-trapped furs has declined. But exotic trout and salmon retain a fiercely loyal following among fishermen on every continent despite the fact that they prey on native frogs, fish, and other creatures.

Many elements of society, then, from the seed and nursery trades to some fish and game managers, are wary of at least some aspects of a general crackdown on biological invaders. That is also true of industries that make no use of imported species but do not want further restrictions imposed on what they move around the world and where and how they do so. Many pests, pathogens, and weeds got where they are because travelers, exporters, or shippers did not take care to prevent hitchhikers from riding along.

A major challenge for nations today is to develop economic tools that shift the risk burden of invasive alien species from the public at large to those industries and individuals who benefit from them or benefit from economic activities that encourage invasions.

The crackdown on invasive alien species is not a "war" that can be won once and for all. Like many of our most serious environmental problems, invasions are chronic and require sustained, often expensive, and decidedly unglamorous solutions.

"I don't think we're ever going to solve the problem of invasive alien species, any more than we're going to solve the problems of crime and public health," says Jeff McNeely, chief scientist at IUCN. "These are perpetual problems that we have to develop an ongoing capacity to deal with."

Even if we were able or willing to do what it takes to intercept all new species at the border, we will always have to deal with the invaders that are already with us and the "sleepers," or time bombs waiting to explode from our gardens, fish tanks, and fields. Established invasions are usually irreversible, and the toll is cumulative. Farmers, gardeners, and land managers certainly know that the task of weeding is never done.

Most environmental and conservation groups have only recently begun to take the issue of bioinvasions seriously. Battling creeping vines or choking weeds is a yawner compared with stopping bulldozers or saving habitat. What's more, solutions are often costly, complex, and unpalatable—using herbicides on weeds, for instance, or poisons, traps, bullets, or disease against feral pigs, rabbits, or cats. Intensive, hands-on management also runs counter to philosophical traditions of wilderness preservation in America. Yet it is an unavoidable fact that in most parts of the world, we cannot declare a piece of land protected, halt the most visible and damaging human encroachments, and assume that the land and its life will remain self-willed into the future. Effective conservation now requires an ongoing commitment to pest and weed control.

No one advocates an attempt to unscramble the world's biota and return it to some historical state, even if that were possible. But not everything that can invade has invaded, and damage from current invaders can be lessened or eliminated. Our ultimate goal must be not to vilify or destroy an invader or set of invaders but to preserve or restore something we value: native biodiversity and the wild places and systems where it can thrive, the look of a landscape, a sense of place, the functioning of an ecosystem, the economic productivity of our working lands and waters, the health of people, animals, and plants.

The challenge before us is how to take full advantage of the world's living riches without diminishing them, how to live in an increasingly fast-paced and borderless world without destroying everything unique, rare, or isolated. The task cannot be left to governments, trade negotiators, and border inspectors, or even farmers and land managers. The solutions will require understanding and participation on the part of consumers, gardeners, travelers, nurserymen, fishermen, pet owners, businesspeople—indeed, all of us who by our very local choices drive global commerce. It is an irony of the human condition that our desire for novelty has become a homogenizing force.

We humans created the problem of invasive alien species, and we must devise the solutions.

TWO Reuniting Pangaea

"But while men slept, his enemy came and sowed tares among the wheat, and went his way."

—Matthew 13:25

"After the wheat, follow the tares that infest it. The weeds that grow among the cereal grains, the pests of the kitchen garden, are the same in America as in Europe. The overturning of a wagon, or any of the thousand accidents which befall the emigrant in his journey across the Western plains, may scatter upon the ground the seeds he designed for his garden, and the herbs which fill so important a place in the rustic materia medica of the Eastern States, spring up along the prairie paths but just opened by the caravan of the settler."

—George Perkins Marsh, *Man and Nature,* 1864

In the Latin Quarter of Paris, on the grounds of the Muséum National d'Histoire Naturelle, you will find the small, old-fashioned Ménagerie in the venerable Jardin des Plantes. It is not the best zoo in Paris. That distinction goes to a modern zoo on the city's eastern outskirts, where lions and mountain goats roam in seminatural enclosures. Here in the Ménagerie, Nubian ibex graze in small corrals and a snow leopard stares through the ornate bars of a circular cat house, a structure little changed in 200 years. The Ménagerie remains an attractive destination for city dwellers, though. On a cool afternoon in late April, I found the shaded walkways lively with mothers and young children ogling chickens and African ostriches with equal fascination. I had made my way to this place to try to recapture an obscure bit of the past that nevertheless haunts us still. Wandering by half a dozen llamas, I paused finally to watch a pair of Tibetan yaks munching hay in a small pen. No one else seemed particularly drawn to these massive, placid oxen, and I tried to conjure a sense of the excitement that reportedly ran through the crowds here in 1854 when the first dozen yaks ever to reach France were placed on display. Those beasts became instant celebrities, not just for their novelty but also for their imagined future in the reinvention of French rural life and industry. Yaks, so the vision went, would one day replace mules, donkeys, and oxen, providing wretched peasants in the Alps with a hardy, multipurpose draft animal.

Such was the bizarre and fleeting dream of the Société Zoologique d'Acclimatation, the French acclimatization society, created only a few weeks before the yaks' debut. News of the animals' imminent arrival after a three-year journey from Tibet had stirred the spirits of 130 naturalists, agronomists, scientists, diplomats, industrialists, and landed gentlemen who had gathered under the leadership of zoologist and Ménagerie director Isidore Geoffroy Saint-Hilaire to found the Société. These gentlemen hoped that with the proper application of science, yaks, as well as llamas, ostriches, kangaroos, and a world of other exotic beasts and plants, could be adapted to a new climate and then turned out to "populate our fields, our forests and our streams with new inhabitants," improving the lot of peasant farmers and launching an agricultural and dietary renaissance in France.[1]

It was the grand opening act in an "acclimatization" movement whose ferment and folly quickly spread throughout Europe and the colonial world. Within the decade, the United Kingdom had launched its own society and the French Société had acquired dozens of affiliates both within France and its colonial empire (including Algeria, Réunion Island, Martinique, and eventually Indochina) and in foreign states from Spain, Italy, and imperial Russia to China, South Africa, Australia, and the United States.

The societies provided ordinary citizens, civil servants, diplomats, and even missionaries an opportunity to join in the importation and exchange of exotic birds, beasts, trees, flowers, and vegetables from across the oceans, an indulgence that from ancient times had been the province of royalty and, later, scientists. For the acclimatizers, the stated goal was not simply to create public spectacles or collect fascinating novelties, but, in Geoffroy Saint-Hilaire's words, to "endow our agriculture, which has languished so long, our industry, our commerce and [our] entire society with blessings which have been neglected or unknown until now."[2] Along with the growing ranks of botanical gardens, zoological societies, and seed merchants, the nineteenth-century acclimatization movement helped fuel an unprecedented rearrangement of the earth's living heritage.

As they reordered the biological world, they refashioned the human sphere as well. The character of the places we cherish today, their peoples, cultures, foods, and landscapes, strongly reflects choices made 150 to 500 years ago, when Christopher Columbus and the explorers who followed opened up new worlds to exploration and exploitation. So, alas, do the diseases, weeds, and pests that plague us and threaten to displace already beleaguered native plants and animals. The spirit of Geoffroy Saint-Hilaire and his followers is still very much with us, and the consequences are increasingly troublesome. The desire to "enrich" or "improve" upon locally available resources—often with ludicrously little knowledge or forethought—set a tone for agriculture, forestry, fisheries, wildlife biology, and species movers in the plant and pet trades that persisted almost unquestioned until very recent decades. We cannot undo history, but there are good reasons to break with some of its traditions.

The story actually begins long before yaks arrived in Paris. The scale on which colonial powers and their citizens could operate and the patina of science they brought to the enterprise were relatively new in mid-nineteenth-century Europe, but humans have undoubtedly been moving other creatures around ever since we stood up and walked. Uncertainty about what they would find to eat combined with a preference for the familiar must have led the earliest migrants to tuck grains and roots, fowl or piglets, into their packs as they ventured into new lands.[3] Thus were domesticated animals and cereal grains spread around the ancient Eurasian world. What's more, for every creature purposely packed along, many others must have stowed away unbidden—pathogens, parasites, rats, mice, roaches, flies, and seeds of weedy plants.

Historian Alfred Crosby points out that by stripping bare the earth at cultivation and harvest, Neolithic grain farmers provided openings for aggressive colonizing plants and so nurtured the creation of weeds as surely as crops. It is no accident that the Old Testament refers often to weeds, tares, and thistles plaguing the fields of the Middle East. And by settling in one place at high enough densities to accumulate garbage heaps, early farmers and villagers also attracted mice, rats, and roaches and "invented the animal equivalent of weeds: varmints." People crowded together in close proximity to varmints, and domesticated animals provided a new arena for parasites and diseases to develop—among them pox and influenza viruses and the morbillivirus family (distemper, rinderpest, measles). These ancient associations would help later generations of Europeans as well as their plant, animal, and microbial camp followers to conquer new worlds together.[4]

Little is recorded of the deliberate trade in plants and animals in the ancient world, but hieroglyphic records tell us that at least 3,000 years before Columbus, the queen of Egypt dispatched ships down the East African coast to collect incense trees for her gardens at Karnak. By Greek and Roman times, ships and caravans were venturing to far outposts of Eurasia to trade in fragrances, medicinal plants, and coveted flowers such as roses, violets, and Chinese peonies.[5] Seafarers from the Malay Archipelago were actively hawking bananas, cardamom, camphor, and other wares to India, China, and the remotest islands of Polynesia. From there, other merchants carried this exotic fare westward to Europe.[6] Medieval crusaders encountered Asian sugar in the Holy Land and developed a sweet tooth. Soon the Moors carried sugarcane westward to North Africa and then to Spain and Portugal.[7] In southern Germany, remnants of early botanical traffic thrive on the crumbling walls of medieval castles, which still harbor exotic lilacs, irises, and dozens of other imports once used for food, spices, medicines, or dyes.[8] By the twelfth century, European rabbits had been imported to England to supply meat and fur.[9]

Along with this deliberate commerce, of course, came the unwanted travelers, some attracting no notice and others changing human destiny. Shipworms and wood-boring crustaceans known as gribbles, along with barnacles, seaweeds, and sea squirts, were hitchhiking on the hulls of wooden ships by the thirteenth century while rats traveled within.[10] The sailors who first battled rats to save their food supplies probably couldn't have imagined there would be worse to come from the spread of these creatures. Then, in the fourteenth century, bubonic plague swept across Europe, carried by infected fleas borne by black rats. Between 1346 and 1350, one-third of the population fell to the Black Death.[11]

It was also in the fourteenth century that European sailors first ventured far enough into the eastern Atlantic Ocean to reach the Canary, Madeira, and Azores Islands. Spanish and Portuguese settlers followed, rapidly transforming these mountainous, well-watered archipelagos. Had anyone been paying heed, that transformation would have served as a chilling preview of the social and ecological upheaval to be repeated a century later in the European colonies of the New World.

In the already peopled Canaries, the Spaniards subdued the natives with horse-mounted soldiers and, unwittingly, pestilence. Throughout the island groups they released sheep, asses, cattle, goats, rabbits, and, incidentally, rats; they chopped and burned off the great forests for which Madeira was named; they planted wheat, dyer's woad, blackberries, wine grapes, and—most precious back in Europe—sugarcane; and they brought in slaves to work the cane fields. Undoubtedly, some native island plants and animals were eliminated before anyone stopped to make a record of their existence, replaced by living commodities in demand in Europe at the time. With some of these ill-considered imports, the colonists also premiered the sort of self-inflicted wounds that would be incurred time and again in colonies to come. Their imported rabbits, for instance, multiplied relentlessly, destroying crops and defeating the first attempt to settle Porto Santo, one of the smaller Madeira Islands. Blackberry brambles also spread out of control; even today, they are among the most widespread weeds in the

Canaries.[12] And the ghastly drama of sugar and slaves would be re-created, with little variation, from the West Indies and Brazil to Fiji and Hawaii.[13]

Finally, in the fifteenth century, Columbus and the explorers who followed crossed beyond the boundaries of the ancient world, driven first by the dream of a direct sea route to the silks, spices, and other biological wealth of the Indies. After months at sea, Columbus and his sailors came ashore on uncharted lands and beheld trees and fruits and wondrous beasts unlike any they had ever encountered. We know today that they glimpsed a high point in the diversity of life on the earth. To understand how biodiversity had reached that high, and why the voyages of Columbus and others who followed him marked a watershed in ecological history, however, we need to look back 200 million years or so to the Jurassic period, when the earliest mammals eked out a living at the feet of dinosaurs.

In the Jurassic, all the dry lands of the earth were huddled together on a single continent known as Pangaea. Then forces within the earth began to drag the great plates of crust apart at the speed our fingernails grow. Slowly, oceans spilled into the widening gaps between landmasses, stranding populations of plants and animals on isolated chunks of island and continent. Left to go their own ways, organisms diverged, evolving into myriad new life-forms and species. When Columbus and other Europeans first ventured across the oceans that now separate the fragments of Pangaea, they encountered the rich array of plant and animal forms that had been forged by millions of years of geologic isolation. At the same time, they opened an era of exploration and commerce that would forever end that isolation.

Even the geologic isolation had never been never complete, of course. Occasionally, plant or animal colonists had drifted between chunks of land on flotsam or blown in on the wind. Geologic forces had rejoined bits of land now and again, sailing Central America into place between North and South America 10 million years ago, for instance. The rise and fall of sea levels during past ice ages opened and closed land bridges, allowing humans to walk into Australia about 40,000 years ago and to cross the Bering Strait from Siberia to Alaska a few tens of millennia later. When rising seas subsequently drowned these connections, human groups, such as

Native Americans and Australian Aborigines, developed in isolation from Old World peoples until Columbus and those who followed reestablished contact.

With every breach of isolation by geologic forces or *Homo sapiens*, species were lost, including giant marsupials, birds, and reptiles in Australia and mammoths, mastodons, and a host of other megafauna in North America, probably felled both by spears and by diseases carried by arriving humans.[14] Yet no past breach of isolation matches that begun in the fifteenth century. The traffic Columbus and other early European explorers initiated across these ancient gaps reestablished the biological connections between the scattered remnants of Pangaea, and the interchange has been accelerating ever since. No natural force, no upheaval of ice or fire or rock in any other 500-year period in history, has rearranged the biology and cultures of the earth with the speed and scope with which humans have in the past five centuries.[15]

Columbus failed to find the spices of the East Indies, but when he landed in what came to be called the West Indies, he and his men became the first Europeans to taste corn and chili peppers. They killed and ate a "serpent"—most likely an iguana. And they loaded up goodies to take home. (Iguanas did not make the cut.) What they and other explorers transported in the subsequent 100 years has been dubbed the Columbian Exchange, a massive exchange of ingredients that eventually transformed menus throughout the world.[16] The New World bestowed upon the Old World corn, potatoes, tomatoes, cassava, chili peppers, sweet potatoes, green beans, chocolate, vanilla, pineapple, pecans, and turkey. The New World, in turn, received wheat, wine grapes, sugarcane, onions, lettuce, walnuts, olives, apples, dates, and domesticated livestock—sheep, goats, chickens, pigs, horses, and cattle, and thus, meat, milk, cheese, and lard. Try to imagine Italian food before tomatoes, French without haricots verts, Mexican without cheese, Thai without hot chilies, African without cassava or corn, and menus anywhere without potatoes, baked, boiled, mashed, or fried.

Unfortunately, a better cuisine is not all the New World got. The experience of the Canaries quickly began to replay itself. Long-isolated peoples in the Americas, Australia, and New Zealand succumbed in uncensused multitudes—some put the number as high as 56 million—to Old World diseases such as smallpox, measles, and influenza that traveled like a wave front ahead of the explorers, missionaries, and soldiers.[17] The ships that brought slaves from Africa to the Caribbean, starting in 1648, delivered more than the obvious cargo of misery—they transported to the New World both the notorious mosquito *Aedes aegypti* and two serious viral infections it transmits, yellow fever and dengue fever.[18]

Explorers were quickly joined on the global oceans by soldiers, merchants, buccaneers, pirates, and whalers. These seafarers dropped off goats, pigs, and rabbits on every remote island they passed, a self-replicating fresh meat supply for future visits. By 1704, when Scottish buccaneer Alexander Selkirk (a.k.a. Robinson Crusoe) began his four-year self-exile on the Juan Fernández Islands off the coast of Chile, he was able to kill and eat feral goats by the hundreds thanks to the release of pigs and goats in the 1560s by the Spanish navigator for whom the islands are named. When he was picked up in 1709, Selkirk moved north to the Galápagos Islands, off Ecuador—which were already crawling with rats more than a century before Charles Darwin's famous visit—and then returned to England, where Daniel Defoe made him a legend. (Defoe's fictionalized castaway took a dog and two cats ashore and soon, as he recounted in a prescient touch of realism, "came to be so pestered with cats that I was forced to kill them like vermin or wild beasts, and to drive them from my house as much as possible."[19])

Some of the earliest explorers even planted seeds in new territories long before any colonists arrived. In 1602, an explorer in what is now Massachusetts reported sowing wheat, barley, oats, and peas, two decades before English Pilgrims landed.[20] In various parts of Australia, early-nineteenth-century explorers not only loosed pigs and chickens but also planted wheat, apples, oranges, lemons, grapes, coffee, and many other food plants for anyone who might follow.[21] When colonists did arrive, they quickly planted more. Dutch settlers carving out the first European colony on South Africa's Cape of Good Hope in 1652, for example, sent orders back to Amsterdam to "send us anything that will grow."[22]

Plant and animal movers, however, almost always got more than they intended. Besides feeding wandering seafarers, feral animals grazed, rooted, trampled, wallowed, preyed on, and spread disease among native creatures and began to transform entire landscapes unaccustomed to the wear and tear of mammals. Without any quarantines, much less knowledge that germs cause disease—a revelation that did not come until the nineteenth century—early settlers and their plants and animals spread pathogens with every move. Weeds dispersed just as readily, clinging to the fur of animals or mingling with crop seeds.

By 1833, when Darwin arrived in Argentina on his now-legendary voyage aboard the HMS *Beagle,* he was able to observe one of the first completely invaded landscapes ever documented. Setting out on horseback to explore the pampas from the central coast north to Buenos Aires, he found that two European thistles—cardoon, or wild artichoke, and giant thistle—blanketed "square leagues of surface almost to the exclusion of every other plant." Even in southern Uruguay, he found an impenetrable mass of thistles covering several hundred square kilometers. Later, in the *Origin of Species,* Darwin expressed doubt "whether any case is on record, of an invasion on so grand a scale of one plant over the aborigines."[23]

Yet other invasive plants—including many that had never been "weedy" at home—were even then establishing beachheads and beginning to transform new lands. By the nineteenth century, the global mix-up of organisms was in its heyday. Yet wounds already visible on foreign landscapes and in the lives of remote peoples—and even those that soon struck Europe—gave little pause to acclimatizers and others busy seeking out and acquiring animal and vegetable curiosities.

Acclimatization Fervor

And that brings us back to the Jardin des Plantes. By the time Geoffroy Saint-Hilaire and his Société set up shop here, the garden had already been in the exotic plant business for 250 years. Before the French Revolution of 1789, when the garden was known as the Jardin du Roi, its staff had accumulated and propagated plants deemed useful for medicine, agriculture, and commerce and made them available to horticulturalists and public gardens at home and in the colonies. Strolling the walkways of the Ménagerie, I found myself shaded by great trees as cosmopolitan as the yaks, bison, and nilgai penned beneath them: Japanese pagoda trees from China, silver lindens from Eurasia, horse chestnuts from Albania, box elders from southern California, and smaller catalpas from Virginia.

By Geoffroy Saint-Hilaire's time, French diet and agriculture, along with that of most of the world, had been changed beyond memory by the swapping of food plants and animals between Old World and New. Initially suspect items such as potatoes and tomatoes had by then been embraced in European cuisine. Indeed, Ireland and Europe were already reeling from the first of the massive self-inflicted wounds caused by this slapdash biological roulette: the potato blight fungus had caught up with the potato and spurred famine and massive emigration from Ireland in the mid-1840s. By the 1850s, when the Société was formed, blight was ravaging potato crops throughout Europe. And less than ten years later, a grapevine-destroying aphid known as phylloxera would arrive from America, eventually to devastate wine production in Europe, Africa, Australia, and America for almost half a century.

One might be tempted to think that by 1854, enough was enough. But Geoffroy Saint-Hilaire had an almost Utopian vision of the progress that might be made by the application of scientific theory to the breeding of exotic plants and animals. Besides, he still believed that the Columbian Exchange had been a bit one-way: "We have given the sheep to Australia; why have we not taken in exchange the kangaroo—a most edible and productive creature?"[24]

Although the term *acclimatization* had a specific scientific meaning to Geoffroy Saint-Hilaire, involving gradual adaptation to a new climate, most of his followers and imitators used the term loosely to mean naturalization in the wild or even domestication—or sometimes little more than importation and release. To drum up effort and enthusiasm, the Société offered medals for successful acclimatization of a whole series of promising creatures: wild asses from Tibet, gray kangaroos, a now-extinct zebra known as the quagga from South Africa, Arabian dromedaries, North American beavers, South American alpacas and rheas, Australian emus, African ostriches, various game birds and fish, new species of silkworms, Chinese yams, bamboos, Japanese mulberries, and quinine trees.

One of the most prolific acquirers of exotics was naturalist, diplomat, and Société member Charles de Montigny, who established the first French consulate in China in 1848. Through a network of Jesuit missionaries, Montigny was able to send back to France and Algeria hundreds of living samples, from cotton, bamboo, and silkworms to those first celebrity yaks.

Accounts by historian Michael Osborne make it clear that the practical aims of the Société never sat well with the academics of the Muséum National d'Histoire Naturelle, despite Geoffroy Saint-Hilaire's prestige. Six years after the Société was founded, Geoffroy Saint-Hilaire and his followers took their yaks and opened their own zoo and gardens, the Jardin Zoologique d'Acclimatation, on the grounds of the Bois de Boulogne, which was then being transformed from a royal hunting preserve into one of the finest public parks in Europe. Visitors could see and buy yaks, llamas, kangaroos, and hundreds of other species, ostensibly ready to be taken to the countryside and turned into profitable livestock—or eaten. The Jardin Zoologique supplied 10,000 exotic birds to the palaces of Constantinople. In 1870, during the siege of Paris, elephants, camels, antelopes, and other animals from the Jardin Zoologique were slaughtered to feed the city.

Société members supported Napoléon III's vision that not just plants and animals but also French peasants themselves could be acclimatized to live and farm exotic crops in Algeria, supplying their homeland with cotton, coffee, sugarcane, olives, and other products at a favorable price. In turn, the government's chief botanist, Auguste Hardy, worked at what might be called acclimatizing the arid Algerian climate to such purposes, planting expanses of Australian eucalyptus and Montigny's Chinese bamboos in hopes of raising the humidity.

All these grand dreams came to naught, alas. Within two decades, it was clear that the Jardin Zoologique had failed to develop any new products, much less revolutionize French agriculture and rural society. Llama and alpaca herds had quickly sickened and died. Yaks proved troublesome, expensive to feed, and not the least bit attractive to conservative French farmers. More important, the Société had proven shortsighted in its social vision. By the second half of the nineteenth century, France was industrializing, her people were flocking to the cities, and attempts to promote rural cottage industries had fallen out of vogue.

The Jardin Zoologique, struggling financially, eventually dropped all pretense at scientific endeavors and parted company with the Société. Today, as the Jardin d'Acclimatation, it thrives as an amusement park for children. Only a small zoo provides any hint of its former mission. The Société itself has persisted, having turned its attention increasingly to conservation until, a century after its founding, it became the Société Nationale de Protection de la Nature et d'Acclimatation de France (the "French Sierra Club").

It was in the European colonies that the biological transformations sought by the acclimatizers and other like-minded purveyors of exotics succeeded to greatest effect, for better or worse, as we'll soon see. Ironically, by actively "globalizing" the production of crops and livestock elsewhere, France and other European powers had undermined incentives for diversifying agriculture at home. Why convince French farmers to husband ostriches and exotic sheep when ostrich plumes, fleeces, and fruit from Britain's Cape Colony in South Africa could reach Europe in two weeks by steamship, and at a better price?[25]

Compared with the French Société, Britain's Acclimatization Society was short-lived and amateurish. It was formed in 1860 at the call of Frank Buckland, an eccentric naturalist who had so enjoyed feasting on a haunch of African eland at a London tavern that he took to heart his host's dream "that we might one day see troops of elands gracefully galloping over our green sward."[26] The British Society members were keen on acclimatization, but they came late to the game and had powerful competition. By the mid-nineteenth century, London was already a world center for the propagation and distribution of exotic plants and animals. The Royal Botanic Gardens, Kew and the Zoological Society of London had long had representatives out scouring the ends of the earth to add to their collections.

Sir Joseph Banks, who had circled the globe with Captain James Cook's first expedition (1768–1771), had built Kew into a world center for economic botany and, with the help of botanical gardens throughout the

Tropics, set the stage for rubber, tea, quinine, and other profitable plants to be dispersed throughout the British Empire. It was Banks who sent one of his plant hunters aboard Captain William Bligh's *Bounty* to grow breadfruit seedlings in Tahiti for delivery to the West Indies as potential food for slaves. After five months lolling in Tahiti waiting for the plants to grow, however, the crew was in no mood to set sail again. During their mutiny on the *Bounty* in 1790, the crew even dumped the breadfruit seedlings overboard. (Bligh succeeded later with a second batch of breadfruit.) In 1798, a decade after the settlement of Australia began, a single plant shipment from Banks to the young colony included "hops, olives, carribs, lemon grass, spring grass, cactis, ginger, strawberries, camphor, vines, apples, pears, peaches, nectarines, mulberries, walnuts, chestnuts, filberts, quinces, pruient plums, oakes, and willows, mint, lavender, tarragin, sage, savoury laurel, cammomile and wormwood."[27] In turn, his collectors brought back to England some of today's most popular flowers. From the Cape Colony alone, the list includes proteas, bird-of-paradise, gladioli, amaryllis, arum lilies, ice plants, aloes, euphorbias, and the progenitors of the ubiquitous garden geranium. By the time of his death in 1820, Banks had been responsible for introducing 7,000 plants.[28]

Naturalist Christopher Lever points out that against this caliber of competition, the British Acclimatization Society managed to acquire no scientific cachet, few members, and no garden or vivarium of its own during its seven-year existence. The British Society claimed its chief purpose was to acquire beasts and plants of utility, yet most members seemed enamored of game birds and shooting sports, and some had an inordinate (and apparently pointless) fondness for breeding up novelties—crossing Indonesian sambar and native red deer, North American prairie chickens and Scottish black grouse, Indian Kalij pheasants and common pheasants. Members' husbandry skills seemed less in evidence when they dispatched a pair of European catfish from Vienna to London packed only in wet moss. The hapless fish, of course, failed to acclimate to breathing moist air! Other schemes were foiled by squabbles with members over the ownership of salmon eggs and various Chinese birds.

Some British scientists publicly challenged the acclimatizers' notion that elands or llamas or giraffes could ever adapt to being turned out in the British countryside. And the press repeatedly ridiculed Society members for their naïve belief that Englishmen could ever be induced to eat wombats and kangaroos. A magazine associated with author Charles Dickens declared in an anonymous article in 1861 that French Société leader Geoffroy Saint-Hilaire was "an intrepid horse-eater who would accord a fair tasting to donkey-flesh, and who appreciates even rats when properly fed."[29] (In fairness, Banks himself had sampled dog, rat, and kangaroo meat while traveling with Cook.)

Ironically, although conservatism in diet helped to doom acclimatization efforts in the Old World, it helped to drive an exuberant transformation of the flora and fauna of Australia, New Zealand, and other colonies. "The British abroad are notoriously unenterprising in experimenting with exotic foods," Lever observes. And so they set about attempting to re-create the motherland in the colonies.[30]

The Colonies Call Home

While the average citizens of Europe were gradually accepting potatoes and other exotic foods into their diet, surprisingly few at first jumped at the chance to pack up and move to remote lands. For the first 300 years after the voyage of Columbus, European settlers were sparse in the colonies. Then, in the early nineteenth century, crowding, famine, wars, and persecution sparked a massive exodus of peoples from the Old World. From 1820 to 1930, more than 50 million Europeans settled in the Americas, Australia, and New Zealand, overwhelming the remaining indigenous peoples.[31] Once safely ensconced in these strange landscapes, however, the settlers began sending home for food plants, livestock, flowers, game animals, songbirds, and other fondly remembered creatures.

Nowhere was that effort more vigorous than in Australia, and particularly in the state of Victoria, where the wealthiest of the colonial acclimatization societies was created in 1861. A gold rush in southeastern Australia in the 1850s had brought miners flocking to the region, but the mines quickly played out. As miners poured into Melbourne looking for work, the population soared sevenfold, to more than half a million. As a possible solution to unemployment, some civic leaders thought miners could be encouraged to resettle on small farms to rear exotic crops and animals.

In 1863, the British Society sent out a questionnaire asking colonials to recommend candidates for acclimatization back home and, in turn, to list species they wished to acquire. The Acclimatization Society of Victoria responded with an astounding wish list. Although the continent was teeming with marsupials, it was unconscionably devoid of hoofed beasts. From the moment the first British settlers arrived in 1788 with seven cattle, the colonials had been working to remedy that lack. Among the mammals the Victoria Society was already working to naturalize were hog deer, axis deer, and rusa deer. But they reported to the British Society their eagerness to try wapiti (elk) and bighorn sheep, as well as snowshoe hares from North America; chinchillas from South America; rock hyraxes, kudu, eland, gemsboks, and spring hares from Africa; and oryx from Arabia. Among birds, they had already purchased and released large shipments of English house sparrows, Chinese tree sparrows, song thrushes, common mynas from India, skylarks, blackbirds, starlings, peafowl, pintails, and a slew of partridges and pheasants. Yet they still perceived a need for "ant-thrushes," woodpeckers, snake-eating African secretary birds, ostriches, black-necked swans, and a dozen more species of pheasant and grouse. For the waters of Victoria, they desired carp, gudgeon, and "other palatable pond-fish," as well as salmon, brown trout, char, "and other principal river fish of Europe," not to mention crabs, lobsters, and various coastal fishes. Then came the plant wish list: new varieties of cotton and tea, cassavas, quinine, pines, firs, walnuts, cork oaks, olives, lemons, tussock grass for dune stabilization, buffalo grass for forage, brambles, bilberries, willows, and on and on.[32]

Individuals and societies in other parts of Australia likewise set about releasing songbirds, trout, furbearers, game animals, camels, llamas, alpacas, and rabbits to remedy perceived deficiencies in the countryside. Not everyone was pleased about this. Farmers quickly complained that imported starlings, house sparrows, and other birds were raiding grain and fruit crops. Supporters, however, maintained that the birds ate enough insects to make up for their pilfering. That conflict has become an old song with an endless refrain: one person's noxious weed or pest is another's profit or delight. And societies traditionally have made little effort to sort out the potential trade-offs beforehand.

Even birds and animals intended for the farm and not the wild soon went AWOL or were turned out into the countryside. Ostriches, initially kept penned on farms, were abandoned to form feral flocks when the plume market collapsed. Today, twenty-six exotic bird species range through Australia, from ostriches, wild turkeys, and California quail to skylarks and mynas. Two dozen mammals have also naturalized, including six exotic deer, three types of rat, water buffalo, dromedaries, horses, and a plague of rabbits.

In Queensland, the local acclimatization society experimented with growing quite a number of commercially valuable crops, from bananas, cotton, grapes, coffee, and tea to sugarcane, which remains an economic staple of the region today. Unfortunately, with the unregulated movement of plant material, the fungal disease sugarcane rust quickly appeared, afflicting the industry. (One of Queensland's most devastating invaders, the prickly pear cactus, had been introduced from Brazil in 1788, along with the cochineal scale insects that feed on it. The bodies of these insects yield cochineal dye, which was wanted in this case to keep the coats of British soldiers red.)[33]

If British colonials thought Australia was wanting, New Zealand must have cried out for enrichment. After 65 million years of isolation, the only resident mammals were three species of bat (one now extinct). Maori settlers had arrived in the thirteenth century accompanied by dogs and Pacific rats and had begun hunting the giant flightless birds known as moa for meat. Within only a few decades, they had eaten a dozen species of moa to extinction, and the rats they brought had apparently eliminated several species of small birds, flightless insects, and reptiles. Overall, some three dozen bird species went extinct after the Maori arrived. Then came the Europeans.[34]

Captain Cook started a trend in 1777 when he stopped by New Zealand on his second round-the-world voyage to plant gardens and release rabbits, pigs, goats, geese, and chickens. Altogether, Europeans since Cook have imported fifty-five mammals to New Zealand, and thirty-five of those have successfully taken up life in the wild. Today you can view the full parade of exotic mammals in the Human Impact section of the Auckland War Memorial Museum, where an impressive taxidermic army of furry invaders marches across a low platform in a high-ceilinged hall: pig, goat, dog, cat, horse, European rabbit, brown hare, hedgehog, various species of rat, deer, and wallaby, Himalayan tahr, chamois, Australian brushtail possum, ferret, stoat, and weasel (the last three imported in a futile effort to control the rabbit). Predators among these have wiped out another ten bird species since Europeans arrived and today threaten several more; grazing animals have so far eliminated three plants that grew only in New Zealand.[35]

More than thirty acclimatization societies formed across New Zealand starting in the mid-1860s and, like most others around the world, showed more enthusiasm than judgment. Their chief preoccupation was importing and releasing the same array of insect-feeding songbirds and game birds that were being turned out in Australia. Most of these introductions succeeded, and many of the small birds are now considered agricultural pests. The societies also helped to stock the country's rivers with various imported trout and salmon. In the 1860s, the societies sought and won legal protection for starlings, song thrushes, blackbirds, trout, and salmon and actively sought to eliminate native hawks, kingfishers, spotted shags, and other creatures that preyed on the exotic imports.

Many of the fish stocks supplied to acclimatizers in New Zealand and Australia in the 1870s came via a group known as the Country Club of San Francisco, an acclimatization group whose main goal was to fill the streams of California with European brown trout (although it also made an unsuccessful attempt to populate San Francisco with nightingales and a few other English songbirds).[36] Across the United States, acclimatization societies largely focused on English songbirds and exotic plants. The leader in this movement—now a legend for his eccentric whimsy—was New York pharmaceutical manufacturer Eugene Schieffelin, a lover of birds and Shakespeare, who is reputed to have taken a notion to import all the birds mentioned in the Bard's plays. On the basis of a single line in *Henry IV*, he made a successful release of forty pairs of starlings into Central Park in 1890. Never mind that this was two decades after starlings had been released into New Zealand and that the birds had already gained a reputation for damaging crops. It was yet another case of heedless folly. Despite other starling releases around the country, it is Schieffelin's birds that are credited with having colonized North America from coast to coast within fifty years.[37] Seven years after Schieffelin released his in New York, Prime Minister Cecil Rhodes introduced the birds into the Cape Colony to enhance South Africa's bird life.[38] Starlings clearly held some appeal a century ago, unfathomable today.

On the Hawaiian Islands, the Hui Manu (bird society) was formed in 1930 to acclimatize exotic birds in a land already reeling with invaders. Polynesians had arrived on the islands about 400 A.D. with rats, dogs, pigs, jungle fowl, and three dozen plants. Archaeological evidence shows that by the time Captain Cook arrived in 1778, half of the native land birds in Hawaii—and, indeed, more than 2,000 species throughout Polynesia—had been eliminated and seabird populations greatly diminished. Cook and the European settlers who followed him brought cattle, goats, sheep, boars, and thousands of plants.[39] Yet the Hui Manu members saw a need for more exotic creatures. They successfully introduced northern mockingbirds, Japanese white-eyes, common cardinals from the U.S. mainland, red-crested cardinals from Brazil, Indian black-headed mannikins, and a number of other birds before the group disbanded in 1968.

A similar story could be told about island after island, country after country. But the tale becomes repetitive, and the favored species form a rather monotonous list. Today, the most visible biological legacy of the acclimatization movement in most countries is the cosmopolitan distribution of so many game animals and European birds, from brown trout to starlings. Just as significant, however, is that these societies helped to ingrain in society and in newly developing applied sciences such as wildlife and fisheries management the ambition to "improve" on local plant and animal communities. The societies were just one of the forces at work in shaping the living world as we inherited it. Commercial enterprises and government agencies were also working enthusiastically to redistribute the world's plant and animal life to their advantage, with no more awareness or concern than the acclimatizers for the species invasions they were setting in motion.

Plantsmen and Presidents

In the depths of a Montana winter, one of the great pleasures I look forward to is the arrival of the first wave of garden catalogs. I leaf through hundreds of pages depicting impossibly lush flowers and perfect tomatoes months before the ground thaws. For years, I assumed that mail-order seed catalogs were a modern amenity unknown to, say, snowbound prairie homesteaders of the nineteenth century. But ecologist Richard Mack of Washington State University, who has spent a great deal of time looking into boxes in the National Agricultural Library in Washington, D.C., and other collections, reports otherwise. More than two centuries ago, before the Lewis and Clark expedition put Montana on the map, seed merchants and nurserymen were circulating catalogs throughout the settled parts of North America and, in fact, the world. The oldest U.S. circular Mack has located is a 1771 fruit tree catalog from a nursery in Flushing, New York. Catalogs from the early 1800s offered mostly food, forage, and fiber crops and home remedies such as horehound, catnip, absinthe, belladonna, digitalis, and opium poppies. After the Civil War, the American economy boomed, and lavishly illustrated seed catalogs offered an increasing array of ornamental plants. The 1868 catalog of German seed merchant Haage & Schmidt offered an astounding 12,471 species to gardeners worldwide.[40] The most influential of all the nineteenth-century seed merchants was Veitch & Sons, a British enterprise that sent its own plant hunters out as far as Patagonia and Burma and became the first to offer—in 1854—seedlings of California's giant sequoia trees to English gardeners.[41]

Mack has identified 139 species sold commercially during the nineteenth century that are now naturalized and growing wild in the continental United States and Hawaii. The list includes some of our worst weeds: Johnson grass and jointed goatgrass (two of the costliest crop weeds worldwide), water hyacinth, watermilfoil, Japanese honeysuckle, barberry, tamarisk, Brazilian pepper tree, casuarina (also called Australian pine, ironwood, or she-oak), strawberry guava, and kudzu—the latter offered in catalogs decades before the U.S. Department of Agriculture officially began promoting it for stabilizing degraded soils.[42]

By the nineteenth century, however, government agencies were as eager as commercial nurseries and acclimatizers to acquire and distribute new species. President Thomas Jefferson embodied the attitude of his time when he proclaimed, "The greatest service which can be rendered to any country to is add [a] useful plant to its culture." One of his successors, President John Quincy Adams, sent instructions to consular officers abroad in 1827 to identify and arrange to send home any "plants of whatever nature whether useful as food for man or the domestic animals, or for purposes connected with manufactures or any of the useful arts." By 1898, that mandate was formalized with the creation of the federal Section of Seed and Plant Introduction. Federal plant hunters joined in the long tradition of tramping about the world's remaining wild places, importing, evaluating, and releasing thousands of new plants, especially forage species.[43]

In Australia, too, as rangelands began to deteriorate under the unaccustomed tromp and bite of exotic cattle and sheep, the government launched a massive effort to introduce new pasture grasses. The caution they employed seems, in retrospect, little better than that shown by the acclimatizers. Plant ecologist Mark Lonsdale of Australia's Commonwealth Scientific and Industrial Research Organisation found that of 463 exotic forage species introduced between 1947 and 1985, only 21 have proven useful. On the other hand, 60 have become naturalized weeds—including 17 of the 21 considered useful for forage.[44]

By the late nineteenth century, governments had also begun to institutionalize the business of "improving" fish and game resources. Even Yellowstone National Park, the world's first national park, did not escape improvement. Although park officials nixed reindeer and exotic game birds, they actively stocked non-native brown, rainbow, and brook trout from 1889 until almost 1960, to the detriment of native Westslope cutthroat trout and Arctic grayling.[45] In 1873, the newly created U.S. Fish Commission fitted out a train car with an aquarium, stocked it with bass, catfish, walleye, and other species from eastern waters, and sent it steaming toward

California. Tragically, the collapse of a railway bridge in Nebraska sent the whole lot plunging into the Elkhorn River. Many subsequent trips succeeded, however, and western settlers were relieved of the monotony of fishing for native salmon and trout. Livingston Stone, who oversaw the transcontinental stocking effort, reported that on every journey, as the train crossed the Mississippi River, he threw in "a few fish for luck."[46]

A Troubling Legacy

The train is still moving somewhere, joined now by containerships, tankers, jets, and tractor-trailer trucks, plus an armada of pleasure boats, cruise ships, military vessels, and private automobiles. The world is in motion as never before. The plant hunters are still poking around remote sites. Botanical gardens, zoos, nurseries, pet dealers, and crop breeders are still importing and distributing exotic wares. Every impulse that drove Columbus, Cook, Banks, Geoffroy Saint-Hilaire, Stone, or the others in this tale still stirs us today: dissatisfaction with things as they are, faith in progress, curiosity, greed, war, carelessness, whimsy, vanity, a love of novelty, a sense of nostalgia, a craving for adventure. Fish and fowl, shrubs and flowers, ants and viruses, sponges and clams, furry animals and turtles are on the move even now, for all these reasons and more. Creatures that missed the Columbian Exchange and its 500-year aftermath—and they are legion—have never had a better shot at expanding their range. Those that have moved just once or twice have myriad opportunities to do so again.

Despite the vaunted complexity of modern life, our world is much simpler in many places, more uniform biologically and culturally, than the one Columbus beheld, and the processes that drive the proliferation of cosmopolitan species the world over are accelerating. In Australia, one ecologist noted: "Man through modern agriculture has managed to achieve in 50 years what has taken over 2,000 years in the Mediterranean Basin. Little remains in its original form, having been replaced largely by fields of wheat and grazing land in all but the driest zones."[47] Much of the credit or blame for this homogenization, of course, goes to plows, axes, livestock, and bulldozers. Yet the global mixing of plants, animals, and microbes, both deliberate and accidental, has teamed with our direct alteration of habitats and landscapes to make any given temperate or tropical place look increasingly like any other. You can now find brown trout and eucalyptus everywhere from New Zealand to Tierra del Fuego, Zimbabwe to California.

This shake-up hasn't been all bad by any means. We all acquired some lovely things, many of which are mainstays of modern economies and cultures. Who would consign potatoes, rice, wheat, tomatoes, corn, sugar, coffee, or chocolate to be grown and eaten only in the regions where they originated? Our most valuable commodities today are grown far beyond their origins—Eurasian wheat in the North American heartland, South American rubber in Malaysia, African coffee in Brazil, sugar and bananas from Southeast Asia in the Caribbean, and Andean potatoes and Eurasian sheep and cattle worldwide.[48] In the United States, 98 percent of food production, valued at $800 billion per year, comes from imported non-native species such as wheat, corn, and cattle.[49] Indeed, these and other widely introduced crop and livestock species provide more than 98 percent of the world food supply.[50]

Yet there are those gaping wounds we created, the vast movement of weeds, pests, and pathogens that interfere with our health, our livelihoods, and the unique diversity of our native plant and animal communities. Public health and agriculture began to tackle this problem early in the twentieth century, and any border controls and import quarantines we encounter around the world today were erected to protect these sectors. Only recently have we begun to realize, however, the burden this biological roulette has placed on the wild places beyond our farms and homes. In some regions, native communities have been overwhelmed. Nearly half of the wild-growing plants on Hawaii have been introduced by humans, as have at least one-quarter of those in Florida.[51] Half of the plants and all the terrestrial mammals living wild in New Zealand today were introduced by people, and four new exotic plants escape gardens and fields and establish themselves there every month.[52] Likewise, on islands from Mauritius to the Galápagos, at least half of the plant life is non-native and was carried there by humans.[53]

Most acclimatizers and like-minded species movers truly believed they were simply enriching, not replacing, the life of their new homes. In many cases, they were. Most exotic species do not survive in new territories without tending, and among those that do, most slip unobtrusively into their new communities. A small but significant number of these imports, however, turn aggressive. They invade, disrupt, and displace, often marginalizing or replacing uniquely local species. A region may end up with the same or even higher head count of organisms, but the new cast is likely to include a striking array of starlings, pigs, thistles, and other creatures that can be found anywhere. This is the new Pangaea, crowded, competitive, a shrinking world connected not by geology but by human commerce. Many unique, rare, or undefended creatures, well suited to isolated realms, will not survive it without our help.

Even more dramatic than the loss of particular species to invaders is the way one or a handful of aggressive exotics can transform whole landscapes, as Darwin first witnessed in Argentina. In fact, if Darwin had ridden through the Central Valley of California instead of Argentina in 1833, he would have seen the native tussock grasslands already dominated by invading wild oats. Thistles were already spreading across Australia as well as Argentina, causing such a problem for farmers that Australia passed the Thistle Act in 1851 to attempt control of the weeds.[54] Today, Darwin would find all the world's temperate grasslands, from South America and Australia to western North America, utterly transformed by invasive plants.[55] If he were to return to the Galápagos, he would find landscapes denuded by goats and pigs and choked with guava and quinine trees and blackberry thickets, with new pests and pathogens arriving continually to threaten the well-being of the remaining giant tortoises, iguanas, finches, and other unique living things that set him to pondering the origins of species. In the South African Cape region today, Darwin would see clots of exotic pines and woody weeds sucking dry the streams; in Tahiti, miconia forests obliterating all other plant life on the slopes; in the southern United States, vast expanses smothered by kudzu; in Africa, lakes and rivers thick with water hyacinth.

Modern societies would never wish to undo many of the events and decisions of the past 500 years. Yet there's much we would undo but cannot. Many bills are just now coming due, generations later, economically and ecologically, as we'll see in the next few chapters. Could we have gotten the benefits of dispersing the world's biological riches without incurring these costs? Perhaps not in centuries past, when organisms were still viewed as singular and interchangeable creations and even learned people had little notion of the web of life beyond village and farm. There is little excuse today, however, for continuing to allow incautious whims and carelessness to govern the fate of our living heritage.

THREE

Wheat and Trout, Weeds and Pestilence

"No one really knows how many species have been spreading from their natural homes, but it must be tens of thousands, and of these some thousands have made a noticeable impact on human life: that is, they have caused the loss of life, or made it more expensive to live. If we look far enough ahead, the eventual state of the biological world will become not more complex but simpler—and poorer."

—Charles Elton, *The Ecology of Invasions by Animals and Plants,* 1958

"Once we were gone, the prairie should have settled back into something like its natural populations in their natural balances, except. Except that we had plowed up two hundred acres of buffalo grass, and had imported Russian thistle—tumbleweed—with our seed wheat. For a season or two, some wheat would volunteer in the fallow fields. Then the tumbleweed would take over, and begin to roll. We homesteaded a semi-arid steppe and left it nearly a desert.

"Not deliberately. We simply didn't know what we were doing. People in new environments seldom do. Their only compulsion is to impose themselves and their needs, their old habits and old crops, upon the new earth. They don't look to see what the new earth is doing naturally; they don't listen to its voice."

—Wallace Stegner, *American Places,* 1981

The arid grassland that stretches from the Rocky Mountains west to the slopes of the Sierra Nevada and the Cascade Range was the last frontier in the American West to be claimed by settlers.

Explorers who passed through in the first half of the nineteenth century saw a landscape dotted with bunchgrass and sagebrush, the bare soil between clumps paved with a living crust of lichens, mosses, and other small organisms. Fires were infrequent and small. Big game was sparse. Occasional herds of pronghorn antelope grazed on the flats, and mule deer browsed in the breaks and coulees, but the vast herds of bison that ranged over the Great Plains to the east of the Rockies were conspicuously absent. Wagon trains rolled through this intermountain region on their way to California and Oregon in the 1840s, but it remained sparsely settled until after the Civil War.

Then, within three decades, aided by the completion of the transcontinental railroad, homesteaders poured in, filling the grasslands with cattle and sheep, wheat and alfalfa. In little more than a generation, settlers brought to a close the saga of pioneer settlement and radically transformed the character of these native grasslands. The bunchgrasses faltered under the unaccustomed grazing pressure of immense herds of cattle and sheep, and the animals broke up the protective soil crust with their trampling and wallowing. Then, with every sack of wheat or alfalfa seeds sent over from Europe, with every shovel of livestock bedding pitched from a train car, weed seeds arrived to take advantage of the injured ground: filaree, wild oats, parasitic dodder, and Russian thistle—a prickly globe of a weed that rolled across thousands of kilometers of degraded range and into cowboy song and legend as the "tumbling tumbleweed." The unsung superstar invader among the new arrivals, however, was cheatgrass.

Sightings of Eurasian cheatgrass patches were first recorded in 1889 in British Columbia and in the 1890s in Washington, Utah, and Oregon. By 1930, a scant forty years later, this annual grass had taken over some 40 million hectares of the western range—an area the size of California. Every few summers, when the grass awns grew sharp enough to damage the eyes and mouths of cattle, vast expanses of tinder-dry cheat fueled unprecedented wildfires that further depressed the native shrubs and perennial bunchgrasses.[1]

The intermountain West was not unique. From the Central Valley of California to southern Australia to the cheat-infested Patagonian steppe, the same forces were at work. Early plant invaders, aided by introduced livestock, have reduced the biodiversity and productive capacity of temperate grasslands worldwide, promoting soil erosion, altering fire cycles, and driving a biological transformation equaled in scale and swiftness only by the depredations of exotic invaders on remote islands.[2] What's more, the process is not over. Indeed, it is accelerating, driven by a whole new cast, many of them plants that arrived more than a century ago but—for unknown reasons—took longer than cheatgrass to begin their aggressive advance. In the decade from 1985 to 1995, for example, noxious broad-leaved weeds such as leafy spurge and yellow starthistle quadrupled their hold on the public rangelands of the American West, and they are now expanding their spread at the rate of 14 percent per year.[3]

All of us who eat the food or wear the fibers that farmers and ranchers produce share the ever-increasing burden of invasive weeds, as well as crop pests and diseases. And agriculture is only one of the many human enterprises and concerns that are disrupted by the myriad plant, animal, and microbial invaders that we continue to spread around the world, wittingly or not. It is the scope of this economic and social toll that we'll examine in the pages that follow.

One crisp fall day, I climbed with Roger Sheley up a hillside east of Bozeman, Montana, to a small fenced plot where he has been monitoring the progress of a weed invasion, trying to document the point of no return—the point at which the native grassland loses the ability to resist and rebound. Sheley is Montana's noxious weed specialist and a rangeland ecologist at Montana State University. That fall, he had been stumping the state, warning ranchers and bureaucrats that a new weed explosion was imminent if they didn't act soon. More than 240,000 hectares of Montana had burned over the summer, from mature forests to stumpy clear-cuts to fields of cheatgrass, all parched by drought and then swept by wildfire. Now the rains had come and the weeds would be the first plants to rebound, he warned—not just cheatgrass but also leafy spurge, yellow starthistle, medusahead, rush skeletonweed, musk thistle, dalmation toadflax, sulfur cinquefoil, and Montana's worst scourge, spotted knapweed.

All these are toxic, thorny, or chemically unpalatable broad-leaved weeds that displace the grasses that cattle and wild grazers prefer and reduce the number of animals the land can support. It was spotted knapweed that we were going to see advancing across Sheley's study plot.

We were walking in the Story Hills, and the muddy road below was near gridlock with pickup trucks and livestock trailers jockeying for access to the loading chutes at the local stockyard. The tinny blare of the auctioneer's bullhorn carried easily up the hill, but the chant of transactions was impossible to make out against the steady chorus of bawling calves—calves the ranchers could not feed through the winter on their drought-, weed-, and fire-diminished range.

Often in spring I had seen these hills awash in soft purple and, like many people in town, failed to recognize the waist-high stalks with their small, daisy-like flowers as spotted knapweed. Spotted knapweed is native from central Europe to western Siberia. Its seeds probably arrived on America's West Coast in dry ship ballast and as contaminants in bags of alfalfa seed. Spotted knapweed was first reported in Victoria, British Columbia, in 1883, several years before cheatgrass was noticed, but unlike cheat, it seems to have bided its time, advancing little until 1920. Then it exploded. Between 1980 and 2000 alone, it spread into every county in Washington, Idaho, Montana, and Wyoming and moved into ten other states as well. Spotted knapweed is now advancing across the West at a rate of 27 percent per year, moving into well-managed lands and degraded range alike. Both elk and cattle tend to avoid knapweed, dining preferentially on grasses and thus giving knapweed another advantage.[4]

Three years earlier, Sheley had walked up this infested hill and noticed, amid the advancing knapweed, a remnant of native grassland. The landowner had agreed to fence off a plot of hillside, unsprayed, where Sheley and his students could track in minute detail a process that is happening across the West on a scale almost too large to grasp.

"I don't know if you're familiar with the Idaho fescue and bluebunch wheatgrass habitat type, but it's gorgeous. It's my favorite," Sheley said as he stepped over the fence and into the plot. "This area right here has a pretty darn good native plant community, with Idaho fescue, some bluebunch, mostly needle-and-thread grass, lots of forbs, a little bit of shrub, some rose." I could see a small, thorny cane sticking up from the ground between clumps of dry grass. "In the summer, this will begin to fill in with mosses and lichens between the plants. This is a really pristine habitat patch."

His enthusiasm was infectious, but this tiny remnant was clearly doomed. Sheley pointed south across the slope at the advancing future. We were standing on the outer rim of a spoke-like series of five transects he and a student, Susan Webb, had marked with metal stakes, each line radiating about twenty meters from a solid stand of knapweed at the hub.

"As you walk along this line toward the dense infestation, you can see the changes in the plant community," he pointed out. With painstaking work on hands and knees, Webb had quantified what we were now seeing as we walked south: for every 1 percent increase in spotted knapweed cover, native grass cover declined by 2 percent and the richness of native species by 3 percent. "When you get to the center, you can see all the native species are gone. Knapweed is there, some cheatgrass, but we basically lose everything, even the mosses. So as you walk along that transect, you're walking forward in time, more or less, and that"—he pointed back to the native bunchgrass patch behind us—"will eventually look like this if we don't do anything about it."

Sheley doesn't intend, of course, to do nothing. That's why he crusades, and that's why he set up these plots and others. Preserving or restoring the grasslands may require a sophisticated combination of tactics, including replanting, changing grazing practices, judiciously spraying with herbicides, and releasing natural enemies for biocontrol of weeds. One fact is clear: the heavy-handed kill-the-weed approach of the twentieth century did little to slow the weed invasion.[5]

"Right now I believe these weeds are moving at their biological potential," he told me. "We're spreading them just about as fast as we try to fight them; therefore, they're doing what they would do if we weren't even around."

If spotted knapweed continues to spread unchecked, this single invader could eventually cost the U.S. livestock industry $150 million per year, a cost that will be reflected either in beef prices at the supermarket or in the financial failure of ranches.[6] And knapweed is only one of many green threats to the productivity and biological integrity of the western range.

Other than the dandelions and thistles that advance on my tomato garden each season, weeds once seemed to me a remote and tedious burden borne by people who make a living from the land. Yet the burden persists like a hidden tax on the global food supply. Worldwide, 67,000 pest species attack crops—9,000 insects and mites, 50,000 pathogens, and 8,000 weeds—and anywhere from 20 to 70 percent of them are non-native in any given region. These pests diminish global crop yields by 35 to 42 percent, a loss of $244 billion each year, despite the enormous labor and expense that go into spraying, tilling, and other control efforts. The United States loses 37 percent of its crop production to pest species.[7] In Australia, the cost of agricultural weeds alone is estimated at $1.7 billion per year, and the vast majority of those weeds are exotic species. Some $22 million of that total is what the Australian grain industry loses to wild oats—a truly cosmopolitan weed that contaminates and competes with crops in the field, reducing world wheat and barley harvests by an amount that could sustain 50 million people.[8] Hidden in the Australian tally also are surprising damages that most of us would not immediately pin on weeds—for example, the loss of 10 percent of the wool clipped from Australia's sheep.[9] One glance at fleeces embedded with the tick-like seeds of Bathurst burr and it's easy to see how such weeds destroy the commercial value of wool even as they hitch rides to infest new pastures.

Production losses, however, are not the end of the story. Pests also attack stored foodstuffs, destroying on average another 20 percent of world crops—a figure that ranges from less than 10 percent in the United States to more than 50 percent in some regions. Combine those production and postharvest losses and you will find that pest species, a large proportion of them introduced, destroy between 50 and 60 percent of global food production.[10]

It is no surprise that our croplands and pasturelands should be extraordinarily vulnerable to invasive weeds and pests, aggressive natives as well as introduced species. Intensive agriculture creates highly modified artificial systems with vast landscapes stripped of all vegetation for part of the year and then planted to a single, usually non-native, species. Many weeds excel at colonizing such regularly disturbed ground, and crop pests and pathogens can spread easily across a field without having to search for their targets amid a diverse community of plants. Furthermore, modern agriculture itself walks a tightrope by enthusiastically importing and promoting potentially lucrative new exotics even as it battles a growing cadre of unwanted invaders. (Many of these exotics considered valuable by agriculture and other sectors have become conservation nightmares, as we'll see in the next chapter.) Thus, the problems of agriculture provide a logical starting point for examining the economic and social toll exacted by invasive plants and animals throughout the human enterprise.

Literally thousands of opportunistic species of plants, animals, and microorganisms have turned aggressive in places far from their origins, and each affects a different sector of our economies or disrupts human life, health, and industry in a unique way. No sector is immune. Invaders not only damage crops and forage, contaminate harvested products, poison livestock, raise operating costs, and damage property values—they also cause power outages (brown tree snakes); clog water intake pipes (zebra mussels); lower water tables (tamarisks, eucalyptus, and pines); block navigation, narrow flood control and irrigation channels, and deplete oxygen (water hyacinth, hydrilla, and giant reed); decimate fisheries (green crabs and comb jellyfish); reduce pollination of crops (bee mites); cause structural damage (Formosan termites and shipworms); enhance urban and wildland fires (eucalyptus and melaleuca trees and buffelgrass); promote erosion and landslides (miconia and spotted knapweed); destroy timber and street trees (gypsy moths, Asian long-horned beetles, and brushtail possums); threaten human safety (fire ants and Africanized honeybees); cause allergic and toxic reactions (parthenium and giant hogweed); spread human diseases (rats and Asian tiger mosquitoes); and cause human diseases (West Nile virus, toxic red tide organisms, food-borne *Shigella*).

Because so many human activities are affected by invading species operating in so many diverse ways, no one has attempted to compute a global bill for damages. But Cornell University ecologist David Pimentel recently made a first cut at estimating the economic and environmental toll caused in the United States alone by the bad actors among the 50,000 or so non-native species that have been introduced into the country. He and his graduate students identified major environmental damage and losses totaling $137 billion per year.[11]

That total includes $34 billion in crop and forage losses and control costs for non-native weeds—and about three-fourths of weed species in U.S. croplands are exotic. In addition, control costs and damage from exotic insect pests (some 40 percent of all U.S. pests) of crops, forests, lawns, gardens, and golf courses came to another $18 billion. The costs of imported pathogens (two-thirds of the country's plant pathogens) in all those settings totaled almost $26 billion.

The rest of the $137 billion toll comes from a diverse array of invaders, and Pimentel's accounting is far from exhaustive. For example, his team calculated damages and control costs for only a few of the 5,000 alien plants now growing wild in natural ecosystems in the United States (purple loosestrife, melaleuca trees, and aquatic weeds such as hydrilla and water hyacinth); only 2 of 97 exotic birds (pigeons and starlings); 1 of the 53 amphibians and reptiles (brown tree snakes); 3 of 88 mollusks (zebra mussels, Asian clams, and shipworms); 6 mammals (mongooses, rats, pigs, horses, dogs, and cats); and only 3 human diseases not native to North America (AIDS, syphilis, and influenza). Clearly, a full list of destructive invasive organisms and their adverse effects would be much higher, although cost figures for most are not available.

Furthermore, the use of pesticides to control invasive species creates an array of additional, though indirect, environmental and social costs, including human health damage, livestock poisoning, destruction of natural enemies of pests, losses of crop pollinators, development of pesticide resistance, contamination of surface water and groundwater, fish and wildlife losses, and government regulatory costs. Although Pimentel did not include these in the $137 billion, he estimated in an earlier study that such indirect costs are double the direct costs of buying and applying the chemicals.[12]

Another large expense not included in the $137 billion is the cost of border inspection, quarantine, and pest and disease exclusion efforts. The U.S. Department of Agriculture (USDA) spent $237 million on such preventive measures in fiscal year 2001.[13]

A similar broad-scale look at invasive species in New Zealand combined quarantine costs, agricultural production losses, and the cost of activities such as clearing gorse and blackberry and controlling possum and rabbit numbers. That study, released by the New Zealand Conservation Authority, put the "directly quantifiable impact of pests on the New Zealand economy" at $354 million per year, about 1 percent of the nation's gross domestic product.[14] A recent study of five other countries by Pimentel's team put the annual costs associated with invasive species at $7 billion for South Africa, $12 billion for the United Kingdom, $13 billion for Australia, $50 billion for Brazil, and $116 billion for India.[15]

Such numbers, rough as they are, largely ignore damage that is not easy to price in hard currency, including the character and pleasure of place. Imagine, for instance, what might happen to the fall foliage tourist season in Vermont—as well as the syrup and timber industries—if the Asian longhorned beetle were to invade north from its beachhead in New York and chew its way through the state's revered maples. Or ignore for a moment the millions of dollars water users around the Great Lakes pay to scrape zebra mussels from their intake pipes, and consider beaches now paved with razor-sharp shells. What price can you put on the lost pleasure of walking barefoot on a beach? Likewise, the cost of pest control and structural damage does not begin to capture what is being lost as Formosan termites turn homes and historic buildings in New Orleans into sawdust.

Weeds and Well-Being

The cold immensity of numbers also cannot capture the full burden of invasions in parts of the world where food is grown for subsistence, not commodity. There, invading weeds, pests, and diseases threaten the livelihood and well-being of families already living on the margin.

A tall grass known as Cogon or alang-alang, for instance—long listed among the world's top ten agricultural weeds[16]—has spread across 60 million hectares of Asia, depleting the soil, building up massive dry fuel loads that carry fires into adjacent forests, preventing tree regeneration, and often increasing the burden of poverty.[17]

Likewise, parthenium, or wild carrot weed, from tropical America, which arrived in India as a contaminant in grain shipments, has spread across 5 million hectares, poisoning the soil and depressing yields of corn and traditional grain sorghums and pulses (jowar and arhar). Clearing the weed can be hazardous because contact with the plant or its pollen can cause dermatitis and allergic reactions.[18] Elsewhere, parthenium exploded across Egypt after a single sowing of contaminated grass seed from Texas in 1960.[19] In Ethiopia, parthenium appeared for the first time in the late 1980s near two food-aid distribution centers, apparently having been transported in contaminated grain delivered for famine relief. The weed is now widespread, is still expanding its range, and—ominously in that stricken land—has acquired a local name that translates to "no-crop."[20]

Lantana is yet another cosmopolitan scourge. A beautiful but uncontrollable ornamental shrub from the Americas, it has become one of the world's most noxious weeds. (See appendix A for a list of 100 of the world's worst invasive alien species, compiled by the World Conservation Union [IUCN].) In India alone, lantana has invaded millions of hectares of cropland and pastureland, forming dense thickets around villages, springing back almost as fast as it is cleared, quickly spreading into native forests, and forcing villagers to walk farther to harvest thatch and other wild-gathered resources or find grass for their cattle. Physicist Vandana Shiva, an outspoken defender of India's biological and cultural heritage, points out that some villagers faced with severe lantana invasions eventually abandon farming.[21]

In the Philippines, entire villages have relocated after pastures and forestry plantations became hopelessly infested by another American import, chromolaena, which is also one of the world's worst invaders. This vine-like shrub can grow as much as six meters high, generating dense masses of dry fuel loaded with oils and other highly flammable chemicals. From West Africa to India and the Philippines, chromolaena rapidly infests forest clearings, preventing tree regeneration and fueling damaging fires.[22]

Aid and development agencies unfortunately have been slow to recognize the full threat presented by weeds, despite the fact that invasive plants are increasingly confounding efforts to relieve rural poverty. "Invasive species are not just an agricultural issue, not just an environmental issue; they are a development issue," ecologist Jeff Waage of Imperial College at Wye points out. "To feed the world, we have to recover degraded land, irrigate, reforest. Yet all these critical goals are being held back by the presence of invaders that prevent recovery of degraded systems. In the Western Ghats range in India, poor people knock down the forests, then chromolaena and lantana move in, and you can't reforest until you do something about that."

Aquatic weeds rank among the worst invasive species problems that developing countries face, Waage says. The World Bank already devotes $45 million each year to battling aquatic weeds such as water hyacinth, another wayward ornamental from South America, in its irrigation and hydropower projects. Water hyacinth has exploded across Africa since the 1950s, from the Nile to the Niger river systems, from Lake Victoria to Lake Malawi, threatening rice cultivation, fisheries, navigation, hydroelectric power generation, tourism, and even human health (by providing habitat for snails and mosquitoes that serve as vectors for schistosomiasis, malaria, and other diseases).[23] Control costs for water hyacinth alone in just seven African countries have been estimated at $71 million per year.[24] Yet governments and aid agencies have only recently begun to act on the threat, helping to fund the release of biocontrol agents that could provide long-term weed suppression at little ongoing cost.

Even in the developed world, floating weeds are an expensive invasion problem. The United States spends an estimated $100 million per year to control water weeds such as hydrilla, watermilfoil, and water hyacinth that block irrigation and drainage canals, restrict navigation and other public use of waterways, lower water quality, and speed the buildup of sediment in reservoirs.[25]

In some parts of the world, weedy trees take an enormous economic and social toll. Many tree species have gone global in the past century, dispersed widely for commercial timber or agroforestry or to provide other valued services in treeless or tree-sparse regions of the world. Too often, however, these exotic trees have escaped cultivation and invaded watercourses, rangelands, and native forests.

In South Africa, more than 100 species of exotic woody plants, from lantana to pines, mesquites, black wattles, and other Australian acacias imported for timber, firewood, erosion control, or ornamental uses, have invaded the treeless bush. More than 8 percent of the country's land area is affected so far, including two-thirds of the remaining Cape fynbos (fine bush), a biologically rich ecological system that has given the gardening world some of its most spectacular and popular plants, from proteas to geraniums. Invading trees not only damage these native communities but also fuel frequent, devastating fires and usurp the country's limited water supplies. Flying over the mountainous Cape, you can see dense green thickets of wattles and pines marching up the watercourses, drying some streams completely and depleting others. And the water loss is growing as exotic forests spread at a rate of 5 percent per year. A massive effort to clear the country's watercourses of alien trees and shrubs is now under way.

In the United States, thickets of Asian tamarisk (saltcedar), imported for erosion control and windbreaks, have invaded watercourses in twenty-three states, including every major river drainage in the Southwest, replacing native willows, cottonwoods, and other riparian vegetation across nearly 500,000 hectares. These thirsty trees consume water 35 percent faster than the native plants, draw down water tables, dry up desert water holes, narrow river channels, exacerbate flooding, and provide poor habitat for native wildlife.[26] Indeed, tamarisks claim more of the arid West's precious water each year than Los Angeles, San Diego, and all the other cities of southern California combined.[27]

Finally, consider miconia, a striking ornamental tree from tropical America that was brought to a private botanical garden on Tahiti by a retired American professor in 1937. Once established there, miconia quickly ran amok in the native forests. Now "Polynesia's green cancer" dominates about 70 percent of the landscape. On steep slopes, solid stands of miconia are blamed for eliminating the soil-gripping understory, promoting landslides and increasing erosion.[28] The tree is still being spread to island groups throughout the Pacific Ocean. Prized by nurserymen for its oval leaves as much as a meter long with dark purple undersides, miconia was imported to Hawaii in the 1960s. In 1992, after state officials were alerted to its devastating environmental effects on Tahiti, miconia was declared a noxious weed, and it has become a veritable poster child for Hawaii's battle against invasive species.[29] State and federal agencies committed $1 million in 2000 to an ongoing and literally uphill battle to keep miconia from destabilizing mountainous watersheds or reaching national parks and protected areas on Maui and Hawaii.[30]

Blights and Beetles

Like weeds, the pests and diseases that afflict our crops and livestock are taking an ever-increasing toll on human livelihoods. In earlier times and in places where people had no money or means to fight them, pests and pathogens have changed the course of history. Think of the parasitic fungus known as potato blight: the famine it provoked in the mid-nineteenth century launched the global diaspora of the Irish people. Or consider the outbreak of rinderpest—"horned beast plague"—that swept from the Horn of Africa to the Cape of Good Hope in the 1890s, wiping out wildebeest and buffalo along with domestic cattle. The loss allowed ungrazed bush and thicket to overtake the open savannas, providing cover for tsetse flies, which became vectors for epidemics of human sleeping sickness and contributed to the devastation of pastoral societies.[31]

Today, thousands of insect species are on the move, from papaya fruit flies and Asian gypsy moths to cassava mealybugs, hibiscus mealybugs, white-spotted tussock moths, and various aphids, thrips, and whiteflies. Thanks to jets and container ships, insects can travel farther and faster than ever before. On Hawaii, an average of twenty new alien invertebrates become established each year, and half of these are known pests.[32] In California, statistics from the state Department of Food and Agriculture show that a new pest species arrives every sixty days.[33] And pest insects, especially leafhoppers and aphids, transmit viruses and other plant pathogens that add to the direct damage the insects themselves inflict. Compared with weeds, insect pests and pathogens can attack with devastating swiftness, one reason agricultural officials have traditionally devoted much greater resources to countering them. Indeed, the battle to protect Florida citrus groves from citrus canker—a wind-borne bacterial disease that originated in Asia and apparently entered the United States on infected fruit—has cost the state and federal governments more than $200 million since it was detected in the mid-1990s.[34]

A recent arrival to California could top that. The invader is a tiny leafhopper from the southeastern part of the country known by the impressive name of glassy-winged sharpshooter. More fearsome than the sharpshooter, though, is the bacterial disease of fruits and nuts it carries, which could devastate the state's multibillion-dollar wine and table grape industries. The USDA declared an agricultural emergency in California in June 2000 thanks to the sharpshooter and joined the state in putting up $40 million to begin combating the new bug.[35]

The sharpshooter joins a diverse roster of recent arrivals in California, from avocado thrips, silver-leaf whiteflies, and tomato bushy stunt virus to the Mediterranean fruit fly, or Medfly, perhaps the single most feared agricultural pest worldwide. From its beginnings in West Africa, the Medfly has spread throughout that continent and into Europe, the Middle East, Australia, South America, and Hawaii, attacking more than 250 types of fruits, vegetables, and nuts. Officials in California, which has been battling Medfly incursions since 1975, figure that if the Medfly, were to settle in permanently statewide, crop losses and postharvest treatment of produce to meet other countries' quarantine requirements could reach $1.6 billion per year.[36] Already on Hawaii, the agricultural sector is losing $300 million per year in potential revenue because infestations of Medflies, and several other alien fruit flies have caused other countries to ban Hawaiian exports. What's more, the relentless pace of new alien pest invasions is limiting the state's agricultural future. Production of sugarcane and pineapples, long the staples of Hawaii's plantation agriculture, has been declining sharply because of global competition, yet the search for profitable new tropical crops is being hampered by the arrival of diseases such as banana bunchy top and papaya ringspot virus.[37]

Recently, international trade in seeds, plants, and produce has also been spreading whiteflies and dozens of new strains of virulent geminiviruses that these flies carry. (These disease agents are called geminiviruses because they are formed of two twinned virus particles.) Until the early 1990s, plant disease specialists thought of geminiviruses as obscure agents that mostly caused trouble for bean producers in the American Tropics. Since then, however, new strains of geminiviruses have shown up in at least thirty-nine countries and inflicted major losses in a wide range of crops, from corn, cotton, and cassava in Africa to tomatoes in the Dominican Republic, Florida, and Central and South America.[38]

Even after crops survive the assaults of pests and pathogens in the field, of course, rodents and insects may attack them in storage. The larger grain borer, which gnaws its way through stored corn and cassava, was accidentally introduced into Africa in the late 1970s, apparently in food aid—another case in which a well-meaning but sloppy aid effort has compounded the region's difficulties in feeding itself. Waage points out that in Tanzania, the larger grain borer is called the "Scania bug" because it arrived with grain delivered by Swedish trucks, although the Swedes did not send the contaminated grain.

Pest invasions, like all bioinvasions, are a cumulative scourge. Once established, pests can rarely be eradicated, and thus the toll continues even in the unlikely event that no new invaders arrive. Some venerable pests, in fact, are getting a second wind. Take the potato blight fungus. Some 150 years after the Irish famine, the blight has experienced a worldwide resurgence thanks to the movement of virulent and chemical-resistant strains. Crop losses to potato blight have soared since the early 1980s, first in Europe and eventually in the Middle East, the Far East, Canada, and the United States.[39]

Dutch elm disease, too, swept across the Northern Hemisphere twice in the past century, each wave caused by a different strain of an Asian (not Dutch, despite its popular name) fungus carried by bark beetles. The first pandemic swept through Europe's elms early in the twentieth century and arrived in North America in 1930 aboard a shipment of elm veneer logs, devastating stately street trees up and down the East Coast. The second, more aggressive strain hit Europe in the 1940s. Now the two pathogens seem to be mingling under the elm bark. Clive Brasier of Britain's Forest Research agency expects to see a new hybrid form emerging any time.[40] "Dutch elm disease illustrates the potential for rapid evolution of introduced pathogens," he says. "It's a real, but until now largely unrecognized, danger in the movement of plant material."

Dutch elm disease is one of an array of introduced pests and pathogens that, along with logging and forest clearing, have transformed the very character of forests in the United States. Another infamous one is the European gypsy moth, which was imported to Massachusetts in the 1860s by a French astronomer who hoped to start a silk enterprise (that was, remember, the age of acclimatization fervor). The moths escaped into the forests, however, and have been stripping trees and advancing south and west ever since.[41] Another fungal disease, chestnut blight, arrived early in the twentieth century, followed by white pine blister rust fungus, which kills some of the country's most valuable timber pines. With blight already eliminating chestnuts from their dominance in the eastern forests, the subsequent arrival of blister rust on nursery stock from Germany prompted the United States to pass its first Plant Quarantine Act in 1912, requiring that imported plant materials be free of pests and diseases.[42]

Most of the quarantine efforts you will see at international borders today, in fact, were launched to prevent the introduction of plant pests and pathogens rather than potentially invasive plants. Yet new pests and diseases keep arriving, and old ones spread. A South African root-rot fungus related to the potato blight fungus (both of which belong to the genus *Phytophthora,* which means "plant destroyer") arrived in Western Australia about the same time the chestnut blight hit New York, and it has been destroying vast tracts of native eucalyptus woodlands ever since.[43] Now it has jumped to the temperate rain forests of western Tasmania.[44] Meanwhile, in California a ferocious killer of coastal live oaks known as "sudden oak death" ("an oak-tree Ebola virus," *Science* magazine dramatically dubbed it) has been identified as yet another species of *Phytophthora* fungus.[45]

And then there is the bug U.S. foresters fear may be the next gypsy moth. The tree-killing Asian long-horned beetle crawled out of a wooden crate from China and infested the urban forests of Chicago and New York sometime in the early 1990s. Thousands of infested trees have been felled to try to halt the beetle's spread. Economists predict damages of as much as $138 billion in lumber, maple sugar, and other industries if the beetle goes nationwide.[46]

As big as those economic losses sound, however, it's not the numbers that have catapulted the Asian long-horned beetle to notoriety. Like the zebra mussel before it, this beetle has consistently won television airtime because it's an in-your-face invader. Instead of setting up housekeeping in obscurity deep in the woods, this beetle walked out of its crate and into our parks and boulevard trees. Urbanites had to watch their neighborhoods being stripped of shade and, thus, pleasure and identity. Arrivals such as the Asian long-horned beetle—and every snake that drops out of a shipping container and onto the docks in snake-free New Zealand—bring the very personal threat of bioinvasions home to city dwellers in a way that whiteflies and knapweed seldom can.

The Cinderella Snail

Most of the pests of agriculture and forestry—at least the small ones we've been talking about—were not deliberately imported to the regions they now exploit. Of course, all those house sparrows, starlings, and songbirds set loose by acclimatization societies are an exception. So are the rabbits that now cost Australian agriculture more than $300 million per year.[47] Ditto the Australian brushtail possums, which outnumber people in New Zealand twenty to one, strip a containership load of foliage from the forests each night, and serve as a reservoir for bovine tuberculosis. But those were mostly nineteenth-century misjudgments. You might imagine that law and experience, not to mention the staggering sums spent in trying to control the damage from previous invasions, would make new blunders like that rare. Not rare enough, however. Take, for example, the Cinderella story of the golden apple snail, as inexplicably charmed a mollusk as you are likely to find.

This snail started its career grazing in the swamps and sluggish waters of Argentina or Paraguay, a seemingly unlikely candidate to inspire a profiteer. But twenty years ago, some Asian entrepreneurs saw virtues in the beast. It eats voraciously, breeds rapidly, and can grow, as its popular name suggests, as fat and round as an apple. Perhaps it could be farmed in backyard ponds in Asia and sold for food in local markets, someone thought, or even to the French for escargot. Boosters smuggled it first into Taiwan around 1979, and then into Japan. The snail acquired various common names that reflected people's hopes or promoters' hypes: mystery snail, miracle snail, golden snail. By 1983, about 500 snail businesses had opened throughout Japan. In the Philippines, agricultural officials embraced the hype and enthusiastically promoted "the golden miracle snail." Soon, others were slipping the snail into China, South Korea, Malaysia, Thailand, Indonesia, Vietnam, Laos, Papua New Guinea, and Hawaii.

No one seems to have checked around, however, to see whether Asian consumers even like the taste of apple snails. Most do not, as it happens. And French health authorities enforce stringent import standards on mollusks like the golden apple snail that harbor rat lungworms, flukes, and other potential human parasites. Worse yet, nobody considered what a snail that enjoys munching greenery in stagnant water would do if it got loose in rice paddies. In fact, had anyone done a bit of checking, they would have found that apple snails have been major pests in the irrigated rice fields of Surinam since the 1950s.

The commercial promise of apple snails quickly fizzled, like many a silkworm-, ostrich-, emu-, nutria-, or chinchilla-farming fad before it. Dumped or neglected, the snails soon found their way into the vital rice paddies and taro fields of Asia and many islands throughout the Pacific, where they spread rapidly. In the Philippines, the snail is now the most important pest of rice. A detailed analysis by Stanford University economist Rosamond Naylor concluded that snail-caused losses to rice farmers in the Philippines in 1990 alone ranged between $28 million and $45 million. On Hawaii, the snail has also become a major pest of taro, the traditional staple crop of native Hawaiians.[48]

In Peninsular Malaysia, apple snails have created yet another headache for officials trying to keep farmers in the field and maintain domestic rice production in the face of rapid urbanization and industrialization. Malaysian authorities first detected the golden apple snail on a prawn production farm in 1991, apparently brought in illegally by a farmer hoping to start a commercial operation. Within a year, the snails began to appear in aquaculture ponds, streams, shipments of water lettuce from Thailand, rice paddies, and even the ponds of the Bird Park, near the heart of Kuala Lumpur. It was there, a decade later, that I got my first glimpse of apple snails.

The Bird Park is a net-topped aviary on a hillside within a larger park called Lake Gardens. Somehow the open netting holds the air below it still and close, exaggerating the sweltering heat of Kuala Lumpur. For a bit of relief, you can retreat to a restaurant at the top of the slope that offers tepid air-conditioning. The building is traditional Malay wood and thatch construction, but there the tradition ends. Inside you will find a Carl's Jr. restaurant, part of a California-based hamburger chain. Behind the counter are teenage employees, the girls wearing discreet head scarves in the Muslim tradition. When they emerge from the counter to deliver Cokes and burgers, however, you will see, from the scarf down, the global teen uniform of T-shirts, baggy techno pants, and running shoes. This embrace of Western pop culture is the most noticeable invasion in the park, but it was not the one I wanted to see.

I left Carl's Jr. and worked my way to the bottom of the aviary, where flamingos and cattle egrets waded in a murky pond. Only a few undistinguished-looking snails were apparent that afternoon. But from a small bridge, I soon began to notice oblong dollops of pink foam everywhere, on rocks, tree trunks, bamboo canes, and the concrete walls of the pond. Female apple snails crawl out of the water at night and lay these bubblegum-pink wads of eggs, a new one every week or so, altogether perhaps 4,000 to 8,000 eggs per year. When these eggs hatch, the snails can reach enormous densities—as high as 150 snails per square meter in rice paddies in the Philippines and 130 per square meter in the taro patches of Hawaii. So far, Malaysia has managed to keep their densities down to 10 to 20 per square meter.

Earlier, I had visited with Asna Booty Othman in an office not far from the park. Part of her job in the Malaysian Department of Agriculture is to direct surveillance efforts for golden apple snails and devise programs to keep their numbers in check. Rice in Malaysia is produced on small, heavily subsidized farms, two-thirds of them less than one hectare in size. The nation's food security goals require that its own farmers produce at least two-thirds of the rice the country needs, yet keeping people on the farm amid an urban economic boom is no easy task.[49] Pests such as the apple snail make farm life even less attractive. Ms. Othman promotes a variety of nonchemical snail-control measures such as filters on water inlets and outlets to keep snails from moving between paddies. Farmers are encouraged to push stakes into the water at the edges of their fields so that snails will crawl up them to lay eggs. Every day, farmers or their children can check for pink egg masses and then turn the stakes over and jam the eggs into the mud, killing them. Some measures can even enhance farm revenues. Her agency supplies farmers with snail-eating catfish and ducklings to rear in the paddies and sell. Such efforts have helped to keep the still-spreading snails at low enough densities that they have not depressed rice yields. But pest control is something farmers must do regularly, forever.

Ms. Othman sighed and looked somewhat exasperated when I asked her how committed farmers are to these efforts. Success, ironically, works against vigilance, and farmers don't see the point of working so hard when snails are not hurting their harvest. "It doesn't cost me money; why should I waste my time?" she intoned in the voice of a complacent farmer. "But we know this is the only way to prevent them from damaging crops. We know that the snails could explode if we don't stay on top of it."

Looking at the sheer number of pink egg masses in the Bird Park pond, where resident birds seemed as disinclined as people to feast on them, I realized that the fairy-tale fortunes of this unlikely creature are still on the rise. Indeed, if anyone needs convincing that people often don't learn from the mistakes of others, consider that in 1995, boosters transported apple snails into yet another country, Cambodia, for culturing in backyard ponds. Within a year, the snails had escaped, and they are now spreading in Cambodia's rice paddies.[50] And that's not the end of it. I later learned from a representative of IUCN in Sri Lanka, Channa Bambaradeniya, that ornamental fish dealers there still give their customers free apple snails as a bonus with purchases of tropical fish.

Whereas the snail seems to have benefited very few, some introduced species do enrich one segment of society while disrupting others. Too often, the costs are borne by the public at large while the profits go to a relatively small group of investors. The large, predatory Nile perch, for example, a fish introduced several decades ago into Africa's Lake Victoria, has become a textbook horror to ecologists because it has helped to drive hundreds of species of small native cichlid fish to apparent extinction.[51] Yet the perch has also dramatically expanded the tonnage of fish caught from the lake and brought prosperity to those with the capital to invest in trawlers and the dozens of factories that now dot the Ugandan and Kenyan lakefronts to process perch fillets for export to Europe. That trade also has brought welcome infusions of foreign currency into both countries. Yet the export trade has driven the price of fish—the cheapest source of animal protein in the region—beyond the means of most local people and cut per capita fish consumption by one-half.[52] On the outskirts of Kisumu, Kenya, I got a vivid snapshot of the collapse of village economies that once depended on men in plank canoes seining for cichlids and other small fish. Many of the men now work in the perch-processing plants. The women, who used to buy fish at the landing beaches to dry and sell in their communities, now lined the edges of the road outside this city, frying and selling perch scraps from the processing plants to passersby.[53]

Scourges Old and New

The litany of threats to our food and fiber supply includes both ancient and emerging diseases of our domestic animals. Worldwide, we maintain some 20 billion domestic animals, 9 billion of them in the United States (all U.S. livestock species are introduced), and we use nearly one-third of the planet's land area to feed them.[54] All their old scourges are with us still. Rinderpest breaks out periodically in parts of East Africa where war or famine interferes with regular vaccination of cattle. And new strains of old diseases keep moving around.

In September 2000, while the Global Invasive Species Programme hosted a meeting in Cape Town, South Africa, to discuss strategies for halting the threat of bioinvasions, a new strain of viral foot-and-mouth disease broke out among pigs in the country's KwaZulu-Natal Province. News reports tagged the likely source as human food wastes that local pig farmers had purchased illegally for swill from an Asian ship docked at Durban. After intensive culling of pigs failed to stem the outbreak, officials in the province began vaccinating more than a million domestic animals and took steps to prevent infection of buffalo, giraffes, and other vulnerable cloven-hoofed wildlife in the region's game reserves.[55] During that year, the identical strain of foot-and-mouth disease struck Japan, Russia, China, South Korea, Taiwan, Mongolia, half a dozen other African countries, and five countries in South America. Then, in February 2001, it began a rampage through the United Kingdom and on to parts of Europe.[56]

Shortly after the foot-and-mouth outbreak in South Africa, the Food and Agriculture Organization of the United Nations (FAO) warned that a number of other livestock diseases were on the move and turning up in new places. Rift Valley fever, for example, made its first known appearance outside Africa that year, not only striking down livestock but also killing 100 people in Yemen. Bluetongue, a viral disease of sheep, broke out for the first time in Bulgaria and Sardinia in 2000, and the United Kingdom confirmed an outbreak of classical swine fever in pigs even as it was battling foot-and-mouth disease.

"All these cases illustrate that trans-boundary animal diseases continue to be a real threat," FAO animal health expert Mark Rweyemamu stated. "No country can claim to be safe from these diseases, and in an increasingly globalized world, veterinary surveillance systems and services are vital to detect these diseases early enough to prepare contingency plants to contain outbreaks."[57]

The figure of 20 billion domestic animals, staggering as it is, fails to capture the growing multitudes of fish and shellfish now raised in tanks, ponds, and net pens around the world for human consumption. Some of today's most difficult animal disease problems involve the movement of pathogens with aquaculture stock, including various plagues, parasites, and viral diseases of crayfish, oysters, shrimp, and salmon. In 2001, infectious salmon anemia, a fatal fish disease that has been moving around the world, began spreading through Maine's fish farms for the first time, causing officials to destroy $12 million worth of salmon to try to stop it.[58] Disease outbreaks such as this not only cause economic losses for industry but also can spread pathogens to wild fish stocks in the region.[59] Conservation officials worry that salmon anemia may spread to already dwindling populations of wild Atlantic salmon. There are certainly precedents. A protozoan parasite that impairs nervous system development in trout and causes a disorder called whirling disease, for example, was introduced into the United States in the 1950s, apparently in infected European brown trout. Release or escape of infected hatchery fish has spread the disease throughout prized fly-fishing streams in eleven western states, devastating some wild trout populations.[60]

Another valuable semidomesticated animal not tallied among the 20 billion is the honeybee, a European native now essential to the pollination, fruiting, and seed development of at least thirty U.S. crops, from almonds to apricots. Invasions of tracheal mites from Mexico in 1984 and *Varroa* mites in 1987 had wiped out as much as 85 percent of feral honeybees in the United States and at least half of the bees in managed colonies by the mid-1990s.[61] In 1997, *Varroa* mites showed up in South Africa, near the Cape Town harbor, and they have since spread countrywide.[62] To protect their own beekeeping industries, New Zealand and Australia enforce strict bans against bringing in bees or honey products. Yet in April 2000, as a New Zealand quarantine official was showing me around some of the strictest border controls in the world, the *New Zealand Herald* reported that *Varroa* mites had been found in a hive in southern Auckland. By July, further surveys had determined the mites were too widespread to eradicate. Beekeepers have had to adjust their practices, including using pesticides in their hives for the first time. Without a government mite control effort, which was expected to cost $17 million in the first two years, total economic loss to the country was projected at $170 million to $380 million.[63]

In the hot, dry summer of 1999, two seemingly unrelated waves of illness appeared in New York City. First, dozens of sick and dying crows started showing up in veterinary offices and on the streets. It was not long before a wide assortment of captive birds, from Chilean flamingos to a bald eagle, began dying at the Bronx Zoo, and dozens of species of wild native birds besides crows—red-tailed hawks, sandhill cranes, mallards, robins, blue jays—were found dying from New Jersey to Connecticut.

The second wave of illness appeared in late summer, when people began showing up in area hospitals with brain inflammations characteristic of viral encephalitis. Eventually, more than sixty were stricken, and seven died. The Centers for Disease Control (CDC) at first attributed the human cases to St. Louis encephalitis, a mosquito-borne disease that first broke out in the United States in 1933 and is now endemic in the southeastern part of the country.

Not until fall, however, when pathologists from the zoo teamed up with colleagues from the U.S. Army Medical Research Institute of Infectious Diseases and the CDC to pin down the deadly pathogen in the birds, did the true identity of the killer of both birds and people begin to emerge. It turned out to be West Nile virus, a mosquito-borne pathogen new to the Americas.[64]

"The introduction of a foreign insect-borne virus never before seen in the Western Hemisphere is a public health threat unprecedented in modern times," Durland Fish later told a United States Senate hearing. "It is reminiscent of the introduction of yellow fever and bubonic plague in past centuries." Fish, a medical entomologist at Yale University School of Medicine, told the senators that despite warnings from the scientific community, the nation had "left [its] guard down against the threat of new diseases."[65]

The West Nile virus was first identified in northern Uganda in 1937. Since then, it has caused disease outbreaks in people, birds, horses, and a few other species in Africa, western Asia, Europe, and the Middle East. As it turns out, at the same time West Nile was making its debut in New York, the virus was also flaring up in the city of Volgograd in southern Russia, where it killed more than forty people.[66]

For anyone concerned with preventing bioinvasions, a central question is how the virus got to a new continent. Research teams found that viral genetic material from a goose that died of West Nile encephalitis in Israel a year earlier matched closely with viral material taken from crows, mosquitoes, a Chilean flamingo, and human victims of the outbreak in New York.[67] So the virus very likely traveled from Israel by airplane in an infected person, a bird, or a mosquito.

"My best guess is that it came in on a mosquito," Fish told me, probably *Culex pipiens,* a bird-biting urban mosquito as cosmopolitan in its distribution as the rat. An infected bird smuggled into the country, bypassing quarantine, or an infected person arriving in New York would have to have spent time outdoors soon afterward and been bitten by local mosquitoes in order to pass on the infection—scenarios that Fish found less plausible.

"There's plenty of evidence that mosquitoes can survive airplane rides. In fact, there's a phenomenon known as 'airport malaria' where mosquitoes infected with malaria parasites get loose from an airplane coming from an endemic area and cause human cases of malaria around airports. There's plenty of air traffic between Israel and New York City. So a mosquito survives the airplane ride, gets out and bites a bird, and infects the bird with virus. Then other mosquitoes that feed on that bird also get infected and bite other birds, and the virus gets established in the bird population." Infected birds serve as a reservoir for West Nile virus, and a mosquito that draws a blood meal from a bird can spread the virus to the next bird—or person—that it bites.

"Durland's guess is as good as any, but I don't think we'll ever know how it got here," John Edman, director of the Center for Vector-Borne Disease Research at the University of California, Davis, told me later. Disease introductions, he believes, are "primarily a game of chance. The more airplanes you have flying back and forth every day from fairly exotic places to other places, the greater the chances of introducing diseases. I think there are introductions all the time, but most just don't take. I can't tell you how many times I've seen mosquitoes flying around on airplanes I've been on." Most people who travel internationally have occasionally endured having flight attendants fumigate the cabin on arrival to meet some country's quarantine requirements—a practice we passengers should probably appreciate rather than protest.

How far will the West Nile virus spread? By late 2001, infected birds, horses, and people had been found from Ontario, Canada, south to Florida, and the virus had already crossed the Mississippi River in its rapid advance westward.[68]

"It's a virus we're going to have to live with; it's entrenched," said Robert McLean, director of the U.S. Geological Survey's National Wildlife Health Center. More than a dozen species of mosquito have been found to be infected, although not all are competent to transmit the virus. Two that are, however, are aggressive recent invaders: *Aedes albopictus,* the Asian tiger mosquito, and its cousin *Aedes japonicus.*

"There's not any place in the United States except the deserts where you don't have a mosquito that can vector this virus," Edman said. "I suspect we're going to see it moving across the entire country over the next few years."

As late as the 1980s, many scientists expressed confidence that vaccines, antibiotics, and modern sanitation had all but conquered infectious disease. The new frontier was a molecular-level assault on disease-causing flaws in our own genetic material. Yet at the millennium, infectious diseases remain the leading cause of death worldwide and the third leading cause of death in the United States.[69] New ones such as AIDS are emerging at an alarming rate even as old scourges such as tuberculosis and malaria develop resistance to drugs and flare out of control. In the 1990s, after a long decline in public health efforts against mosquitoes, yellow fever reemerged in Peru and dengue fever flared throughout Latin America.[70]

While I was poking around the Kuala Lumpur Bird Park looking for apple snails, a previously unknown encephalitis virus—eventually named the Nipah virus—was raging among pig farmers and farmworkers farther south on the Malay Peninsula. Nipah encephalitis, it turned out, is spread not by mosquitoes but by contact with infected tissue. Yet like West Nile encephalitis, which hit a few months later in New York, Nipah encephalitis was misidentified at first, in this case as mosquito-borne Japanese encephalitis—a mistake that greatly hampered efforts to stop its spread. Agencies were still spraying to kill mosquitoes even as workers slaughtered a million pigs without taking precautions against direct infection. By the time the 1999 outbreak was quelled, 106 people had died, and many survivors were left with brain damage. Fruit bats are the suspected natural host for the Nipah virus, although scientists are still investigating how the disease emerged in domestic pigs.[71]

Animal-borne diseases are called zoonoses, and fully half of the viruses, bacteria, fungi, protozoa, worms, and other disease-causing organisms known to plague humans arose from zoonotic diseases. Among the 156 infectious diseases such as Nipah encephalitis that are known to be emerging today, three-fourths are from animals.[72] The more we move animals, disease vectors, and people around, or disturb or intrude on wild habitat, the greater the rate of new disease emergence and spread.[73]

For instance, the continual movement of indigenous people and goods in the Torres Strait Islands, a special treaty zone in the narrow passage between Queensland, Australia, and Papua New Guinea, presents a high risk for the transfer of pests and diseases between those two lands. So do the monsoon winds that annually carry fruit flies, mosquitoes, and other introduced pests south across the strait. Japanese encephalitis, caused by a mosquito-borne virus now widespread in Papua New Guinea, resulted in two deaths in 1995 in these closely watched islands on Australia's "Top End" and in 1998 reached the mainland for the first time. Several dengue fever outbreaks on the islands and in northern Queensland during the 1990s were attributed to the same source.[74]

Worldwide, however, indigenous travelers in small boats represent a tiny fraction of the people on the move each day. Most of us in the major cities of the Americas, Europe, Asia, Africa, and Australia can fly to almost any other part of the globe in thirty hours or so. Two million people each day cross international borders separating countries. Every week, a million people travel between industrialized countries and the developing world. Between 1990 and 1995, the number of refugees and people displaced from their homes by economic or environmental crises or war increased from 30 million to 48 million.[75] Thus, an infectious disease outbreak anywhere in the world should concern us all.

The global movement of produce is also helping to drive a rapid increase in food-borne bacterial infections in industrialized countries.[76] And the ships that carry most of this produce also move millions of tons of ballast water from port to port, water teeming with microorganisms that include human pathogens. A recent sampling of ships arriving in Chesapeake Bay from foreign ports, for instance, found cholera organisms among the plankton samples taken from all of them.[77] There is evidence, in fact, that the return of epidemic cholera to Peru in 1991—the first outbreak in the Western Hemisphere in a century—was caused by cholera bacteria delivered from Asia in ballast water.[78]

The staggering economic damage caused by aggressive invasive species runs into hundreds of billions of dollars per year worldwide and affects nearly every aspect of human life, health, industry, and economy. Yet many people complacently reason that this musical chairs game with weeds, pests, pathogens, and nuisance animals has gone on for so long that there is little point in trying to halt it now. That complacency is dangerous, for the problem of invasions is accelerating. New nuisance species are arriving continually, compounding the threats to the health and well-being of people almost everywhere. And there are many more invasions to come. Only a fraction of the species that can move have yet done so. By one estimate, the 4,000 weeds that have been interchanged between regions so far represent less than 15 percent of the potentially invasive plants from the global stock.[79] Every country knows of species it does not want, and there are thousands of other problematic species out there, not on any warning list, that most would undoubtedly regret acquiring. The economic and social stakes are too high and the threats to our natural heritage too severe to continue letting living organisms be moved around the world by whim or accident.

FOUR Elbowing Out the Natives

"From prehistory to the present time, the mindless horsemen of the environmental apocalypse have been overkill, habitat destruction, introduction of animals such as rats and goats, and diseases carried by these exotic animals. . . . Each agent strengthens the others in a tightening net of destruction."

—Edward O. Wilson, *The Diversity of Life,* 1992

"An ecological explosion means the enormous increase in numbers of some kind of living organism—it may be an infectious virus like influenza, or a bacterium like bubonic plague, or a fungus like that of the potato disease, a green plant like the prickly pear, or an animal like the gray squirrel. I use the word 'explosion' deliberately, because it means the bursting out from control of forces that were previously held in restraint by other forces."

—Charles Elton, *The Ecology of Invasions by Animals and Plants,* 1958

The moisture-laden trade winds that sweep into the Hawaiian Islands from the northeast can foil the usual pattern of clouds around mountains, creating an inversion layer that prevents the damp air from rising up the volcanic slopes. Thus, on Maui and on Hawaii—the Big Island, as it is commonly known—the highest places are often driest. For someone living in the northern Rocky Mountains, where clouds often envelop the peaks, it was surprising one May morning to drive out of Hilo on the Big Island in a gray drizzle and emerge—nearly fifty kilometers later and more than a kilometer higher—into sunshine on the rim of the Kilauea caldera. The only haze in the sky came from sulfurous fumes escaping Kilauea's craters and vents.

Some 1.8 million visitors drive up there each year to marvel at two of the earth's most active volcanoes. Most tourists are content to circle the Kilauea caldera and perhaps drop down to the coast on Chain of Craters Road, where lava has been spilling from the southeastern flank of Kilauea and into the sea since 1983. Less than 1 percent of visitors ever leave the roads, venturing into the forests or deserts or hiking up to the windswept alpine communities of Mauna Loa, more than four kilometers above the sea, to glimpse the surprisingly varied life of this harsh land. Instead, most are back on the beaches along the Kona coast by evening, enjoying the lush tropical vegetation and bird life of hotel grounds and golf courses and feasting on roast pig at hotel luaus.

Without some training, few of us would recognize the flora and fauna of those delightful settings as pure artifice. Yet most of the birds of the coastal lowlands were brought there from Asia, Europe, or South America. Most of the spectacular flowers and greenery were introduced by humans, too, including the plumeria leis tourists wear over their Hawaiian shirts. Perhaps few would care, if told. After all, little that blooms in Miami or London or Sydney is native to those places. We are comfortable with landscaped environments, delighted by gardens assembled from pleasing bits and pieces of the world's bounty. That's why I surround my Montana home with Asian lilacs, peonies, roses, and poppies. Our familiarity with the exotic makes it hard to grasp how profoundly Hawaii's native plants and animals are threatened, not by the molten violence that pours from the earth but by some of those very ornamental plants, exotic birds, and barnyard animals that help create the Hollywood image of paradise that draws us there. It was an attempt to understand the threat behind the lowland façade that took me up to Kilauea that day.

Among people who work to conserve the earth's living heritage, Hawaii is considered a frontline battleground, the extinction capital of the United States. One-third of all the endangered plants and birds in the country live only there, although the state represents less than 0.2 percent of the U.S.

land area. These islands rose from the sea more than 4,000 kilometers from the nearest continent, and more than 90 percent of their native plants and animals are originals—endemics—unique offshoots of the few hundred seeds, snails, birds, and insects that successfully rode the winds or currents out there during the few million years since the main islands emerged, then adapted to fill a wide array of habitats, from beaches to snowy peaks. No snakes, ants, or wasps made it on their own, and prior to *Homo sapiens,* only one land mammal—a bat.

In the 1,600 years since humans invaded these islands, hunting and land clearing by Polynesian and European settlers alike have taken a massive toll on the long-isolated life of Hawaii. But that toll has been matched or exceeded by the damage done by rats, dogs, pigs, cattle, goats, sheep, cats, mongooses, and thousands of exotic plants and insects that people have introduced, intentionally or accidentally. Today, exotic invaders present the greatest threat to the survival of the remaining native plants and animals of Hawaii as well as to the character and functioning of its landscapes.[1] The situation is so dire that some disheartened conservationists speak of their work in Hawaii as "hospice ecology," merely easing the inevitable decline of faltering species into extinction.

Tim Tunison is chief of resources management at Hawaii Volcanoes National Park, and he knows the numbers better than most. Eighty percent of his budget goes toward combating alien pests, a percentage that's typical for protected lands on other islands in the Hawaiian chain as well.[2] Yet he seems constitutionally unable to remain glum. That May morning, as I stepped over a thigh-high pig-proof fence and followed him into a patch of ʻOlaʻa Forest, Tunison sounded more like a man in charge of a recovery ward than of a hospice. Most of his patients remain grievously impaired, but he takes delight in any progress.

"This place is changing so much, I come in here and I can't find where I used to take people just a couple of years ago," he said with obvious satisfaction as we threaded our way between tall native tree ferns. "We fenced this section in 1991 and got the pigs out soon after. My crew gets really stoked working in here because all of these tree ferns have come back. It used to be such a wasteland."

We paused under a canopy of tree ferns and older ʻohiʻa trees, the red-blossomed relatives of bottlebrush that dominate Hawaii's forests. At our feet, Tunison pointed out native peperomia, half a dozen types of fern, an endemic lobelia shrub called *Clermontia,* and a nettleless nettle—one of the many plants that dropped their defenses after spending eons in a land with no mammals to munch them.

ʻOlaʻa Forest is a 4,000-hectare rain-forest tract designated as wilderness and disconnected from the rest of the park. The short drive from the park gate to ʻOlaʻa cuts through cleared fields and pastures, remnants of the now-declining sugarcane, cattle, and pineapple economy that long ago stripped away most of the low-elevation forests on the islands. Until this section of ʻOlaʻa was fenced, feral pigs rototilled the ground in their search for exotic earthworms and other edibles, destroying the native shrub and fern layer, killing native tree seedlings, and depositing with their dung the seeds of lovely but highly invasive ornamental plants such as kahili ginger, strawberry guava trees, flowering passion fruit vines known as banana poka, and Himalayan raspberries.[3]

The park itself had been fenced and invasive feral goats eliminated by hunting by the late 1970s, but pigs present a more difficult challenge. There is little hope of eliminating pigs throughout the rugged terrain of the park, especially given that pig hunting remains a fiercely valued tradition among native Hawaiians, and state agencies promote thriving pig populations on many lands outside the park. Instead, Tunison and his staff focus their efforts on defending only the "most intact, most representative, and biologically richest sites." They erect hog-wire fences—pigs can dig under or push through a merely goat-proof fence—around these sites, hunt out the pigs, and pull and spray the weeds. We were standing in one of the chosen sites, a "special ecological area" in park parlance.

To one side, I spotted a fallen tree-fern trunk with the girth of a telephone pole, hollowed out like a toy canoe in the making. "That's been carved up, troughed, by a pig," Tunison pointed out. His crew inspects the fences once each month, but a pig gets through every year or so. "There's a big, starchy center to the plant that pigs love. The outside is fibrous roots, but the inside is like a cream-colored refrigerator full of food. Those cavities they make fill with rainwater, and that's where *Culex* mosquitoes breed, the ones that carry avian malaria."

What Tunison described in shorthand is an example of the insidious and unpredictable synergy that often occurs among invaders. Pigs, nourished by alien earthworms and exotic guavas as well as native tree-fern innards, not only damage the forest directly, create openings for invasive plants, and spread the seeds of those plants but also threaten forest birds by scooping out breeding pools for the introduced southern house mosquito, which carries a bird malaria parasite brought to the islands by imported birds. Thus, the effects of multiple invaders combine and enhance one another to create what ecologist Dan Simberloff calls a juggernaut that devastates native plant and animal communities.

"You can hear our native crickets, but no birds in here," Tunison continued. "It's an avian desert. There were plenty of birds 200 years ago, but now this is the mosquito and malaria belt. If you go up to the forests at 5,500 feet [1,675 meters], above where mosquitoes live, you'd start running into birds again." Even here at 1,200 meters, if flowering ʻohiʻa trees were dense enough to provide abundant nectar, we might see the fairly malaria-resistant ʻapapane, or Hawaiian honeycreeper. But the ʻohiʻa have not yet started to regenerate in their fenced refuge.

"So it's pretty bleak in terms of birds," he finished. "About half the birds that were around when the Polynesians first arrived are gone, and about half those that remain are endangered species."

We walked across the road and stepped over another hog-wire fence into a section of ʻOlaʻa Forest that has been pig-free for almost twenty years. It was a veritable jungle of tree ferns, close and damp in the deep shade below the lacy frond canopy. As we made our way along the spongy ground, we stepped across numerous fern logs lying untroughed, carpeted with moss, small ferns, and ʻohiʻa seedlings. "Nurse logs," scientists call them. This patch of forest is largely native now, I learned, but it's not what it was pre-pig.

"It's tree-fern heaven," Tunison explained. "One-third of the vascular plant species in here are ferns, which is a very high proportion. And probably 80 percent of the biomass in here is fern biomass. So this is pretty darn unusual. I think it's a little bit artificial." He said this almost like a question because there are very few undamaged rain forests to use for reference. "I think it's a response to all the years of pig activity and then removal of the pigs." The tree ferns grew in quickly and now they dominate, perhaps because of the thick layer of litter they shed. "Their fronds fall down and crush and bury seedlings. So it's probably kind of a skewed composition. I think what's missing is a native shrub layer."

Eventually, the ʻohiʻa and other native trees and shrubs may regain prominent places in this community. Tunison also hopes to propagate and replant native orchids and other rare and threatened understory species that have been depleted or eliminated by invading pigs and weeds. But mostly he plans to keep the fences intact, knock down the weeds periodically, and let what is left of the native community sort itself out. There are too many other critically stricken patients on his ward, many that require hands-on care to survive, not just a safe haven. And a bevy of ills besides pigs threatens them. For starters, there are fire-promoting grasses invading dry forests; faya, or fire, trees advancing into mesic areas and colonizing young lava flows; formerly innocuous roadside weeds such as common mullein and knotweed suddenly spreading aggressively; yellow jacket wasps and ants threatening native pollinators; mouflon sheep jumping the fences on Mauna Loa to threaten recovery of the park's agave-like silversword plants; and rats, feral cats, and mongooses plundering the nests of endangered native sea turtles, dark-rumped petrels, and Hawaiian geese (nene).

"Most of us out here in Hawaii consider it to be a test case of our ability to stem the tide of invasives and restore what is left," Tunison told me. "The way invasions are going, continental areas will have all of Hawaii's problems in twenty-five, fifty, one hundred years. If we fail out here, the continental areas will probably fail, too."

Biodiversity Loss on the New Pangaea

Hawaii represents an extreme, a region hit particularly hard by the range of problems that exotic invaders pose for native species and natural areas worldwide. Yet the state's problems differ only in degree. Today there are few, if any, places so remote or well monitored that they remain unthreatened by the biological baggage introduced by humans.

Penguins in the Antarctic carry evidence of exposure to potentially fatal domestic poultry viruses, apparently the legacy of tainted chicken dinners eaten by tourists or researchers visiting the nesting colonies. Antarctic seals in the South Shetland Islands show exposure to the cattle disease brucellosis. Endangered sea turtles find it impossible to scoop out nests on some Florida beaches amid the dense roots of Australian casuarina trees that have taken over the sand. Western snowy plovers search in vain for open stretches of sand in which to nest on West Coast beaches now carpeted with European grasses. If you hike up to the alpine tree line in the remote mountains of Glacier National Park where Montana meets Canada, you will find yourself, as I have, amid a silver ghost forest of whitebark pine trees, killed by a Eurasian blister rust fungus brought to North America on overseas nursery stock. At high elevations in Great Smoky Mountains National Park in North Carolina and Tennessee, you will find similar ghost stands of Fraser firs where European aphids known as balsam wooly adelgids have finished the destruction begun by acid rain. Below these firs, the ground may be rototilled by pigs just as on the forest floor in ʻOlaʻa Forest. In Africa, Mexican poppies are advancing along the roadsides in Tanzania's Ngorongoro Crater, and more than one-third of the lions in neighboring Serengeti National Park have died of canine distemper brought in by domestic dogs. In the mountains of eastern Europe, wolves are threatened by breeding with feral dogs. In the beech forests of New Zealand, exotic wasps usurp the honeydew that once sustained threatened forest parrots. North American beavers gnaw at the forests of Tierra del Fuego while South American nutrias (coypu) devastate wetlands as far away as Spain and Kenya.

As these few examples indicate, the damage caused by invading plants and animals takes many forms. Some invaders directly threaten the survival of individual species by displacing, disrupting, consuming, mating with, or spreading disease among their populations. The most damaging invaders operate on a larger scale, transforming the character of habitats and landscapes—turning open beaches to forest, forest to grassland, or wetlands to dry lands. In doing so, they inevitably alter ecological processes such as the cleansing and cycling of water, the renewal of soils, the natural checks on pest and disease outbreaks, and the frequency and intensity of wildfires. When invaders with this power arrive, they change the terms of life for every other species in the community. And that's the last thing most creatures need.

That the lavish array of life on the earth is under siege is no secret. The trend is clear, though the details of this biodiversity crisis remain fuzzy. Just how rapidly we are losing species, populations, and even whole habitat types such as rain forests and wetlands remains a subject of fierce contention. To begin with, we have no firm count of how many species of plants, animals, and microorganisms live on the earth, even to the nearest tens of millions. Formal listings of named species put the number at about 1.4 million, but virtually every time specialists explore some previously neglected spot on land or sea, they discover new species. The 1995 Global Biodiversity Assessment released by the United Nations Environment Programme (UNEP) estimated that there are between 7 million and 20 million species on the earth and suggested 13 million to 14 million as a good working estimate.[4] Biologists Paul Ehrlich of Stanford University and Edward O. Wilson of Harvard University estimate there may be as many as 100 million species on the earth, many of them endemic or uniquely local species such as the birds, snails, and plants that evolved in isolation on Hawaii. Because the Tropics are so rich in endemics, Ehrlich and Wilson speculate that one-quarter or more of the species on the earth could be eliminated in the first half of the twenty-first century if tropical deforestation continues at its current pace.[5]

Deforestation, bulldozing, plowing, draining, dredging, trawling, damming, polluting, paving, and other activities that destroy and fragment habitat remain the biggest threats to our natural heritage. Humans have already transformed one-third to one-half of the land surface of the earth in these ways.[6] In a world where so many species are already hard-pressed to survive in shrinking and degraded habitats, invaders too often provide the final shove to the brink. Indeed, loosing a pig or goat into a forest can be as destructive as sending in a bulldozer. Smothering vines and fire-promoting grasses are slower and less dramatic than pigs—and too subtle for most of us without botanical training to notice at first amid a pleasing blur of green—but ultimately just as damaging. "Extinction by habitat destruction is like death in an automobile accident: easy to see and assess,"

Wilson wrote. "Extinction by the invasion of exotic species is like death by disease: gradual, insidious, requiring scientific methods to diagnose."[7]

The diagnoses that we have so far are not encouraging. The World Conservation Union (IUCN) 1996 Red List concluded that 11 percent of all birds and 25 percent of known mammals are threatened with extinction.[8] Invaders have a significant hand in this threat: UNEP's Global Biodiversity Assessment found that nearly 20 percent of the world's endangered animals are imperiled in some way by non-native invaders.[9] In a 1997 Red List report, IUCN warned that more than one in ten of the world's plants face extinction, the two chief threats to them being land clearing and exotic invaders.[10]

Data from the United States confirm the threat invaders pose: a 1998 study found that 85 percent of the country's imperiled species were threatened by habitat loss or degradation and 49 percent were either preyed upon by alien invaders or forced to compete with aliens for resources. That latter figure included two-thirds of birds, more than half of fishes and plants, and one-third of imperiled butterflies and reptiles, all under threat from invading species.

Those numbers serve to confirm what conservation biologists have long suspected: invasions now rank second only to habitat loss as threats to the survival of native species in many regions. And the authors of the U.S. study added a sobering note: "The discovery that nearly half of the imperiled species in the United States are threatened by alien species—combined with the growing numbers of alien species—suggests that this particular threat may be far more serious than many people have heretofore believed."[11]

To make matters worse, habitat degradation and invasions are not independent threats. Not only do many invaders such as pigs create novel disturbances, but also many others—including the ginger and guava the pigs spread in their dung—specialize in colonizing disturbed sites. To use a medical analogy, "wounded" ground is quickly "infected" with weedy invaders. Conservation biologist Michael Soulé points out that as the expanse of undisturbed field and forest, beach and marsh continues to shrink, invasive exotics are likely to become the number one "engines of ecological disintegration."[12]

Consider, then, the most extreme outcome of all this. How extensive could the loss of species be if we allowed modern trade and transport to commingle freely all of the earth's creatures, effectively reuniting the isolated landmasses that were joined 200 million years ago in the single supercontinent of Pangaea? Ecologist James Brown of the University of New Mexico conducted just such a thought experiment using a formula known as the species–area relationship, which correlates the number of species with the land area of a continent.

"Imagine that continental drift could be reversed and all of the earth's land could be reassembled in a single giant continent," he wrote. "[F]rom 35 percent to 70 percent of existing species would be lost," he estimated. The aggressive species would be free to invade, altering habitats, preying on their longtime inhabitants, competing with the natives for food, shelter, light, and resources. It would be equivalent to moving Hawaii to the continent, a move that "would just accelerate the replacements of endemic species that have been occurring for the last century because of the transportation of continental species to Hawaii." Indeed, Brown's extrapolations predicted a loss of two-thirds of the earth's land mammals, nearly half of the birds, one-third of the butterflies, and nearly three-quarters of flowering plants. (And that loss estimate is for a hypothetical supercontinent without bulldozers, chain saws, plows, or guns.)[13]

As Brown indicates, those most vulnerable in this great "cosmopolitization" of the earth's living things are the island species. Long isolated and often undefended, they are least likely to survive unassisted in the global melting pot. Ninety percent of the imperiled plants on IUCN's Red List are endemic to a single country, many to island countries.[14] Of course, continental species are also at risk, some because they live on natural habitat "islands" such as isolated mountaintops and lakes, but many others because we are gradually trapping them in islands of habitat surrounded by human developments and besieged from within by imported enemies.

One paradox of the accelerating trend toward commingling of the earth's biota, however, is that even after massive extinctions of native species have occurred, the number of plants and animals in a given place may actually be larger than before. Hawaii, Tahiti, Mauritius, New

Zealand, and many other islands certainly have more species now than ever before, although they have traded a suite of unique organisms for a set of animals and plants you can now see in almost any tropical country you visit. This phenomenon is sometimes derided as biological "McDonaldization": it may create more variety locally, but the menu is monotonously the same the world over. The global menu loses an item with every extinction but gains none with an introduction.

In most cases, biologists have good reason to regard "artificial diversity" in the form of exotic species as an indicator of general degradation and loss of ecological integrity in a community.[15] There are a few cases in which human-caused additions of species to natural ecosystems appear to have enriched the life of a region without causing any losses from the original community—at least, no losses of charismatic species. An often-cited example is Lake Nakuru in Kenya, a formerly fishless saline lake where more than thirty fish-eating bird species have taken up residence since 1961, when tilapia were released into its waters to control mosquito larvae.[16] Whether the fish and birds have driven any insects or amphibians extinct apparently remains unstudied.

Extinction, of course, is merely the final act in a prolonged drama. Especially on continents, it may take decades or centuries for the last individuals of a faltering plant or animal species to vanish. In the meantime, however, as its numbers shrink, a species becomes increasingly marginalized and ecologically irrelevant. The canopy of a formerly abundant tree ceases to define the character of the landscape; its roots and detritus no longer influence the chemistry of the soil; its flammability, the fire cycle; its fruits and flowers, the livelihood of birds and beasts. An animal in decline no longer shapes the destiny of other species on which it feeds, the plants whose seeds it disperses, the flowers it pollinates, the soil it churns, the trees it topples. By the time such a marginalized plant or animal merits headlines because it teeters on the brink of extinction, its influence in the living community has already faded to insignificance. By that point, something else probably dominates the system and sets the rules—often the very species that pushed the imperiled one to the brink—and that species is likely to be an invader.

Low Probability, Large Impact

In fairness, most exotic species, whether accidentally or deliberately moved to new territories, do not survive at all without human help. Of those that manage to survive, most plants, at least, settle into their new communities with little fanfare.[17] However, the few bad actors that do make their mark can be devastating. As ecologist Mark Williamson of the University of York put it, damaging invasions are "events of low probability but high impact."[18] Within Hawaii Volcanoes National Park, the 400 surviving native ferns and flowering plant species are now rivaled by 400 naturalized exotics, but it's only the three dozen most disruptive ones (along with nearly all the exotic mammals) that worry Tunison and his crews, for example.

There is good reason, however, to keep an eye on any exotic as a potential invader and to deal with it while only a few scattered infestations exist in unwanted places. Delays of years to centuries can occur between the arrival of a new species and its sudden, explosive spread across a landscape. Ecologists describe docile exotics such as the common mullein and knotweed, which have recently turned aggressive in Hawaii, as "sleepers," or incipient weeds. Brazilian pepper trees brought to Florida as ornamentals in the nineteenth century also behaved well at first. Not until the early 1960s did land managers and scientists begin to notice dense stands of pepper trees. Now Brazilian peppers blanket some 280,000 hectares, often to the exclusion of all native plants.[19] In Germany, researchers who surveyed 184 non-native tree and shrub species introduced into the Berlin and Brandenburg areas found it had taken anywhere from 8 to 388 years for those plants to start "spontaneous spreading," an average lag time of 147 years.[20]

Michael Soulé suggests that any of three factors might be at work in these lags. First, the perceived delay might simply reflect the time it takes for a small number of initial colonists to multiply and build up to noticeable population sizes. Second, some species may adapt genetically over time to their new settings and only then begin to flourish—a difficult change to detect. Third, some change in the environment may enable a previously limited species to explode.[21] Sometimes the enabling environmental change involves the arrival of another invader. Avian malaria clearly could not have taken hold on Hawaii until its mosquito vector became established. And barnyard pigs may have found life in the Hawaiian forests too meager until imported guava trees and earthworms spread.

Whatever the delay, a percentage of newly established species do become invasive. Mark Williamson has put forward a statistical generalization known as the "tens rule": roughly 10 percent of free-living introduced species become established, and roughly 10 percent of those that are established become invasive pests. "Roughly 10 percent" means somewhere in the range of 5 to 20 percent.[22] The congressional Office of Technology Assessment concluded in a 1993 report that at least 4,500 non-native species occur in a "free-ranging condition" in the United States—an admittedly conservative estimate—and approximately 20 percent of these cause serious economic or ecological damage.[23]

Others believe, however, that the tens rule will falter and the proportion of invaders will increase with time as the sleepers among them awake.[24] Time, Soulé points out, is on the side of the invader. And, of course, as new alien species arrive, the cumulative total of harmful invaders and possible sleepers grows. Even among troublesome invaders, however, some clearly do much greater harm to native communities than others. It's hard to imagine that knotweed and mullein will ever rival the power of pigs and ginger in Hawaii, for example, even though mullein is already a serious invader atop Mauna Kea on the Big Island, as well as on the mountains of Réunion Island in the Indian Ocean.

The types of invaders that cause the greatest damage fall into a handful of categories. One is the "transformers," including metastatic species such as the "green cancers" that grow out of control and swamp whole ecosystems and landscapes. A second group is disruptive animals, which may be predators, grazers, or occasionally even the animal equivalents of transformers that overrun whole landscapes or seascapes. Third are the disease-causing agents and the vectors that carry them. Finally, there are the plant and animal invaders that work more subtly, competing for resources or interbreeding with native species.[25] And inevitably, whenever an invader alters one element of a community, the disruption sets up chain reactions that affect other elements.

Green Transformers

As Tim Tunison and I hiked along a cinder trail on the rim of the Kilauea crater that May morning, we veered off the track and into the deep shade of a stand of small trees. The crunch of our footsteps was quickly muffled by a thick duff of soggy leaves on the barren floor of the woods. There appeared to be only one species present, and Tunison told me it was fire tree, an exotic invader that has proven a notorious transformer on the Big Island. The fire tree, an aromatic evergreen related to American bayberry and wax myrtle, had been brought to Hawaii from the Azores and Canary Islands by Portuguese immigrants in the 1890s. During the 1920s and 1930s, it was planted widely throughout the islands to reclaim deforested watersheds—a practice that stopped when forestry officials noticed its habit of aggressively invading pastures. Yet the damage was done. The trees spread, abetted by introduced birds, such as Japanese white-eyes and house finches, that eat the fire tree's berries and disperse its seeds. In Hawaii Volcanoes National Park, the first fire trees were noticed in 1961. Intensive control efforts were started but soon abandoned, and by 1992 the trees had advanced across 16,000 hectares of the park, forming solid stands like the one we had entered.

On this site some forty years earlier, Tunison explained, a volcanic eruption had rained lava cinders, thinning out the ʻohiʻa and tree-fern forest. Later, pigs had added to the disruption. That's when fire trees moved in. Fire trees are nitrogen fixers, meaning that the symbiotic bacteria living in their root tissues can capture gaseous nitrogen from the air and convert it into a form of fertilizer. The trait gives these trees an edge in colonizing cinder falls and the nitrogen-poor soils on young lava flows and allows young fire trees to expand in girth ten to fifteen times faster than ʻohiʻa trees. Once fire trees take over a site, they profoundly alter the fate of the system by quadrupling the amount of nitrogen available in the soil and undoubtedly biasing the types of plants that will come next. Just what those plants will be—native or invader—remains an open question.[26]

Tunison's team, searching for ways to give natives the edge, has found that on this particular site, killing the fire trees slowly and laboriously by girdling them and daubing on an herbicide opens up gaps in the canopy that admit light and allow tree ferns to spring back. He has great hopes for this place. Unfortunately, he recently found a plant that doesn't wait for the fire tree canopy to be breached. In the deep shade, Tunison pointed out a spot where a low ground cover was spreading out across the leafy duff. "This is one of our sleepers I was telling you about, knotweed," he said. "Now we're finding it in a broad range of habitats, even under fire trees, where nothing else grows. It's depressing."

Fire trees are classic transformers, dominating the communities they invade, overrunning whole ecosystems, and displacing native species. By definition, transformers also change the character and physical structure of the landscape, creating new types of habitats and destroying old ones, and altering the processes of earth, water, or fire. In doing so, they necessarily rewrite the rules for every other creature in the community. Any type of plant can be a transformer: trees, grasses, shrubs, climbing vines, herbaceous plants, floating plants, seaweeds. Some notorious transformers have been moved around accidentally, but many others, such as the fire tree, were deliberately imported for use in gardens, pastures, forestry plantations, erosion control efforts, or aquariums. Then they escaped.

Among trees, the miconia now blanketing two-thirds of Tahiti is certainly a transformer. It is directly endangering forty to fifty endemic plants, especially lobeliads, by overtopping, shading, and crowding them from the forested slopes.[27] On nearby Raiatea Island, a veritable army of schoolchildren and French soldiers has chopped and pulled 800,000 miconia trees since 1992 in an effort to avoid repeating Tahiti's fate, but miconia seeds are still sprouting.[28] Equally unstoppable are the guava trees that have commandeered vast tracts on islands worldwide, from Hawaii to the Galápagos Islands to Mauritius.[29] Ironically, in parts of the Galápagos, guava has teamed up with invading quinine trees to overrun a native miconia, a relative of Tahiti's "green cancer."[30]

Such tit-for-tat ironies are legion in this scrambled world of transformers. Australian melaleuca trees are turning treeless saw grass marsh and prairie in Florida's Everglades to dense forest while along the northern coast of Queensland, shrubby pond apple trees from the southeastern

United States are aggressively displacing native melaleuca woodlands.[31] Along the California coast, native Monterey pines are succumbing to pine pitch canker in their few remaining strongholds. Yet this pine has become the world's most widely planted non-native timber tree and an aggressive invader in many parts of the Southern Hemisphere, beyond the natural range of pines.[32] Tamarisk invasions in the southwestern United States have dried up desert springs and oases, threatening native creatures from pupfishes to bighorn sheep.[33] Yet in Pakistan, native tamarisks and shrubs are being pushed out by dense stands of mesquite trees introduced from the American Southwest a century ago to afforest the deserts.[34]

Many invading alien grasses become notorious transformers by altering the wildfire cycle. Eurasian cheatgrass has greatly accelerated wildfire frequency on the western plains of the United States. To the south in the Sonoran Desert, invading African buffelgrass and Mediterranean red brome grass have increased the frequency of summer fires, to the peril of desert tortoises and saguaro cacti.[35]

In the drier ʻohiʻa woodlands of Hawaii, a wide array of non-native grasses such as molasses grass, beardgrass, and broomsedge have invaded, fueling frequent fires in a land where wildfires were once rare. In the lower-elevation woodlands on the southwestern flank of Kilauea, only nine fires flared between 1920 and 1968, and they burned a mere 2.3 hectares in total. Then, in the late 1960s, grasses began to invade. From 1969 to 1988, some thirty-two fires swept across 7,800 hectares on these slopes, knocking back the native trees and shrubs and allowing the grasses to take more ground.[36] Even in the wet tropical forests of Hawaii and Australia, invaders such as molasses grass fuel hotter, more frequent fires.[37]

When parts of a tropical forest are carved away for agriculture, the sunny perimeters of the farmed lands frequently become colonized not only by alien grasses but also by invasive shrubs and vines, which can smother or crowd out trees at the forest edge, gradually shrinking and degrading the remnant forest. In Queensland, the transforming shrubs and vines include rubbervine, blue thunbergia, potatovine, pond apple tree, lantana, and privet. Indeed, rubbervine, imported to Australia from Madagascar in the 1860s as a garden plant, now infests 350,000 square kilometers of Queensland, an area larger than the state of Victoria.[38]

Anyone who has seen rubbervine, bridal creeper, potatovine, and other smothering vines at work in Australia does not need to guess what an infestation of kudzu looks like as it overtops and obliterates trees, telephone lines, barns, derelict vehicles, farmland, and stream banks and transforms another 49,000 hectares each year in the southeastern United States. In 2000, kudzu even showed up in Oregon.[39]

Some of the most notorious and widely traveled transforming shrubs include raspberries and blackberries (both *Rubus* species), lantana, chromolaena, mimosa, clidemia, privet, Japanese barberry, gorse, and broom. Although some are introduced accidentally, most invasive shrubs, as well as most weedy annuals and perennials, are imported deliberately as garden and landscaping plants. Invariably, they escape from cultivation. Impenetrable thickets of South American prickly pear cactus—long a scourge in the arid lands of Africa and Australia—now infest 35,000 hectares of savanna inside South Africa's Kruger National Park. Like lantana, chromolaena, privet, oleander, pampas grass, wandering jew, and many other exotic plants invading Kruger, the cacti escaped from gardens in the park's own rest camps and employee villages.[40] South of Kruger, in Africa's largest coastal estuary, dense infestations of chromolaena even threaten the future of South Africa's Nile crocodiles. By overgrowing and shading once-sunny nesting sites in Greater St. Lucia Wetland Park, the shrubs cool the crocodile eggs to the point that too many embryos develop as females, skewing the sex balance among hatchlings.[41]

Many of the world's wetlands, freshwater lakes, and rivers are choked by transforming waterweeds, such as giant reeds, water hyacinth, and salvinia ferns, that destroy open-water habitats and alter water chemistry and productivity. Indeed, even the seafloor is not secure from such a takeover. A lush, fern-like tropical seaweed called caulerpa was dumped into the Mediterranean Sea from a public aquarium in Monaco in the mid-1980s, and within fifteen years it had carpeted—some say AstroTurfed—4,000 hectares of the seabed off France, Spain, Italy, and Croatia, replacing much of the native life of the ocean floor. There is little to stop it from colonizing the shallows of the entire Mediterranean.[42] Patches of the same seaweed showed up half a world away in the year 2000, on the bottom of several southern California lagoons, prompting a multimillion-dollar eradication and monitoring effort.[43] The likely source was people dumping the contents of their home aquariums.

Animal Engineers and Enemies

Swamping a landscape or seascape is usually the work of aggressive plants. But in recent years, a few invading animals have taken on the role of transformer, too. These are sometimes small creatures that may seem laughably benign by ones and twos, but when their populations explode, they can overwhelm vast domains. Two of them in the seas—the zebra mussel and the comb jelly—have caused such severe economic damage that they have helped thrust the problem of alien invasive species into the spotlight of international policy making. Their ecological effects will be equally severe, but much less visible.

Zebra mussels are fingernail-sized mollusks native to the Black, Aral, and Caspian Seas of eastern Europe and western Asia. In 1988, they showed up in the North American Great Lakes, near Detroit, apparently having been carried in the ballast tanks of a freighter across the Atlantic Ocean, through the locks of the Saint Lawrence Seaway, and into the lakes. The mussels spread quickly and aggressively, invading twenty states and reaching the mouth of the Mississippi River at the Gulf of Mexico within a decade. In 1999, they were spotted in the Missouri River system for the first time. This creature's claim to fame—or infamy—is its habit of encrusting and fouling anything solid in the water, from rocks, boat hulls, and pilings to water intake pipes and filtration equipment, cementing mussel upon mussel into clumps as dense as 700,000 per square meter. The ten-year bill for controlling and cleaning up after the zebra mussel has been estimated at $3.1 billion. And with each mussel filtering as much as a liter of water per day, vacuuming up the phytoplankton that sustains the lake's food web, these invaders are profoundly altering the fate of other aquatic creatures.[44] Predictions are that the zebra mussel could eventually drive ninety species of native freshwater mussels extinct in the Mississippi River basin alone.[45]

Shortly before the zebra mussel made its way west, a predatory blob native to the coastal waters of North and South America was riding eastward in the ballast tank of another ship. The comb jelly was first spotted in the Black Sea, far from its native waters, in 1982, but its explosive growth began in the late 1980s, just as the zebra mussel was making its move into the Great Lakes. The two- to five-centimeter jellies reached astounding densities, 300 to 500 per cubic meter, a mass of gelatinous life that would have weighed in at ten times the tonnage of the entire world's fish catch in 1988. But the jellies did more than dominate that landlocked and degraded sea. They consumed vast quantities of young fish and other tiny animals, simultaneously competing with and preying on the native fish and driving the already beleaguered fisheries in the region to near collapse.[46]

This jelly and the zebra mussel are anomalies among invading animals, however. The vast majority wield their influence not by physically dominating space and resources through sheer numbers but through their behavior and appetites. Animal engineers such as beavers and nutrias, for example, may change the fate of the communities they invade by reshaping the landscape.[47] But the major adverse effect of animal invaders, as Mark Williamson writes, comes from their "being an enemy of the native biota," consuming plants or preying on other animals.[48] The majority of invader-caused extinctions, in fact, involve predation.

The roster of invasive vertebrate animals includes a number of notorious repeat offenders, animals that create upheaval in natural ecosystems almost everywhere they are allowed to run wild: goats, pigs, feral dogs and house cats, rats, mongooses, rabbits, foxes, horses, donkeys, camels, water buffalo, and mosquito fish, to name a few. Spineless predators also wreak havoc, from northern Pacific seastars and green crabs to ants, gypsy moths, wasps, and even exotic scale insects.

All of these creatures wield their greatest influence in communities that have never seen anything like them before—that is, places where they bring a novel talent or role, create a novel disturbance, or add a whole new link to the food chain.[49] The hooves of cattle and goats that stepped ashore with the First Fleet in 1788 were the first that ever trampled the plants and soil of Australia, for example. The first pigs and goats that arrived on Hawaii and other islands formerly free of large herbivores encountered nettleless nettles, mintless mints, and other plants that had felt no pressure to evolve thorns or unpalatable chemistry. Cats, rats, mongooses, and brown tree snakes arrived on previously predator-free islands to discover petrels, nene, flightless rails, and other tasty birds nesting heedlessly on the ground. Indeed, sometimes a large consumer introduced into such a setting becomes what ecologists call a "keystone species"—an animal that wields power disproportionate to its size or numbers over the structure and destiny of a community.[50]

The first creature to be given that label was a native seastar that makes its living eating mostly mussels along the rocky shallows of the Pacific Northwest coast of the United States. Ecologist Robert Paine, in a now-classic experiment in the 1960s, showed that the seastar's appetite keeps the mussel population in check and allows a wide assortment of other animals to find space in the community. Removal of the seastar led to an explosion in the mussel population and a drop in species diversity in the community.[51] It is easy to see that anytime a new player usurps the position of keystone species, the face of a community can change drastically.

"Our keystone species now really is the pig," Tim Tunison had told me as we poked around 'Ola'a Forest. "The pigs and the weeds are the great ecosystem modifiers." In both Hawaii Volcanoes National Park and Haleakala National Park on Maui, in fact, pigs bear the major responsibility for preparing the ground for invasion and spreading weeds that subsequently overwhelm the native vegetation.[52] Until their control is broken, it is pointless to try to repatriate the orchids and other rare plants that the park and conservation groups are propagating in protective custody.

Animals do not have to be large to upset the community or threaten other species. Remember that it is exotic earthworms that help nourish the keystone pigs in Hawaii, though not by choice, of course. And invading worms themselves can proliferate to such great densities that their work also directly affects the survival of native plants. That phenomenon is not limited to Hawaii. In the previously earthworm-free northern hardwood forests of the United States and Canada, for instance, European earthworms introduced long ago in dry ship ballast, in the soil of imported potted plants, or by fishermen dumping unused bait are speeding the decay of the deep duff layers on the forest floor. The bare ground left by the worms in many areas no longer supports trillium, yellow violets, Solomon's seal, and other spring flowers and instead invites colonization by exotics such as garlic mustard.[53] In Sri Lanka, the rise of organic gardening has led to the importation of earthworms, which can now be found in the rain forests. No one knows yet what ecological consequences they will have.[54]

The ranks of noxious small invaders also include slugs and a wide array of ants. As if Hawaii's plants did not have enough troubles with pigs and goats, common garden slugs have spread into remote rain forests on Maui, damaging the spectacular greenswords and rare lobeliads.[55] What's more, exotic ants—about forty species of them, including the big-headed ant, crazy ant, and Argentine ant—are preying on many of the native insects that pollinate Hawaii's besieged plants.[56] In South Africa, one of these same species, the now-global Argentine ant, is likewise driving out native ants that disperse the seeds of rare proteas and other endemic plants. As a result, large-seeded plants are growing sparser.[57]

In a few odd cases, a predator may create as much threat to the native fauna in the role of prey as it creates by preying. Take the cane toad, a whopping South American amphibian introduced into Australia and many of the Pacific Islands, including the Hawaiian chain, in a horribly misguided effort to control beetles in sugarcane fields. Cane toads secrete a deadly poison through their skin. In the fragmented rain forests of Queensland, cane toads have hopped in on logging roads and now pose a deadly temptation to owls, pythons, and native marsupials called spotted-tailed quolls or tiger cats, which evolved without any lessons about poisonous prey.[58]

Imperiled plants and animals usually face multiple threats, so it can be difficult to prove a direct link between the decline or extinction of a species and the activities of a specific invader, whether the suspect is a pig, an ant, or a toad. Daily acts of eating or being eaten or the rooting up of plants deep in the woods are not as obvious or easy to track as the presence of transforming vines spreading over a landscape. Thus, it can be hard to prove that a plant is growing ever more rare because goats are browsing it or that feral cats are the major cause of a bird's decline. Feral cats and outdoor domestic cats kill some half a billion birds in the United States each year, but do they kill devastating numbers of the species in question?[59] Supportable answers to questions like this are often vital when an invader has a devoted constituency—whether it is hunters who want to keep populations of an introduced game animal high or animal rights groups who object to the killing of any animal. (Pigs in Hawaii have strong advocates from both camps.) The clearest demonstration of an invader's effects comes indirectly, when that creature is finally removed or excluded from a region and native species or communities recover.

One of the most spectacular successes has been the recovery of Aleutian Canada geese, which were among the first animals to be listed under the United States' Endangered Species Act of 1973. The geese winter on the coast of Oregon and California but nest in summer on remote islands in the Aleutian Islands chain in the Bering Sea off Alaska. Starting in 1750 and continuing until 1932, fur farmers and trappers introduced foxes on nearly 200 of these islands. The stocking hit its high mark in the early part of the twentieth century, when fur markets were booming. For almost thirty years after that, biologists thought the goose had gone extinct as a result of fox predation. Then a small population was discovered on a remote, fox-free island. In 1949, the U.S. Fish and Wildlife Service began shooting and trapping foxes on certain islands and relocating birds to those islands. By the early 1990s, some twenty-one islands were fox-free, and the goose population had grown to 32,000.[60]

Most other successful predator removal projects also have taken place on islands. Rodents, for instance, have been eliminated from more than eighty islands worldwide, leading to significant improvements in the status of native plants and animals.[61] Now even the island continent of Australia is trying to match that success to allow recovery or reintroduction of such imperiled native marsupials as wallabies, bilbies, and bettongs. It's a project of unprecedented geographic scale.

Starting in 1992, Flinders Ranges National Park in South Australia launched Bounceback, a program that has succeeded in nearly eliminating foxes, greatly suppressing feral cats, and dramatically reducing goat and rabbit grazing. The best measure of success, however, has been recovery of the little yellow-footed rock wallaby as its predators declined.[62] Another high-profile effort in Western Australia called Project Eden has fenced off the entrance to Peron Peninsula, which juts into Shark Bay 725 kilometers north of Perth. After years of feral cat and fox removal on the peninsula, native vegetation and animals such as the nocturnal echidna have rebounded. In 1997, scientists began reintroducing malleefowl and rabbit-sized woylies, or brush-tailed bettongs, and planning for the return of bilbies, bandicoots, and wallabies that had earlier been wiped out by exotic predators. Project Eden itself is part of a larger wildlife recovery effort called Western Shield that aims to control exotic predators across 3.5 million hectares of Western Australia.[63]

These and many other protected places outside Australia will forever require active human management, whether they are fenced or not. The sturdiest fences will not keep birds or winds from depositing seeds of invasive plants, much less halt incursions by ants, earthworms, mosquitoes, mongooses, rats, or cats. And not all endangered animals can be fenced in for their own protection. To help the fewer than 200 nene in Hawaii Volcanoes National Park breed successfully, Tim Tunison's crew must control mongooses, pigs, and feral cats wherever the geese choose to nest. On Floreana Island in the Galápagos, park personnel must selectively poison rats and chop back lantana bushes each season around the beaches where petrels and sea turtles nest.

Sometimes, fencing an area and removing the worst invaders are enough to restore a community to its self-directed path with only minimal human upkeep. In many cases, however, natural areas may be too degraded to recover their former structure, even with extensive help. In the pig-fenced sections of 'Ola'a Forest, removing pigs and knocking back a handful of key weeds has been enough to start recovery of the native plant community. But in drier areas of Hawaii Volcanoes National Park, where grasses and fire trees have taken over, wresting control from the invaders and restoring power to the natives has been less successful. So far, even active intervention has done little good. Part of the problem is that—as often happens—a new invader has arrived to complicate the picture. This time, it's a plant-eating insect called a two-spotted leafhopper. The good news is that the leafhopper is causing a massive dieback of fire trees in these dry areas. The bad news is that it also kills mature 'ohi'a trees. What remains is molasses grass and a scrubby ghost forest of flammable leafhopper-killed trees.

"So our strategy for now is to give up on the former 'ohi'a community there," Tunison told me with a rueful laugh. "We'll do some prescribed burning of the molasses grass and plant seeds of fire-tolerant native trees and shrubs." If it can't be a native forest for now, perhaps it can be a native shrubland.

Pathogen Pollution

Another invader that fences cannot shut out is disease. Pathogens, like larger predators, hit hardest in populations and communities that have never encountered anything like them before. A pathogen thus stands its best chance of encountering naïve and inexperienced hosts when it invades a territory for the first time. If a new host turns out to be a dominant or keystone plant or animal, then a pathogen can indirectly alter the workings of an entire ecological community. The Asian chestnut blight fungus, introduced into New York City around 1900, was able to eliminate the American chestnut from its dominant place in the forests of the eastern United States within a few decades. The loss of the huge chestnut crop, fed on by bears and many other animals, must have sent ripple effects through the woodland food chain, but unfortunately, no one was chronicling the spin-offs of that epidemic.[64] We know that rinderpest transformed life for humans and animals alike on the African savanna when it swept the length of the continent in a decade, killing as much as 90 percent of the buffalo and wildebeest as well as domestic cattle.

Emerging diseases in wildlife are gaining increased attention today because of the threat they pose to biodiversity as well as to humans and domestic animals. Diseases emerge when we introduce animals into new settings, move ourselves and our pets and farm animals into wildlife habitat, or unwittingly transport pathogens to new places—a phenomenon parasitologist Peter Daszak calls "pathogen pollution."[65]

Daszak and his colleagues are finding, for instance, that a chytrid fungus disease blamed for many mass die-offs of frogs worldwide, and possibly three extinctions, may have moved between continents and regions with the amphibian trade. The chytrid fungus has been found in Europe and the United States in frogs and other amphibians shipped internationally for pets, outdoor ponds, zoo display, and laboratory use, as well as in widely introduced species such as cane toads and bullfrogs.[66]

The mass die-offs of frogs that made headlines in the United States, Australia, and elsewhere during the 1990s gained scientific notice only because researchers happened to be in the field studying these frog populations when the declines began. The same goes for outbreaks of canine distemper virus in Tanzania's Serengeti National Park, blamed for eliminating the last of the area's wild dogs in 1991 and then jumping from canines to felines to kill more than one-third of the park's lions in 1994. Unfortunately, in most places and at most times, diseases in wild animals go unnoticed and undiagnosed—that is, unless they spill over into the human population.[67]

West Nile encephalitis made headlines in 1999 when it invaded the New World for the first time, but the focus, understandably, was on human deaths and illness and not the deaths of thousands of birds. However, this virus is likely to be a far more serious threat to birds and other wildlife than to humans as it moves across the United States, encountering a continent full of creatures that have never had an opportunity to develop immune defenses against it. Some sixty bird species have tested positive for the virus, along with a dozen mammals, including horses, cats, bats, raccoons, rabbits, squirrels, chipmunks, and skunks.[68]

Subtle Struggles

If the effects of pathogens invading wild populations are hard to spot, other harmful interactions between newcomers and locals can be even more cryptic. Competing subtly for space, services, or resources, or even interbreeding with native species, can be more insidious and subtle than transforming a landscape or preying on keystone species. It often takes painstaking scrutiny to pin down who mates with whom or to monitor who manages to claim the best nourishment, the most attention from pollinators, the safest nesting sites, or the sunniest patch of ground.

One example is the surprising competition between alien wasps and native forest parrots called kaka in the lowland beech forests of New Zealand's South Island. Native scale insects naturally infest the beech trees and excrete abundant honeydew, a sweet bounty that supports an array of insects and other birds as well as the threatened kaka. Several wasps, including the common wasp, have now invaded these forests, however, and in autumn, when their populations reach a peak, these tiny invaders claim 95 percent of the honeydew and cause the kaka to abandon the forest. By one estimate, the sheer mass of wasps in these forests now exceeds the combined mass of all the native birds plus the furry enemies that also plague them—nest predators such as stoats and rats.[69]

Competition in the water is even harder to sort out. In the Pacific Northwest, where predatory eastern bullfrogs and bass have been dumped into marshes and lakes, native red-legged frogs decline. Both bullfrogs and bass eat red-legged frogs, but that's not the whole story. It turns out, after some painstaking scientific detective work, that the intimidated red-legged frogs are forced to retreat to cooler waters, which are not optimal for nurturing frog eggs. In this stressful setting, their tadpoles weigh less and die at higher rates, and the survivors take two weeks longer to metamorphose into frogs.[70]

That is certainly not the sort of interaction you would notice on a casual visit to the marsh. Yet it is just the sort of thing Edward O. Wilson had in mind when he said that extinction by invasion is "gradual, insidious, requiring scientific methods to diagnose."[71] The process goes on all around us yet often remains largely invisible until it's too late.

Hybridization between an invader and a closely related native can be equally cryptic, sometimes requiring genetic testing to confirm. Such evidence shows that European wolves are already breeding with feral dogs in Romania and western Russia, and the wolf population in the Italian Apennines is also considered at risk of interbreeding with feral dogs.[72] The now-cosmopolitan North American mallard is breeding with both the New Zealand gray duck and the Hawaiian duck, threatening the survival of both as distinct species.[73] The survival of European white-headed ducks—the object of a remarkably successful twenty-year conservation effort in Spain that has brought their numbers back from 22 to more than 1,000—is now compromised by the wooing of aggressive North American ruddy ducks. Ruddy ducks were imported to England in the 1950s, thrived, and have flown across the English Channel to invade France and Spain, where they readily mate with their white-headed relatives.[74]

More difficult still is documenting the domino effects that follow when a new plant or animal—or, worse, a whole cadre of invaders—sweeps into a community. Who would have imagined that honeysuckle vines and buckthorn invading the forests of the eastern United States might set up an "ecological trap," encouraging robins, wood thrushes, and other songbirds to nest in sites more accessible to marauding predators?[75] Or think of the sailors who first turned goats out on the Galápagos Islands. There, the scrubby trees of the highlands, festooned with liverworts and mosses, comb moisture from the seasonal garua mists, or marine clouds, to create a lush, dripping habitat where giant tortoises and other animals thrive. Could those early sailors have imagined that voracious goats stripping bare the trees would shut off the drip and turn the pools to dust?[76]

Harder by far to foresee are impromptu alliances such as those that link pigs, earthworms, transforming weeds, exotic birds, mosquitoes, and avian malaria in Hawaii. Each organism in this alliance arrived separately, for different reasons, and by different routes. Each had to establish a beachhead before it could begin to advance across the islands and take advantage of the talents of other invaders. We will have to understand, and do a better job of intervening in, those early stages of invasion if we are to conserve remnants of every region's unique living heritage.

Whether invaders act alone or in concert, as transformers, predators, or subtle competitors, it seems clear that the threats they pose to native biodiversity are far more serious than we once imagined. If we continue to allow the commingling of the earth's species to proceed unchecked, greater losses among native plants and animals and further transformations of natural landscapes and ecological processes are inevitable. The pace of global trade and travel is accelerating, and all the economic and social motives that brought the current crop of alien invasive species to every shore are still at work. Like pollution, overexploitation, and habitat destruction, species invasions are at root a human problem, and only a change in our attitudes and actions can halt further degradation of our natural heritage.

FIVE The Good, the Bad, the Fuzzy

"It may have begun with Noah, but, wherever it started, the whole idea of rearranging the earth's wild creatures still seems irresistible. Man, the supreme meddler, has never been quite satisfied with the world as he found it, and as he has dabbled in rearranging it to his own design, he has frequently created surprising and frightening situations for himself. . . . Release of wildlife into territory foreign to it involves, not a calculated risk, but a risk too great to calculate."

—George Laycock, *The Alien Animals,* 1966

"When I climbed I almost always carried seeds with me in my pocket. Often I liked to carry sunflower seeds, acorns, or any queer 'sticktight' that had a way of gripping fur or boot tops. . . . I have carried such seeds up the sheer walls of mesas and I have never had illusions that I was any different to them from a grizzly's back or a puma's paw. . . . You can call it a hobby if you like. In a small way I, too, am a world-changer and hopefully tampering with the planetary axis. Most of my experiments with the future will come to nothing but some may not."

—Loren Eiseley, *The Night Country,* 1971

As snakes go, the brown tree snake is hardly a standout: not particularly colorful, only mildly venomous, a one- to three-meter-long tree dweller and nighttime hunter that few people ever encounter. Sometime around the end of World War II, a few brown tree snakes apparently stowed away on derelict military equipment that was being salvaged from the jungles of New Guinea and barged to the small Micronesian island of Guam for disposal. On arrival on Guam, the snakes slipped off into the degraded forests without attracting much notice.

By the 1960s, however, biologists had begun to observe other changes in the forest. The southern one-third of the island was completely devoid of birdsong by then, and the silence in the forest was advancing steadily northward. Lizards and mammals, some themselves invaders, were disappearing along with the birds. Closer scrutiny in the 1980s revealed a startling development. The brown tree snake population had exploded, reaching densities as high as 100 per hectare, 10 to 100 times normal for a snake its size. By the mid-1990s, the snake had eliminated nine of Guam's thirteen native forest birds and pushed the others to the brink.

Although the struggle in the forest has been taking place largely out of sight of the human residents of Guam, the snake is affecting their lives, too. Brown tree snakes crawling up power poles spark frequent power outages that cost millions of dollars, disrupt traffic, and cause food to spoil and computers to fail. As forest creatures decline, the snakes attack poultry, pet birds, and even puppies. More horrifying, however, is the snake's habit of entering island homes at night and—possibly because of what researchers describe as its "poor judgment with regard to the size of potential prey"—biting sleeping infants and sending a number to the hospital each year.[1]

Between its bird-destroying and baby-biting behaviors, the brown tree snake has earned notoriety among invasive species worldwide. Special search teams on serpent-free Hawaii now inspect all planes and cargo arriving from Guam for hitchhiking snakes and have intercepted seven so far. The U.S. Fish and Wildlife Service (USFWS) has added the snake to its short list of injurious wildlife banned from importation into the United States. This unlikely marauder has also garnered more television, magazine, and newspaper coverage than any rattlesnake or python. It was from one such article, in a magazine for reptile fanciers, that bird-watcher Harriett Peuler in southern Louisiana learned about a creature she never expected to see.

One day in the fall of 1998, Peuler and a friend drove out to a wildlife reserve in the Atchafalaya River basin west of Baton Rouge to watch swallow-tailed kites soaring over the forested wetlands. For some reason, instead of taking the interstate highway back to the city as she usually did, Peuler drove home along back roads. It was there that she chanced upon a small roadside snake farm in Port Allen, one of those now-fading public attractions of the rural South that she remembered visiting as a child.

"I said, 'Oh, I haven't been there in years; let's stop,'" she recounted. "So we went in and began looking at all the snakes." That's when she saw the Brown Tree Snake sign. Peuler was skeptical. She had taught classes, even produced an educational CD-ROM, on native snake identification. "People will call everything a brown snake, a green snake, or a black snake." But a closer look revealed a label for *Boiga irregularis* and a note saying that the three caged snakes inside had come from Guam.

"I thought, 'Oh, lord,' and I said, 'This is illegal,'" she recalled. But the proprietor assured her the snakes were legal and didn't seem to share her concern about their reputation. "He just liked snakes, and it was a new snake for him, one that was from far away, and people like to come and see exotic snakes in his place. But I'm thinking, 'What if these things got loose into the environment?' Because I've had snakes, and they do get out. They're Houdinis. I'm a bird-watcher, and I was concerned."

As soon as she got home, Peuler e-mailed state biologist Jeff Boundy, who in turn contacted the local USFWS law enforcement agent, Downie Wolfe. Unknown to her, Peuler's find set in motion a quick investigation that, as Wolfe recalled, "sent a shiver up and down the Service" all the way to Washington, D.C.

Boundy, who knew the snake farm owner, quickly confiscated the snakes and learned that they had been passed along by at least one other reptile collector before arriving at the roadside display. At that point, the matter was passed to Wolfe: "I had to assume they had come in through some illegal means." His investigation led to an importer in Miami who had sold the snakes several months earlier. Wolfe notified Special Agent Eddie McKissick in Miami.

"I went through our records, and I found the importer's declaration,"

McKissick recounted. What he found is what set off the administrative shivers. The snakes had arrived clearly labeled as *Boiga irregularis,* yet they had been cleared for import. More than 12,000 shipments of wildlife come through the port of Miami each year, and USFWS inspectors attempt to visually examine 25 percent of those. But they check the import paperwork on all shipments. "It's just one of those things that slipped through the cracks," McKissick said.

Without mentioning what he had found, McKissick called the importer on his cell phone and asked permission to drop in even as he drove toward the place. "We can't come unannounced," he told me. "We have to give notification." The importer opened his cages, his books, even his computer to McKissick, who found *Boiga irregularis* openly listed. The snakes had been shipped by a reptile dealer in Indonesia as part of an order for an unspecified assortment of mangrove snakes that a Louisiana collector had requested. The price of a brown tree snake: $35. Not the sort of financial incentive that usually drives illegal trade.

"Essentially, because I really did not see any malice intent in the importation, he was just given a $1,000 fine," McKissick said. No charges were brought against the Louisiana collector or the snake farmer. The snakes are now pickled in jars, one at Louisiana State University and two adorning Downie Wolfe's back office near Baton Rouge, although he, too, planned to send his booty to the university at some point. At the port of Miami, USFWS inspectors are now "very aware" of *Boiga irregularis.* So far, that has been the only interception of a deliberately imported brown tree snake in the United States.

What, I wondered, would an invasion by escaped brown tree snakes mean in the swamps of Louisiana or even southern Florida? Wolfe and Boundy are confident that the abundance of raccoons, skunks, coyotes, and other predators; competitors such as rat snakes; and cold winters would keep such a newcomer from playing a dominant role, at least in Louisiana. Harriett Peuler is not so sure: "There are too many species that people thought wouldn't do well, and they ended up taking over." It was hard not to wonder about the fate of all those other assorted "mangrove snakes" still out there—well caged and tended, I hoped—along with the millions of other exotic reptiles that people on every continent have been buying in unprecedented numbers.

The United States dominates both imports and exports in the booming international reptile trade. U.S. residents bring in more than 2.5 million live reptiles each year, almost half of them green iguanas that sell for as little as $8 in pet shops—a price that encourages impulse purchasers to release their pets "into the wild" when they outgrow their cages.[2] Feral colonies of green iguanas, boa constrictors, caimans, cane toads, and thirty-two other exotic reptiles and amphibians have already set up housekeeping in Florida.[3]

In turn, the United States ships out more than 9.5 million reptiles per year, including 8.4 million red-eared slider turtles from the swamps and turtle farms of southern Louisiana.[4] The large sliders go to Asia to be eaten; the small ones mostly go to Europe to be sold as pets. (It's currently illegal to sell the sliders as pets in the United States because of the frequency with which they pass salmonella infections to their young owners.) Odd and sad as it seems, these little sliders, which die so quickly in the hands of most children, thrive to become nuisances when they are dumped or escape into waterways and wetlands around the globe. The little omnivores gobble up insects, crayfish, shrimp, small fish, frogs, worms, snails, and aquatic plants and compete with native turtles. Sliders have established themselves in the wild from Malaysia to South Africa and earned a place with *Boiga irregularis* on the list of the world's worst invaders.

Reptiles are only one part of a burgeoning $10 billion per year world trade in wildlife.[5] And even this trade represents but a small fraction of all the living things we are moving around the world these days, intentionally or otherwise. The human factors that put species in motion have changed surprisingly little over the past 500 years, but with 6 billion people on the planet now and more to come, the pace of invasions is inevitably accelerating.

The Conveyor Belt of Global Commerce

As global market forces batter away at trade barriers, industries worldwide pander to our desires for exotic pets, livestock, game animals, flowers, ornamental shrubs and trees, and food products, as well as a dizzying array of consumer electronics, clothing, and other goods. The paradox, of course, is that by design and by default, our appetite for novelty and variety helps drive homogenization—not just of global cuisine, clothing, entertainment, and cultures but also of our landscapes, our plant and animal communities, even our diseases. For better or worse, one place is becoming more like any other today, not just culturally but also biologically.

World trade drives the rearrangement of the living world in two ways, one intentional and the other accidental. The exotic plants and animals we import or move intentionally beyond their natural ranges—pets, flowering plants, boutique crops, plants for restoring degraded lands, animals destined for the table or for release into woods, fields, and streams as game animals—provide an increasing reservoir of potential invaders.

The organisms such as snakes and turtles and nursery plants that we import intentionally, however, pale in numbers beside the masses of smaller living things we set in motion incidentally. These are the hitchhikers that ride in ballast tanks or aboard the hulls of ships, in airplane cargo holds and cabins, in the nooks and crannies of shipping crates and containers, or mingled with grain, fruits, vegetables, cut flowers, timber, minerals, soils, and the other goods bought, sold, and shipped worldwide. The global infrastructure of trade and transport is linked to a growing web of roads stretching from teeming cities into the remotest jungles. These corridors not only foster more land clearing and resource exploitation but also bring pathogens, pests, and weeds into contact with the global transport network. The pace of this biological traffic has been accelerating along with the astounding growth of world trade in recent years. The value of goods traded worldwide rose from $192 billion in 1965 to $6.2 trillion in 2000, for example, thanks to liberalization of trade barriers. Economic downturns slowed the growth rate of world trade in 2001, but the overall trend is clear.[6]

Globalization means that retailers everywhere can buy their clothing, fruit, or other goods from countries that turn them out most cheaply or out of season. The "cool chain" industry, for instance, produces fruit, vegetables, cut flowers, shrimp, and other farmed fish and shellfish relatively cheaply in tropical countries and ships them via refrigerated carriers to booming markets in North America and Europe.[7] When winter blankets Montana, I expect to find peaches, plums, and grapes from Chile and fresh-cut bundles of roses, carnations, and alstroemerias from Colombia in my supermarket. Americans like me are the world's best customers, buying a record 18.9 percent of all global imports in 2000.[8]

The vast majority of goods bought and sold worldwide moves by ship. After fifteen straight years of increases, world seaborne trade in 2000 reached a record 5.33 billion metric tons. That equates to more than 120 million fully loaded eighteen-wheel trucks lined up bumper to bumper for some 2.25 million kilometers, a caravan of material goods that would circle the earth nearly sixty times. About two-thirds of that volume consists of bulk commodities such as grain, oil, and minerals, but world container traffic continues to expand, with 192 million six-meter units in motion in 1999.[9] Worldwide air cargo comes in a distant second, totaling about 30 million metric tons in 2000.[10]

People as well as goods are on the move in record numbers. There were 664 million international tourists traveling in the world in 1999, up from a mere 25 million at midcentury. And we are traveling more widely than ever. In 1950, most international tourism was concentrated in fifteen countries in Europe and the Americas. By 1999, some seventy countries and territories were hosting more than 1 million foreign tourists, with Asia and the Pacific region the newest stars of the tourism industry. Package tours now allow the least intrepid among us to sample in relative comfort the remotest reaches of the earth, from tropical jungles to Antarctic ice shelves.[11] (A sharp drop in airline travel after the September 11, 2001, terrorist attacks in the United States did damp the growth in world tourism for 2001.[12])

Often we travel to experience unique places, peoples, cuisines, and cultures, yet we find them more alike than ever, from shopping mall to hotel to fast-food counter. When we do find something new and foreign to delight us amid the increasingly homogenized products, we not only expect to be able to buy it and bring it home but also expect our local stores to begin importing and offering it for sale. We no longer accept the limitations of geography and season, and the wants of consumers in wealthy nations drive the global market. Even in much of the developing world, television signals arriving by satellite keep people clued in to what the rest of the world is eating, wearing, wanting, listening to, and driving. And so we put more things in motion. The reasons, the trade routes, the means of conveyance, the goods involved, and the associated packaging all matter in determining what sorts of living organisms are moving, too. Most of the world's pests, pathogens, and marine invaders travel uninvited. A small percentage of plants and larger animals do, too, but most of these by far travel as wanted cargo.

Hitchhikers and Stowaways

Water pumped into the ballast tanks of ships to stabilize and balance them carries with it myriad living things, from small fish and larval clams to red tide organisms and cholera bacteria. Jim Carlton of Williams College–Mystic Seaport in Connecticut estimates that anywhere from 3,000 to more than 10,000 species are on the move in ballast tanks each day on the global oceans, most of them collected from coastal bays and estuaries, hauled across otherwise impassable oceans, and released along new coastlines. The range in Carlton's estimates is vast, but he will tell you that the actual number is less important than the staggering scale of the process, which far exceeds any other means of species movement.[13] In U.S. ports alone, some 62,000 commercial vessels arrive each year, dumping 79 million metric tons of foreign ballast water at a rate of 9 million liters per hour.[14] Most aquatic invertebrates that travel the world do so in ballast, as do most seaweeds and marine plankton. It was ballast water that brought the zebra mussel to the Great Lakes, voracious northern Pacific seastars to southern Australia, and comb jellies to the Black Sea.

But there are other significant pathways for marine organisms as well. Nearly 300 species, from fish to mollusks, have invaded the eastern Mediterranean region from the Red Sea since the Suez Canal opened in 1869, and the process is far from over. In fact, the number of invaders moving through the canal is increasing exponentially.[15] Similarly, the seaway and canal system that opened the Great Lakes to oceangoing ships and their ballast also provides open passage for such invaders as the sea lamprey and alewife. Canals from Lake Michigan now provide passage to the Mississippi River system for invaders such as these as well as zebra mussels and other ballast arrivals.[16] A growing number of aquatic invaders also stow away in aquaculture materials such as fish stocks and eggs that are shipped around the globe.

Another commonly overlooked route by which we are redistributing marine creatures is that of discarded plastic bottles, bags, and other trash that litters the oceans. For eons, floating mats of seaweed, driftwood, pumice, and other debris have provided natural rafts that occasionally carry both land and marine creatures to new homes, seeding newly emerged islands such as Hawaii and the Galápagos. Today, human garbage greatly speeds that process. Biologists examining floating rafts of plastic have found it crusted with barnacles, tube worms, bryozoans, mollusks, and corals and colonized by crabs, shrimps, limpets, and many other marine animals and algae.[17]

Small land-based creatures are also on the move as never before. Insects, for instance, overwhelmingly arrive at new locations stowed away in cargo or riding the vessels that carry it. In Japan, entomologist Keizi Kiritani has identified four different waves of exotic insect invasions that reflect shifts in that country's commodity imports: from 1868, when Japan emerged from isolation, until World War II, a rush to import fruit tree stock brought scales and mealybugs. From 1945 to 1965, grain imports to relieve postwar food shortages prompted a massive invasion of stored product pests, especially beetles. The 1966–1985 period was dominated by arrivals of various weevils that infest "upland crops, turf, vegetables, and ornamental trees." The fourth wave began in 1986 along with a massive increase in cut-flower imports, which rose from fewer than 1 million flowers in 1970 to 1.5 billion in 1998. These imports have carried into Japan greenhouse pests such as thrips, aphids, whiteflies, scales, mealybugs, and leaf miners.[18]

Invading microorganisms often arrive in shipments of plants, animals, and produce, and their identities reflect trends in the goods being traded. A number of new powdery mildews of poinsettias, petunias, impatiens, and other ornamental flowers have recently entered the United States, for instance, as trade burgeons in cuttings and nursery stock.[19] And global movements of fresh fruit and vegetables are spreading food-borne diseases. An exotic intestinal bug, *Cyclospora cayetanensis,* that sickened U.S. consumers in 1996 was shipped in on Guatemalan raspberries. More recently, outbreaks of shigella in California and Chicago were linked to imports of parsley and cilantro from Mexico.

Many weeds have been moved around the world as contaminants in seed shipments and other materials, as we've seen, and unwanted plants are still moving in that way. Seeds of mimosa were first brought into Malaysia in the early 1980s in trucks of fill dirt and sand from Thailand during a construction boom.[20] Toxic fireweed, known locally as the "pony killer," invaded pastures on the Hawaiian island of Maui in recent years after arriving from Australia in hydroseeding material (a mixture of seed, fertilizer, mulch, and glue-like substances that is sprayed onto bare ground to promote fast turf growth). Seeds can also travel clinging to a vehicle rather than in cargo: more than 18,500 seedlings of 260 species of plants were germinated from mud washed from vehicles at a single car wash in Canberra, Australia.[21] Like plants, some mammals such as rats and mice have been carried around the world for centuries as unwanted hitchhikers. Some still travel that way. Bats recently rode a truckload of oak planks from France to England, for example, and a raccoon made headlines when it journeyed from Texas to France in a shipping container.[22] But these creatures are the exception. By far, most plants and most vertebrate animals have been, and still are, shipped around the world deliberately.

"Dirty" Wars

We've been exploring how various living organisms are moved around the world as an unintentional side effect of normal commerce. It is also worth noting that wartime movements of people, ships, planes, equipment, and foodstuffs have historically caused both deliberate and accidental movements of species. Just as the brown tree snake hitchhiked to Guam, Formosan termites apparently rode to Hawaii and the southeastern United States, and eventually spread to Sri Lanka and South Africa, in infested wooden pallets and packing material aboard returning World War II vessels.[23] Also during that war, parthenium entered Australia with machinery parts, mice crawled onto Pitcairn Island from cases carrying radio equipment, German wasps arrived in New Zealand in crates of airplane parts, chromolaena spread widely through Asia and the Pacific islands, and all manner of barnacles, jellyfishes, and crabs and hundreds of other marine species were redistributed around the world's bays and harbors by ship traffic.[24]

Today's military operations are no "cleaner" in terms of species invasions than those of World War II. For instance, parthenium moved into Sri Lanka along with Indian peacekeeping forces in the late 1980s, apparently carried by goats brought in to feed the troops.[25] And the western corn rootworm showed up in the war-ravaged Balkans in the early 1990s, its first appearance outside North America. Scientists suspect that American military transport planes, while being loaded around the clock on lighted airfields in Texas, probably attracted the adult beetles inside and then delivered them to eastern Europe. The beetle and its worm-like larvae were first spotted around Belgrade, and by the end of the decade they had spread as far as southern Switzerland. There are neither geographic nor quarantine barriers to stop the rootworm's advance across the cornfields of Europe.[26]

Some species have been introduced deliberately by military units for logistic purposes. During World War II, the Japanese Imperial Army introduced the fast-growing, weedy tree leucaena on many Pacific islands for camouflage, and the U.S. military planted thorny barricades of gorse around gun emplacements on Washington State's Puget Sound.[27] Even mesquite, which now forms dense thickets on remote Ascension Island in the South Atlantic Ocean and threatens to overgrow nesting beaches used by green turtles, is believed to have been brought in deliberately—although perhaps more for barbecuing than for military logistics—during the Falklands War in 1982, when the island served as a staging post for British forces en route to and from the Falkland Islands.[28]

In an effort to break this pattern of war and weeds, Australian quarantine scientists in 1999 helped the nation's peacekeeping force in East Timor develop a strategy for moving goods, equipment, and people back and forth while minimizing the risk of bringing chromolaena, Asian honeybees, exotic mosquitoes, giant African snails, or *Varroa* mites (bee mites) back to Australia.[29]

Degus, Emus, Swamp Eels, and Sika Deer

Let's leave war, though, and return to the economic and consumer sphere to look more closely at what is motivating deliberate movements of plants and animals that result in species invasions. Although intentional imports represent a relatively small proportion of the alien species arriving in most countries today, dwarfed by the number of insects, fungi, microbes, and other small organisms hitchhiking in the nooks and crannies of commerce, they present a troubling dilemma. It would seem that snakes and birds, seeds and saplings arriving openly at a country's borders ought to be far easier to control than myriad bugs stowed away in crates and containers or foreign ballast water. But it turns out that regulating deliberate importation of garden plants, agricultural breeding stock, pets, and other wanted organisms is politically far more difficult because each import has a constituency, someone who stands to benefit immediately from its arrival. In most cases, when such an organism proves economically or ecologically destructive, the harm does not accrue to the same person or industry that reaps the benefits. Too often, in fact, the public at large bears the cost.

The things we move intentionally are the things we value. These importations are driven fundamentally by our choices as consumers, travelers, businesspeople, nursery owners, gardeners, pet keepers, farmers, foresters, researchers, and even foreign aid and development workers. Our motivations have changed little over the centuries, although the favored species shift with need or fashion.

One wet fall night in a rain forest in southern Queensland, I walked slowly along a dirt track with a dozen other people, all of us trying to dodge unseen puddles while staring high into the eucalyptus trees. We were following a man with a powerful spotlight who was scanning the dark canopy for the gleam of tiny eyes. His hope was to show us a rare giant glider, but most of us were just as delighted when the beam settled on a sugar glider, a tiny marsupial possum that looks a bit like a flying squirrel. Sugar gliders live in small colonies in the forests of Australia and New Guinea and glide through the treetops at night, feeding on insects, nectar, and tree sap. However, I need not have traveled halfway around the world to see one outside a zoo. With a credit card and an Internet connection, I could buy a hand-tamed baby sugar glider for $125, I recently learned, and have it shipped right to my home from breeders in the United States or Canada.

Dogs and cats, though still dominant in the pet trade, are passé in many quarters. The lucrative growth in the pet industry now centers on exotic reptiles, aquarium fish, and furry "pocket pets"—small, reputedly low-maintenance companions for today's fast-paced lifestyles and cramped apartment and condominium living. A quick search of the Internet shows that trendy companions range in price from $5 for hamsters to $5,500 for capuchin monkeys. The offerings include sugar gliders, bush babies, capuchins, marmoset monkeys, chinchillas in eight colors, degus ("the poor man's chinchilla"), chipmunks, coatimundis, African dormice, duprasi (fat-tailed gerbils), fennec foxes, ferrets, three varieties of hamster in eighteen coat colors, African pygmy hedgehogs, kinkajous, possums, pygmy mice, spiny mice, jerboas, prairie dogs, and southern flying squirrels. By one estimate, there were 14 million of these animals in U.S. homes in the mid-1990s.[30] The increasing popularity of small exotic mammals prompted the U.S. Department of Agriculture (USDA) in 1998 to begin licensing retailers as well as breeders.

Many states, including Hawaii, California, and Florida, prohibit sales of some or all of these creatures, especially small exotic predators, such as ferrets and foxes, that have a bad track record when introduced into the wild. The rise of the Internet marketplace, however, as well as enormous trade shows and pet fairs displaying exotic wares, have made sales of these animals across state lines and even international borders hard to control. This is especially true where urban customers do not know or simply do not believe or care about the risks these creatures could pose to native wildlife. Ferret enthusiasts, for example, have waged passionate but unsuccessful battles for years to overturn California's ferret ban.[31]

The tropical fish trade is also booming. According to an industry group, there are more than 10 million fish tanks in U.S. homes holding 200 million ornamental fish of 3,000 varieties. Some 95 percent of those fish come from 300 Florida breeding farms that sell fish and aquatic plants worth $60 million per year.[32]

The desire to own exotic pets is not a phenomenon limited to the United States, as the volume of U.S. reptile exports to Europe and Asia indicates. South African conservation authorities are dealing with increasing requests to import exotic reptiles. France has cracked down on ownership of pit bull terriers only to find that youths in search of fearsome pets are turning to the illegal trade in Barbary apes.[33] The booming economy in Chile since the 1980s has sparked a parallel boom in the illegal pet trade, with authorities confiscating thousands of birds, monkeys, and land turtles from would-be importers.[34] Affluent people in Sri Lanka who once filled their ornamental ponds with small guppies, platys, and swordtails now seek large, carnivorous knifefish, catfish, and piranhas—and, as we saw earlier, get free golden apple snails as a bonus.[35]

When exotic pets grow too large, lose their cute appeal or novelty, or become expensive to feed or troublesome, a few of the luckier ones end up in sanctuaries. Many others are simply turned loose. In Hawaii, despite the most stringent laws in the United States prohibiting the importation and release of wildlife, about thirty species of reptiles and amphibians are found on the loose each year, most of them originally smuggled in as pets.[36] U.S. Geological Survey fisheries biologist Pam Fuller has noticed a recurring seasonal increase in tropical fish in the waters around Gainesville as students leave the University of Florida for the summer and apparently "liberate" their aquarium pets on their way out the door. She noticed the same thing around campus ponds in Colorado. Mollies and swordtails cannot survive winter in the Rocky Mountains, but many exotic fish—just like exotic reptiles—do survive to form breeding populations in the waters of Florida. Of the 100 foreign fish species established in Florida, most were released or escaped from aquariums or breeding farms. Nationally, though 50 percent of the exotic fish in inland waters of the United States were stocked intentionally for sport, another 25 percent were dumped from aquariums. Former pets have become economic or ecological nuisances around the globe, from golden apple snails in Sri Lanka to monk parakeets in central Chile to crab-eating macaques in Irian Jaya and, as noted, red-eared sliders just about everywhere they are sold.

In some situations, people who buy live animals do not intend to keep them. Customers in San Francisco can buy live frogs and turtles in some Asian food shops to release in the wild as religious gestures or to create favorable karma. Similarly, conservation officials in Sri Lanka round up feral water buffalo to keep them from destroying forests, only to find that locals buy them from the slaughterhouses and return them to the forests for luck or karma.

Although the pet trade dominates world traffic in animals today, there are a number of other motivations for continuing to ship live animals around the globe. One, of course, is for display and captive breeding in zoos, public aquariums, butterfly houses, and safari parks. The problem of escape or release into the wild is negligible for most such facilities, except when war or social upheaval leaves them damaged or neglected.

Another motivation for importing animals is to ranch "alternative livestock" or to stock aquaculture facilities. Emus, ostriches, and llamas all went through boom-and-bust cycles in the United States during the 1990s while the ranching of exotic hoof stock, from red deer to African antelopes, continued to grow.[37] Even in Africa, with its spectacular native wildlife, game ranchers have felt the need to import exotics. Since the 1950s, South African game ranchers have brought in European fallow deer, red deer, sambar, sika deer, hog deer, nilgai, llamas, dromedaries, addaxes, mouflon sheep, scimitar-horned oryx, Indian water buffalo, and others.[38]

Aquaculture, too, is on the rise, and production of farmed fish, shrimp, clams, and oysters has more than doubled in weight and value since the late 1980s. Farming of high-value salmon in particular has expanded rapidly, and the dominant species farmed, from Norway to Chile, is Atlantic salmon from the eastern United States.[39] One need only think back to the salmon escapes and the spread of the golden apple snail described in earlier chapters to realize that aquaculture is a "leaky" business when it comes to spreading exotics. Likewise, North American bullfrogs have escaped from frog-leg "ranches" from Ecuador to Italy to Taiwan, to become one of the world's worst invaders. These elephants of the frog world, weighing up to half a kilogram, are major predators of other frogs as well as fish, salamanders, and even mice and birds.[40]

Sometimes it is difficult to tell just why a species has turned up in a new location. Predatory Asian swamp eels, for instance, appeared at several sites in Florida and Georgia in the late 1990s, including waters bordering Everglades National Park. The fish (not true eels) are a popular menu item in Asia but are also sometimes kept in aquariums in the United States.[41] Did a number of hobbyists dump their aquariums, or did someone release live eels in hopes of starting a new local fishery? A similar unanswerable question involves the Chinese mitten crab, which appeared in San Francisco Bay in 1994. It could have arrived in ballast—one more invader in a bay already dominated by alien species. But it could just as easily have been imported and set loose by a private party hoping to create a new crab fishery.[42]

The tradition of stocking game animals, particularly birds and sport fish, continues at a strong pace, although it has slowed from its heyday in the 1950s as wildlife and fisheries scientists have become more aware of the adverse effects exotics have had in reducing habitat and resources for native animals and in spreading disease.[43] State fish and game officials in the United States still engage in an informal but frenetic wildlife bazaar among themselves, swapping everything from bass and pike to moose and chipmunks to "enrich" their regions.[44] Even efforts to boost populations of declining native wildlife can be taken to unlikely extremes. Aficionados of North American wild turkeys, admittedly magnificent birds, have not only reintroduced them throughout their native range since the 1970s but also set them loose in Hawaii amid a dozen other exotic game birds, from pheasants to peafowl.[45]

A large gap in values and outlook persists between fisheries biologists and fishermen, who often consider trout an improvement to any waterway, and ecologists alarmed at the damage large predators such as exotic trout and game fish inflict on native fish, frogs, salamanders, and other creatures. And sport fishermen are a powerful lobby, from North America to southern Africa. The number of fish species introduced into U.S. waters for all purposes in the past half century is nearly triple the number introduced prior to 1950. It includes such foreign fish as bass-like peacock cichlids for sport and grass carp and black carp for control of weeds and snail-borne parasites, respectively, in fish farm ponds.[46] Even in places where wildlife agencies have ceased stocking exotics and are working to restore native species, fishermen sometimes take matters into their own hands with "bait-bucket" introductions. That is the suspected route by which predatory lake trout showed up in Yellowstone Lake in 1994, an act of "environmental vandalism" that park officials believe threatens the lake's native cutthroat trout and a host of animals from grizzly bears to bald eagles that fatten each spring on the stream-spawning cutthroats.[47] In a less deliberate variation, fishermen dumping their live bait at the end of the day are responsible for widespread introductions of exotic crayfish and minnows. The state of Illinois, worried about the fate of four endangered native crayfish, banned the importation and sale of non-native rusty crayfish for bait in 1990.[48]

Sometimes, of course, the danger comes not from the newcomer itself but from pathogens it has picked up in its travels. The release of live red shiner bait minnows spread Asian tapeworms to endangered fish in the southwestern United States. The bait minnows had picked up the imported parasite from exotic grass carp used for weed duty in bait farm ponds.[49]

The grass carp represents another category of exotic animals released into the wild deliberately—those introduced for biocontrol of troublesome pests or weeds. Most biocontrol agents released these days are insects or pathogens that have been extensively tested to ensure that they attack only a specific target weed or pest, a topic we'll come back to later. But the disastrous days of importing and releasing mongooses, stoats, weasels, ferrets, predatory snails, and cane toads are not as far behind us as many would like. The rosy wolf snail, introduced on the Pacific island of Moorea in 1977 in a failed effort to control exotic giant African snails, instead wiped out at least seven unique local snail species. Indeed, the rosy wolf snail is imperiling native snails on islands throughout the Pacific.[50] Similarly, mongooses were introduced on Japan's Amami-Oshima Island in 1979 in hopes they would kill rats and snakes. Instead, the ferocious little predators did what they have done for a century now on islands from Jamaica to Hawaii and began preying on endangered animals. The government is now considering the costly prospect of trying to bring them under control.[51]

One little guppy-sized topminnow from the southeastern United States was endowed a century ago with the name "mosquito fish," and on the basis of that undeserved name has become the most widely distributed freshwater fish on the earth, introduced on every continent but Antarctica and most of the world's islands for mosquito control. The fishes' influence on mosquitoes has been minimal, but the damage they inflict on native fish and other aquatic species has earned them a spot on the list of the world's worst invaders. Yet "mosquito fish" continue to be widely marketed.[52]

When Good Greenery Turns Bad

In many parts of the world, the most serious invasion threat comes not from animals but from continuing deliberate importation of new species of exotic plants. The threat is exacerbated by the fact that even conservationists often find it hard to think of plants as potential thugs and invaders.

Early in 2000, a membership promotion from the National Wildlife Federation (NWF) arrived in 9,000 mailboxes in Hawaii. Inside each was a packet of North American "butterfly garden" wildflower seeds. Probably only a relative handful of Hawaiians recognized the incongruity of a major environmental group sending exotic seeds from the continent into these critically invaded islands. The ones who did, however, were very vocal. From members of the Maui Invasive Species Committee to scientists at the Bishop Museum in Honolulu, incensed citizens e-mailed, wrote, and called the organization's headquarters in Virginia. Clearly embarrassed, the acting president of NWF reported in a memorandum to the federation's directors and trustees that a recall notice had been sent "to everyone in Hawaii who received the seeds, enclosing a postage-paid envelope for return of the seeds, and asking that any plants grown from the seeds be destroyed." The memorandum also stressed that "in no way should this mailing mistake be construed as an expression of NWF's policy on invasive species or indicative of a lack of commitment on our part to conservation of native species."

Neither the mailings to Hawaii nor the 2.4 million seed packets sent to mainland addresses violated any laws. Indeed, such promotions are becoming commonplace. Recently, I've received seed packets in mailings from a lifestyle magazine, a low-income housing charity, and a group raising money for medical research. Advertising and marketing agencies in many parts of the world seem to regard generic "wildflower" seeds as the ultimate "green" promotion these days.

Editors at *New Zealand Geographic,* for instance, were embarrassed to find that barley seeds had been attached to a beer advertisement in magazines mailed overseas, including to Australia. (Australian authorities were less worried about the barley itself than about fungal diseases the seeds might carry.) In turn, Banrock Station, an Australian winery that prides itself on its environmentally friendly image, recently shipped bottles with neck tags containing golden wattle seeds to the United States and bottles with tags holding red bottlebrush seeds to New Zealand. Both were accompanied by a brochure that read, "Celebrate the Good Earth, Plant a Tree."

A tree, any tree. A flower, any flower. The concept that not everything green is benign and interchangeable, or that anything green could cause ecological degradation, is still a hard sell in many quarters. Being "green" in the sense of being environmentally aware does not yet seem to require an understanding of the threat of plant invasions. What's more, it is a message the horticultural world has only recently begun to acknowledge.

The garden trades, like the pet trade, are on a roll. In the United States, sales of horticultural products reached $10.6 billion in 1998, more than double the 1988 figure of $4.8 billion. The majority of that comes from sales of bedding and garden plants, along with potted foliage and flowering plants.[53] With the volume comes variety. At least 59,000 species and varieties of plants are for sale today in North American seed and nursery catalogs, and nearly twice that number are being traded worldwide.[54] Even plants listed on many state noxious weed lists in the United States, such as St. John's wort, or Klamath weed, or purple loosestrife, are readily available in catalogs and on seed racks. Many known invaders of forests and wildlands that have not been legally declared weeds are also widely sold and planted, including oriental bittersweet, capeweed, Japanese barberry,

Japanese honeysuckle, English ivy, periwinkle, privet, tree of heaven, and Chinese tallow tree. Any tourist or resident of Hawaii can buy packets of kahili ginger seeds to plant, even as Tim Tunison's crew at Hawaii Volcanoes National Park and other land managers all over the islands battle to clear ginger from the forests. A significant part of the plant trade, just like the pet trade, goes on informally and largely unregulated at fairs and farmers' markets and on the Internet, through both commercial sales and private seed-swapping sites. Gardeners the world over, it seems, love exotics.

"Gardeners in South Africa are used to exotics," Dave Richardson of the University of Cape Town explained. "They want oaks and flame trees. In every category—perennials, shrubs, vines, whatever—exotics are much more widely used than native plants." And this is in a region that gave the world proteas, bird-of-paradise, geraniums, and so many other spectacular plants. Even, as we saw earlier, in the staff gardens at South Africa's Kruger National Park, people want to retreat from their work on the savannas to homes landscaped with privet, lantana, ginger, and pampas grass.[55]

But is our fascination really with "exotics"? Are we insisting that nurseries offer us something new and different every year? As I think about the gardens of Cape Town and Kruger, Honolulu and Paris and Sydney, it seems to me that the attraction of many of our cosmopolitan garden plants may be their familiarity rather than their novelty. Yes, we snap up the new varieties and hybrids that the nursery trades headline each season, but these are mostly variations on a theme. Most of us plant the gardens our parents and grandparents planted. Last century's exotics have become our heirlooms, no more strangers in their adopted lands than we are. We do not plant for novelty so much as for nostalgia, for the comfort of familiar personal landscapes.

"Nursery people will tell me, 'The public demands new species,'" Sarah Reichard said one summer day as we browsed through a garden center in Missoula, Montana, during a break from a meeting. Before entering graduate school and starting on the plant-hunting venture that led to her research on invasiveness, Reichard managed a nursery for four years. "Most people come into a nursery and say, 'I've got this bare spot in my garden.' And they'll look around and say, 'Oh, look at these,'" she said as we wandered past a table loaded with flats of blue-flowering forget-me-nots. "They'll pick this up and say, 'Forget-me-nots! My grandmother had forget-me-nots. I'm going to buy some of these.'"

Few of us today even know the origins of most of the greenery and flowers that surround us. Half of the earth's population now lives in cities. We are an urban species, surrounded by artificial constructions and landscapes that frequently offer more variety than nature ever crowded into one spot. I see no reason why we shouldn't assemble our cityscapes to please ourselves. What is still hard for us to recognize, however, is that some of our imported plants are not constrained by the boundaries of city gardens and boulevards, even when they are carefully pruned and trained. Carried by birds, riding the wind, clinging to feet, tires, or clothing, seeds and pollen escape into the semiwild places beyond the concrete and crop fields. This is more than an occasional occurrence.

At least two-thirds of the imported plants now growing wild in the United States, Britain, Scandinavia, and Australia were imported deliberately, most for gardens. Among woody invaders in the United States, 82 percent have a background as landscape plants. Of the top eighty-four weeds of natural areas in South Africa, the majority were introduced as ornamentals (although forestry brought in some of the most damaging invaders). Likewise, most invasive waterweeds—water hyacinth, salvinia, Eurasian watermilfoil, hydrilla, Brazilian elodea, and even the turf-forming seaweed caulerpa—have been widely sold to decorate water gardens and oxygenate aquariums throughout the world.

Horticulture is a valuable industry and gardening is a treasured activity worldwide, and an increasing number of people such as Reichard are working to make both "greener." One key step, they believe, is to convince botanical gardens to lead the way in setting conservation standards for the horticultural profession.

There are 1,700 botanical gardens around the world today, nurturing 4 million specimens representing 80,000 species—almost one-third of the world's known plant species. As they have for centuries, botanical gardens, as well as commercial nurseries, are still sponsoring plant hunters to explore for new species. Most gardens not only display and conserve the plants they collect but also make many available to the public at plant sales and share their plants with other botanical gardens worldwide through international seed exchanges. Unfortunately, scrutinizing their own offerings for weedy potential is seldom part of the process. As a result, a number of destructive alien plant invaders, from miconia trees in Tahiti to the passion fruit vines strangling forests in Java, have gotten their start in botanical gardens.[56]

In recent years, plant hunters and botanical gardens have at least become aware of this "dark cloud over their shoulder," says ecologist Peter White, director of the North Carolina Botanical Garden in Chapel Hill. In 1996, White's garden became the first in North America, if not the world, to enact an anti-invasive species policy, including a decision not to distribute seeds outside its region. White and Reichard are leading an effort to make such policies part of the conservation ethic of gardens worldwide.

Since the 1980s, botanical gardens have greatly increased their involvement in conservation activities such as promoting off-site propagation and seed banking for endangered plant species, discouraging collection of rare plants from the wild, and protecting and restoring native plants and natural areas within their own grounds. Some 500 gardens now belong to Botanic Gardens Conservation International, which was formed in 1987 to coordinate these conservation efforts.[57] But gardens have been slow to acknowledge the role of plant invasions in endangering plant biodiversity.

"The piece that is most troubling and most recent is the invasive plant issue," White says. "There are many gardens that talk a conservation line in all ways but that. That's always the hidden thing that contradicts the conservation mission, even if they have jumped on the bandwagon in many other ways."

When White—one of the few ecologists to direct a botanical garden—first began raising the issue in the mid-1990s, he was surprised to find that "the initial reactions were pretty hostile. Back then, it created great angst in the horticultural world because horticulturalists tend to think of themselves as inherently pro-environment, doing a good thing, making the world beautiful. A lot of gardening is more emotive than intellectual, and they were being dragged into an academic frame of reference with this issue. Gardeners tend to think of plants as dabs of paint on a palette, and they forget that these plants have developed their characteristics as a result of evolution in an environment full of lots of other species, friends and enemies."

But attitudes are evolving quickly, White noted, and "the old plant-anything attitude has become the extremist position. The middle ground in the horticultural world is now willing to admit there's a problem and the horticultural community has to participate in the solutions." The new battleground, he said, is deciding "which are the true bad guys and which are the good guys" in any given region. That's a constructive struggle that we can expect to last for quite a while.

Horticulture is not alone in coming slowly to the recognition that some plants can cause economic and ecological damage when moved to new settings. There is also no tradition of checking plants for weediness before they are released into working landscapes, as we saw earlier with the pasture grass trials in Australia that produced far more weeds than forage crops (and even forage crops that largely doubled as weeds).

Invasive plants such as kudzu, multiflora roses, and Russian olive trees were originally introduced into the United States as ornamentals but were then widely dispersed by what is now the Natural Resource Conservation Service (NRCS) to prevent soil erosion. The service, for instance, distributed 85 million cuttings of kudzu to southern farmers and paid them a bonus per acre to plant their fields with them.[58] In New Zealand, seeds of North American lodgepole pine were sown over native grasslands and shrublands by helicopter in the name of soil stabilization, and imported willows were planted to stabilize riverbanks. Both the willows and pines are now regarded as major invaders. Despite this history, soil conservation agencies "seem to be trapped on a devil's merry-go-round," as ecologist Francis Harty put it, continuing to promote plantings of imported dogwood, for instance, over any of the eleven native species of shrubby dogwood that could provide equivalent service in stabilizing soil and supplying fruit for wildlife in the eastern United States.[59] At least seven of the twenty-two exotic plant species that the NRCS promoted during the 1980s were potentially invasive.[60]

Since the 1980s, with financial support from the USDA, seed companies in Oregon have promoted the export of multiple varieties of ryegrass, fescue, and orchard grass to China for lawns and erosion control on sites such as the Three Gorges Dam and highway cuts and to help eliminate the dust storms that plague Beijing. From 1994 to 1999, Oregon grass seed exports to China grew from 39,000 to nearly 3 million kilograms, and the Oregon Seed Council projects that figure could rise to more than 45 million kilograms annually in five to ten years. Scientists in China have just begun to push for investigation of the invasive potential of these vast plantings. Meanwhile, the Chinese Academy of Agricultural Sciences has collected and preserved more than 1,000 native grass species, many of which could fill the same need as the exotics.[61]

Aid and development agencies likewise frequently rely on a few widely husbanded exotic species rather than investigate the potential of native resources to provide sustainable food and fiber crops. Thus, many well-intentioned agencies are still spreading known invasive species such as mesquite and leucaena trees, exotic fodder grasses, crayfish, and tilapia. Such species can cause long-term economic and ecological problems that eventually outweigh the short-term benefits to local people. Still, agencies continue to search for purported "wonder trees" like the drumstick tree, or moringa, and other exotic species that seem to promise a quick fix to overwhelming problems such as desertification, erosion, and deforestation.

Even the status of apparently useful exotic species can change as local land-use patterns shift. The introduction of electrical service, for example, may reduce the need for firewood, the cutting of which once kept potentially invasive trees in check. Or a change in livestock grazing may allow a species to grow out of control. On the island of Vanuatu during the 1980s and 1990s, aid programs promoted the development of numerous small timber plantations of laurel trees from Central America. Unfortunately, the laurel trees do not grow as well in the South Pacific as they do when used to shade coffee plantations back home, and the timber ventures have largely failed. The trees do, however, sprout profusely in overgrazed pastures. When cattle numbers are reduced, the laurels quickly form dense thickets.[62] Even in southern Australia, plant ecologist Mark Lonsdale points out that *Lolium* ryegrass imported to nurture sheep has become one of the most noxious weeds of grainfields now that sheep farming is on the decline.

Just as botanical gardens serve an increasingly vital role in conserving the world's plant diversity through their collections and seed banks, agricultural germ plasm collections store and maintain seeds of wild and cultivated plants that form the basis of the world's food supply. The National Plant Germplasm System, for instance, has nearly half a million varieties stored in dozens of refrigerated vaults around the United States. Yet little consideration has usually been given to weediness when these are handed out to crop breeders or planted to generate fresh seeds. Indeed, the system is an outgrowth of the Section of Seed and Plant Introduction, created in 1898, which introduced tens of thousands of new plants into the country.

Nor does invasive potential seem to be an integral concern at the USDA Agricultural Research Service (ARS), which recently had researchers evaluating rubbervine, the kudzu of Queensland, as a potential alternative energy crop. The ARS report makes no mention of weediness but does give rubbervine positive marks for its "potential to flourish in marginal arid lands" and its "rapid growth rate suitable for annual harvesting."[63] Imagine what could happen to those fields of vines if the market for biomass fuels collapsed and made harvesting the crop unprofitable! But rubbervine would not be the first noxious weed that farmers eagerly spread across their fields when market demand was high. St. John's wort was once a rangeland scourge, as spotted knapweed is today. In the 1940s, this toxic weed infested some 2 million hectares in the northwestern United States and western Canada and also plagued cattle ranchers in Hawaii, Australia, New Zealand, Chile, and South Africa until a successful biocontrol effort knocked the weed back dramatically.[64] St. John's wort remains on many state noxious weed lists in the western United States, but its sudden popularity as an herbal antidepressant in the 1990s created a lucrative market. As a result, farmers in Washington successfully petitioned for state permission to plant their fields in St. John's wort.

Finally, even as nations commit increasing resources to controlling the damaging side effects of the great Columbian Exchange, there are researchers promoting wider use of the plants and animals that missed that first great mixing. The United States' National Research Council, through its Board on Science and Technology for International Development, has published a number of studies with titles such as *Lost Crops of the Incas: Little-Known Plants of the Andes with Promise for Worldwide Cultivation* and *Microlivestock: Little-Known Small Animals with a Promising Economic Future.*[65] To its credit, the latter volume warns that "if misunderstood, this book is potentially dangerous" because of the risk that some of the species it recommends—including such rodents as the agouti, capybara, hutia, and mara and the widely notorious invader nutria—could become serious pests if introduced into new locales. There is one recommendation in the series that would have particularly stirred the approval of Isidore Geoffroy Saint-Hilaire and his followers in the French Société Zoologique d'Acclimatation—the suggestion that wider use might still be made of yaks.[66]

Invasions almost always result, directly or indirectly, from economic activity. That makes controlling invasions a profoundly human problem, not just a biological one. We move species to new locales, nurture them, open up ground for them, spread them around for reasons that range from primal to frivolous. It is our behavior, our wants, needs, and values, that creates most invasions. Jeff McNeely, chief scientist for the World Conservation Union (IUCN), puts it this way: "Ecologists can't solve the problem of invasive alien species any more than doctors can solve [that of] AIDS. They can help with the symptoms, but it's people who are the drivers." If we are to slow the rising tide of invaders, we must be willing to apply more judgment and restraint in decisions about which plants and creatures to move as well as in efforts to prevent the accidental movement of harmful species. For that, we will need a better understanding of how to spot potentially aggressive and troublesome species before we decide to introduce them.

SIX The Making of a Pest

"Changes will occur whether we intend them or not. We cannot leave one footprint in a new country, pass through it with horses or mules, careen our *Golden Hind* on its empty beach, without bringing in our luggage or our pockets, in our infested hair or clothes, in our garbage, in the dung of our animals and the sputum of our sick and the very dirt under our fingernails, things which were not there before, and whose compatibility with the native flora and fauna is utterly unknown."

—Wallace Stegner, *American Places,* 1981

"The gardening definition of a weed is 'a plant growing in the wrong place.' I can't agree with that. A chrysanthemum may grow in the wrong place, but I can dig it up and be done with it. If weeds were so amenable to interference, they would not be weeds. A weed is a plant that is not only in the wrong place, but intends to stay."

—Sara Stein, *My Weeds,* 1988

In the austral summer of 1988, Sarah Reichard spent months hunting for plants in the Chilean Andes, searching subalpine forests and roadsides alike. Her main target was a small evergreen tree called *Drimys winteri,* or Winter's bark, a member of one of the most primitive of the flowering plant families. Most of us would find her purpose esoteric at best. She was doing the sort of thing that systematists or taxonomists have done for 250 years, gathering evidence that could help her figure out the family tree of relationships among three similar forms of *Drimys* that occur at middle to high altitudes in Chile and Argentina. That was to be the topic of her master's thesis at the University of Washington and the prelude to an intended career in systematics. Like most plant-hunting systematists over the centuries, however, Reichard had a second, less scientific mission. The university arboretum had paid her airfare to Chile and asked in return that she keep an eye out for "anything pretty" that might grow in the gardens of Seattle, a cool region of wet winters and dry summers like much of the Chilean Andes.

By day, as Reichard hiked the fields and dry riverbeds, she gathered seeds from a little *Calceolaria* with an unusual spike of tiny yellow flowers; a local species of *Buddleia,* or butterfly bush; a lovely purple-flowered vine; and more than a dozen other pleasing garden prospects. At night, however, as she sat in village hotels cleaning the day's seed collections, Reichard began to wonder about things traditional plant hunters never considered.

"I thought, 'Wait a minute; what am I doing? I could be bringing back a terrible weed.' This was before the consciousness about invasive species really hit, but I'd taken ecology classes where people talked about invasions," she recalled. "So I thought, 'When I get back to Washington, I'll go to the library and look up how you do an evaluation for invasive potential, and I'll evaluate all these species. If there's one I think is really going to be bad, I just won't turn it over.'"

It was a good plan, but there was a hitch, as Reichard found when she returned to Washington. "I read that there was no way to predict invasive potential, but that just didn't make sense to me. We're seeing the same things invading over and over again and others never invading, so there has to be some reason. It's just a matter of figuring out what it is."

Figuring it out became so compelling that after finishing her master's degree, Reichard plunged full-time into the topic of invasiveness. Now, as an assistant professor at the University of Washington, she remains hooked, one of a growing number of ecologists who have recently found themselves drawn into the search for what has been called the Holy Grail of invasion biology.

The search for a singular object like a Holy Grail, however, seems to me the wrong metaphor to describe the pursuit of such a dynamic and elusive quality as invasiveness. Instead, the search for reasons why a rather small proportion of the earth's plants and animals succeed at invading new territories—and, in turn, why some communities are more often breached than others—is more like a quest for the secrets of successful entrepreneurs. It is necessarily an after-the-fact exercise, an attempt to discern a formula for success by scrutinizing the traits and behaviors of the successful. The result is more often an explanation for success than a formula to predict it. Identify the traits that propelled Bill Gates, say, or Sam Walton or Ted Turner into the ranks of billionaires and you are still unlikely to come up with a general profile that will let you predict with any confidence which earnest young person with a business plan will become tomorrow's corporate mogul. The invasive potential of plants, animals, and microbes can be similarly cryptic, especially when we have nothing to go on but their behavior in their home communities.

Take the house sparrow, a rather sedentary bird that fledges three to five chicks each year in its European homeland. What formula could have warned the acclimatizers and their like—had they cared—that this sparrow would rapidly take much of the New World by storm? Yet nineteenth-century observers reported sparrow pairs producing twenty-four fledglings per year as the birds exploded across North America, and thirty-one fledglings per year in New Zealand. The birds essentially bred continuously, with the young of each clutch brooding the eggs of the next, and so on, year-round—an unpredictable behavior, given that it had never been known to happen in Europe.[1] In a similar surprise conquest, shrubby longleafed wattle trees upped their seed output by more than a thousandfold when they were moved from Australia to South Africa and began spreading along watercourses in the South African Cape.[2] Some introduced plants and animals explode aggressively in certain settings yet remain unobtrusive in others. Eurasian cheatgrass has transformed the American West, yet in New Zealand it is little more than a roadside weed. And then there are the exotic plants such as Brazilian pepper trees in Florida that appear benign for decades or even centuries before their populations explode and they begin to expand their territories.

Despite such complexities, the field of invasion biology is booming. Reichard and other ecologists are searching as never before for patterns and clues in the traits of successful invaders, the circumstances of successful invasions, and also the characteristics of heavily invaded habitats. The power to spot potential troublemakers before they begin to run amok could help scientists, policy makers, and managers craft reasoned decisions about which exotic species to admit deliberately into new territories—for agriculture, horticulture, forestry, aquaculture, the pet and aquarium trade, fish and game stocking, economic development, or biological pest control—and which accidental arrivals to battle most vigorously.

Talent, Luck, or a Helping Hand

Colonizing or expanding into new territories is a natural process of plants and animals, a necessary process, and the identities of species in any habitat change over time as environmental conditions shift and various species thrive or falter. Today, however, this process has been blindingly accelerated by human activities. The first person to bring that fact to wide public attention and lay the foundation for a science of invasion biology was pioneering British ecologist Charles Elton. Elton first pulled together his thoughts on bioinvasions for a series of three radio broadcasts he delivered for the British Broadcasting Corporation in 1957. A year later, he published the material in an accessible and enduring classic titled *The Ecology of Invasions by Animals and Plants.*[3] Elton recognized both the dramatic environmental effects invasive species were already having and the likelihood that those effects would increase with more human trade and travel. Although geologic forces had occasionally connected continents and allowed some mixing of previously isolated species, he wrote, "we are artificially stepping up the whole business, and feeling the manifold consequences. . . . We must make no mistake: we are seeing one of the great historical convulsions in the world's fauna and flora."[4]

Elton chronicled the depredations of many invaders, such as chestnut blight, starlings, muskrats, cordgrass (spartina grass), sea lampreys, Chinese mitten crabs, Argentine ants, and gypsy moths, that continue to cause severe economic and ecological losses nearly half a century later. In his writings, Elton likened himself to a war correspondent, "recounting the quiet infiltration of commando forces, the surprise attacks, the successive waves of later reinforcements after the first spearhead fails to get a foothold, attack and counter attack, and the eventual expansion and occupation of territory from which they are unlikely to be ousted again."[5] Elton clearly recognized that the vast majority of those infiltrators that reach new territory quickly perish, and he lamented our inability to predict which ones would beat those odds and go on to invade.

Invasion is a multistage process from arrival to establishment to spread, and accidentally introduced immigrant species face long odds at each step. Some 90 percent of those that survive each step are winnowed out at the next.[6] A potential invader must first survive some type of journey, perhaps sloshing in ballast water, sealed in a shipping container, mingled in grain shipments, or nestled in a load of fruit, and then establish a beachhead wherever it arrives. It is impossible to say how many living things perish on such journeys or soon after arrival. Even among species known to be invasive, most individuals fail this first step (unless, of course, they are imported deliberately and cared for during and after the journey). The survivors, their numbers perhaps bolstered by continuing arrivals, then face new and different hurdles. They must persist and reproduce successfully until their descendants establish a self-sustaining population—that is, naturalize—in their new community. Finally, a few years, decades, or even centuries later, some of these naturalized species will begin to proliferate wildly and disperse across the landscape.[7] These are the "ecological explosions" Elton described, the militant invasions that threaten economic activities, ecological services, or native species.

Plants and animals seem to vary dramatically in their mastery of the invasion process. Water hyacinths, Cogon grass, guavas, lantanas, wild oats, starlings, rats, and a few hundred others have had such spectacular and repeated success around the globe that it's easy to see why some ecologists suspect they possess extraordinary talents. All the while, many other species have never made a mark outside their native range.

Biologists have long pondered the remarkable success, for instance, of the Old World plants and animals that conquered the New World in the wake of the European conquistadores and colonists—a case of "ecological imperialism," as historian Alfred Crosby called it, or, in ecologist Dan Simberloff's words, a "synergistic juggernaut" of humans and their "biological allies."[8] Had these species become preadapted by enduring thousands of years of human influence in Eurasia to thrive and compete well in the kinds of situations we create? Or did human history simply provide them with exceptional opportunities?

Early on, biologists tried to formulate broad general laws that would explain invasion success, but these are plagued with too many exceptions to be useful. One seemingly plausible idea, for example, is to assume guilt by association. Some of the world's most notorious flowering plant invaders are concentrated in a few singularly weedy families such as the aster or composite family (which includes the knapweeds, thistles, parthenium, ragwort, cockleburs, groundsel, and dandelions) and the meadow grasses of the Gramineae family (including jointed goatgrass, quackgrass, wild oats, cheatgrass, crabgrass, and Johnson grass). Yet these are huge families with many "benign" members, so family ties give us unreliable clues about invasive potential. Furthermore, some of the world's worst invaders, such as water hyacinth, are black sheep in their families and have no close relatives of similarly bad character.[9]

Another idea, which began with Charles Darwin, suggests that alien plants are more likely to invade successfully in places where they have no close relatives. Certainly, pines are among the most invasive trees in the temperate regions of the Southern Hemisphere, where there are no native pines. And some of the worst North American invaders, including melaleuca, tamarisk, kudzu, knapweed, hydrilla, and water hyacinth, had no close native relatives there. One possible advantage of being the first of your kind in a new land is that you're less likely to meet up with pests that coevolved with native relatives.[10]

Many immigrant species also have the advantage of having left behind their natural enemies, including parasites, pathogens, predators, and competitors. House sparrows in North America are burdened by half as many parasites as those in Europe, and the birds have acquired two-thirds of those since they arrived. That means they had a chance to colonize while relatively unencumbered until the local parasites learned to target them.[11] (Like the marathon breeding mentioned earlier, relative freedom from enemies may or may not have been the sparrows' secret to success.) The Australian brush-tail possum has 80 percent fewer parasites and less competition for food and shelter in the forests of New Zealand than it does back home, and it reaches population densities ten times greater. The idea that invaders proliferate wildly because they escape their natural enemies provides the rationale for biocontrol efforts that seek out enemies back home and import them to try to control invasive species.[12] Yet if escape from enemies were the key to all invasions, nearly every exotic arrival ought to run amok. Even biocontrol is complicated by the fact that leaf-munching insects, fungal blights, and other natural enemies may profoundly damage some individuals of a target species without effectively reducing the population.

Finally, you might expect that the climate in a species' home range would help to predict the areas where it could invade. Yet this sort of climate matching is fraught with exceptions and limitations. Species may be more adaptable than one might conclude by looking at conditions in their original range. Also, species evolve. European rabbits in the interior of Australia now thrive in hotter, more arid climates than their ancestors experienced in Europe. Monterey pines were once confined largely to the foggy coastline of northern California, yet they now proliferate in a wide array of conditions in the Southern Hemisphere.[13] Birds from temperate regions introduced on tropical Hawaii have invaded just as successfully as birds imported from other tropical areas.[14]

With such an unreliable set of generalities to go on, it was reasonable for ecologists to ask whether there is something special about aggressive species themselves. That is, are there certain biological attributes or talents within any given group of plants or animals that mark certain species for success? One of the best-known efforts to profile the "ideal weed," for instance, was made in the 1960s and early 1970s by Herbert Baker, a biologist at the University of California, Berkeley. The characteristics he came up with include continuous and prolific seed production and no fussy requirements for seed germination or specialized pollinators. Many known invaders do not fit the profile, however, whereas many meek species do. Furthermore, Baker defined weeds as plants that grow only in crop fields and other areas that have been "markedly disturbed by man" and have no detected human value. Many transforming invaders of forests and other natural communities, including shrubs and vines that dominate the shady understory, bear little resemblance to Baker's "ideal weed."[15] What's more, many of our worst weeds, from herbs such as St. John's wort to pines to ornamentals such as English ivy and lantana, have economic value to some parts of society. Ecologist Mark Williamson of the University of York believes that trying to apply Baker's characteristics in risk assessment is "pointless" (although some companies do just that to argue that their genetically engineered plants will not be weedy).[16]

In recent years, a number of scientific teams have turned to statistical approaches to try to link certain traits of trees and shrubs with invasiveness. (A word of caution: the studies I will describe here use the term *invasive* as ecologists often do, to mean the ability to naturalize and spread in places where they are not native, even if they cause no obvious impacts.[17] Elsewhere in this book, *invasive* means not only proliferating but also causing economic or ecological harm.)

In one such effort, Marcel Rejmánek of the University of California, Davis, teamed up with Dave Richardson of the University of Cape Town to examine the life history traits of two dozen pines from Europe and North and Central America that have been widely planted throughout the Southern Hemisphere. Half of these pines have spread to some extent outside of cultivation—including the five most troublesome invaders: Monterey, lodgepole, Aleppo, cluster, and Mexican weeping pines. The other dozen species have not spread. All the pines that do spread share three traits, the researchers found: they mature to reproductive age quickly, they produce large seed crops at short intervals, and they bear small seeds. When Rejmánek and Richardson checked these same three traits in woody plants other than pines, they found they could separate out reasonably well the species known to spread in disturbed landscapes. Other factors, however, can complicate any simple predictions based on these traits. An otherwise noninvasive shrub or tree may be more likely to spread, for instance, if it produces small-seeded fruits and berries that are eaten and spread by animals such as birds, bats, monkeys, or squirrels. European stone pines introduced into South Africa stayed where they were planted for two centuries, until the American gray squirrel showed up to disperse their seeds.[18]

Reichard, too, has probed the attributes of woody plants, working with Clement Hamilton at the University of Washington to assemble a "decision tree" that policy makers, quarantine officials, or even plant hunters or sellers might use to decide whether to allow the importation or sale of new plant species—a potentially valuable tool, given that demand for new plant introductions for horticulture and agriculture continues to increase and, in the United States, remains largely unregulated. Reichard and Hamilton found, as many others have, that the strongest predictor of invasiveness is a history of invasions elsewhere in the world.[19] Looking back to the analogy of human entrepreneurs, this means your best bet at picking the next winner is to put your money on someone with a track record of success. For biological bets, you can be fairly certain about the result if you unleash water hyacinths, wild oats, or rats in a new setting.

Indeed, an invasive track record is a key bit of information that quarantine officials in Australia and New Zealand look for in determining whether to permit importation of a new plant species of any type—trees, shrubs, grasses, annual flowers, vines. Invasion history is also the basis for an early warning system and computerized database developed by the World Conservation Union (IUCN) as part of the Global Invasive Species Programme (GISP).[20]

Even for woody plants without such a track record, Reichard and Hamilton found they could predict invasive potential with 76 percent accuracy by looking for such red flags as the ability to reproduce or spread by means of vegetative growth such as stolons or rhizomes rather than seeds, the ability to grow rapidly in the juvenile stage or reach maturity relatively quickly, or the ability to produce seeds that sprout without any pretreatment. The team found it harder to predict plants that would not invade than those that would, meaning that the decision tree tars the name of many plants that would be likely to behave well in North America.

And the researchers did not attempt to predict which of the plants that spread beyond cultivation would produce harmful ecological or economic effects.[21]

If invasive plant prediction is still in its infancy, the situation is worse for other species: only recently have similar predictive schemes been attempted for mammals, birds, insects, other types of animals, or nonwoody plants. University of Notre Dame biologist David Lodge and graduate student Cindy Kolar have used the same approach as Reichard to attempt to predict which fish species that arrive in the Great Lakes in ballast water are likely to become established. Most important, they found, is how many individuals are delivered—the more, the better chance of success. This factor is called "propagule pressure," and many studies of other organisms have pinpointed it as a strong predictor of invasion success. The other factors of importance for fish invasions are how fast a species grows, its ability to survive the water temperature in the lakes, and—not surprisingly—its history of invasion elsewhere. Lodge and Kolar used those criteria to assess the invasion risk posed by species from the Ponto-Caspian Basin (composed of the Black Sea, the Caspian Sea, and the Sea of Azov), which has supplied 70 percent of the invaders discovered in the lakes since 1985, including the zebra mussel, ruffe, and round goby.[22] Twenty-three of the sixty-eight species screened could invade, they concluded. Lodge can envision using such an approach to help set ballasting restrictions for ships visiting ports infested with high-risk fish species.[23]

So far, the traits linked to invasiveness appear to differ from one study to the next and from one group of taxa to the next. Although Rejmánek and Richardson found that small seeds mark woody invaders, others have found that big seeds are the key or, like Reichard and Hamilton, that seed size is irrelevant. This apparent inconsistency has led Williamson to conclude that predicting the "behavior of a species in a new environment may be effectively impossible."[24]

Lodge, Richardson, Reichard, and others, however, are not surprised or troubled that various scientific teams are coming up with different lists of predictive traits. "There is no single Holy Grail," Lodge contends. "Invasiveness is too broad a topic. We have to hone our questions more narrowly." Predictive tools can be developed, but each will have to be specific to a single group of organisms entering a specific ecosystem. Furthermore, Lodge and Kolar say, researchers must specify what stage in the dynamic process of invasion they want to predict. The probability that a fish will become established in the Great Lakes may be associated with different traits from those that indicate its potential for rapid spread. And the analysis developed for the Great Lakes would not necessarily serve in other waterways.[25]

For now, prediction certainly seems largely impractical on any basis but invasion history. Some traits that have been linked with invasiveness—relatively small genome size and high "specific leaf area," for instance[26]—require technical information not easily available to plant importers or quarantine officials.

You can find a wide range of opinions about the practical value of pressing on with such efforts. At the Commonwealth Scientific and Industrial Research Organisation (CSIRO) in Canberra, Australia, plant ecologist Mark Lonsdale describes the approach as "barking up the wrong tree" because predictions of fundamentally rare events such as invasions will always yield mostly false positives—that is, will finger more innocent species than actual troublemakers.[27] Simberloff has been equally pessimistic about pursuing trait-based prediction schemes for the opposite reason, fearing they could create "a false sense of security among managers and policy makers" that the species they clear for importation will behave.[28]

Note that the researchers who pursue predictions always speak of likelihood, probability, and potential rather than declaring that a species will or will not invade. That's because timing, chance, and historical circumstance play a big role in determining the success of any immigrant population, no matter how talented at invasion the species has proven to be. Think once again about human entrepreneurs. No matter how great their talents or ideas, any number of chance events or quirks of timing can cause their enterprises to fail—political events that shake the financial markets, a shift in styles or fashions, a breakthrough in technology, or even a fire at a distant factory that supplies critical parts.

As Lodge and others have found, the fate of plant and animal immigrants hinges at least as much on how many individuals arrive and the circumstances that greet them as on their own aptitudes. The larger the population that arrives, and the more often, the greater the chance that the species will succeed if it has at least the minimum qualifications to survive. Numbers count most in the perilous period when only a limited number of individuals has established a beachhead in a new land and their fate is buffeted by random events—what scientists call stochastic events—such as bad weather, predators, poor reproductive success, or sparse food or nutrient supplies. Yet humans can do a great deal to buffer the vagaries of chance and timing and to enhance the likelihood that a newly arrived species will become established—or, for an unwanted species, the likelihood that it will fail.

Cultivation versus Chance

Washington State University ecologist Richard Mack remembers being under the impression in the mid-1980s that most of the exotic plants that had naturalized or settled into the native communities of the world had arrived by accident, as seed contaminants or hitchhikers. His perception began to shift, however, as he pored over catalogs from eighteenth- and nineteenth-century seed merchants and nurserymen. As we saw earlier, he found that many of the worst exotic weeds in the United States were once sold commercially. Mack began to wonder about the origins of other exotic plants, both weedy and benign, and how the circumstances of their arrival had affected their fate. Since the mid-1990s, he and a number of colleagues have been examining records of thousands of non-native plants now growing wild in various regions.

"We've got solid numbers now for four representative sections of the United States—the Northeast, Florida, north-central Texas, and Hawaii—and conservatively, two-thirds of these naturalized floras were deliberately introduced," he told me. "Probably many of the remaining third were deliberately introduced, too, but we just don't have the records. That two-thirds figure is also true for Britain and for South Australia."[29] Among trees and shrubs, the vast majority of exotics in every region of the world were intentionally introduced.[30]

What difference does it make whether seeds arrived from overseas in a packet mailed to a gardener or clinging to the pant leg of a traveler? "It's not much of a leap to figure out that the deliberate imports had cultivation lavished upon them," Mack pointed out. And cultivation is exactly the sort of help a small founder population requires to survive environmental ups and downs and begin to grow in size until it is no longer so subject to the vagaries of the environment. The deliberately introduced species weren't pulling themselves up by their bootstraps. Someone was watering them, fertilizing them, spraying for pests, pulling the weeds that competed with or shaded them, perhaps covering them on frosty nights. The garden plants, the street and timber trees, even weeds like cheatgrass that found their way into cultivated fields, all benefited from human nurture. Likewise, pigs, goats, cats, and many other animals often had the benefit of our husbandry to get them off to a good start.

"The traits of a species are clearly important—you can't grow palm trees in the Arctic. But traits don't mean anything unless the population survives long enough to reproduce," Mack commented. "The best of all possible worlds for naturalizing species is to bring in a huge number and deliberately cultivate them generation after generation." Instead of probing the traits of successful invaders, Mack and his students are now investigating the minimum levels of human assistance that different-sized populations of a species need to become established.

In the rolling hills west of the university's Pullman, Washington, campus, on a narrow strip of floodplain along the Snake River, graduate student Mark Minton has been providing various levels of help to plantings of four weedy crop species: Japanese millet, buckwheat, native sunflower, and crimson clover. As I watched on a blazing summer day, sprinkler heads clicked and whirred like locusts as they sprayed the field plots with water from the sluggish, dam-tamed Snake. Various plantings, from 50 to 1,000 seeds each, were receiving irrigation water; some were caged to protect them from rodents and birds; some were defended with pesticides.

"We know most introductions consist of small populations, and we think chance is a big factor, maybe the most important factor, in whether they establish," Minton said as we ducked beneath the rotating sprinkler head. "Often it takes lots of introductions before even a species that is known to do well, like sunflowers or millet, takes hold." After three summers, Minton's results show that, indeed, populations of his experimental plants increase as more cultivation is applied. Irrigating the plants and excluding pests and grazers upped the number of survivors twofold in each generation. Also, as theory predicts, the populations that started with fewer seeds experienced more radical swings in size from one generation to the next than did the populations with a larger number of founders.

The farm at the U.S. Department of Agriculture (USDA) Central Ferry Research Station where Minton conducted some of his experiments is part of the USDA's National Plant Germplasm System, one of a group of plant introduction facilities around the country where stocks of exotic seeds are stored, and periodically grown out and reharvested, to provide a continuing supply of novel genetic material for plant breeders. At Central Ferry alone, the cold storage unit holds thousands of strains of warm-season grasses from throughout the world. Minton can imagine a day when any plant, before it is introduced into agricultural or ornamental use, has to go through some sort of cultivation trial like the one he conducted rather than an invasiveness analysis based on traits alone. (Currently, the USDA requires no test of any sort for invasiveness, although it does require field testing for genetically modified organisms.)

"I would like to develop a test that we could use to screen plants before we introduce them into new areas," Minton said. "If there's a high probability that a species is going to become naturalized without any cultivation, let's don't introduce it. The facilities are here to do the tests; it just requires a change in attitudes and protocols and emphasis."

As I listened to his ideas for a screening system, it struck me that many of the traits that would cause an ecologist to reject a plant are the very same ones that would cause an agriculturalist or nurseryman to embrace it. I remembered Mark Lonsdale's tally of pasture grass introductions into the savannas of northern Australia: of 463 exotic forage species introduced, only 21 had proven useful, and 60 had become weeds of crops and natural areas—including 17 of the 21 useful ones. The vast majority of invasive plants, it seems, were introduced because someone found them useful or pleasing. And we gardeners as well as farmers have a special fondness for hardy, vigorous plants that thrive with little tending. Thus, in many cases, we have unwittingly selected for potential invaders.

Invasion-Prone or Invulnerable

The traits of a species and the conditions of its introduction are clearly important for invasion success, but so is the response of the community. This brings us to the second long-standing question in invasion biology: Do some communities repel or accept invaders more readily than others? And if so, why? This is one aspect of invasion biology in which Elton's ideas may have been off track. When he looked at such heavily invaded habitats as islands such as Hawaii or weedy roadsides, crop fields, and other sites greatly altered by humans, Elton hypothesized that these areas are inherently more invasible because they are poor in biodiversity. In other words, they contain a smaller number of native species, either because humans have eliminated them (as on roadsides and in plowed fields) or because certain types of creatures, such as mammals, ants, and snakes, never made it to such places on their own.

Elton proposed that a community's resistance to invasions goes up with its species count because a large complement of species is more likely to fill all the available habitat niches. A newcomer arriving in such a community would have to oust a native to have a chance at becoming established: "They will find themselves entering a highly complex community of different populations, they will search for breeding sites and find them occupied, for food that other species are already eating, for cover that other animals are sheltering in, and they will bump into them and be bumped into—and often be bumped off. . . . The shortest way of describing this situation . . . is to say that (the invader) is meeting *ecological resistance.*"[31]

Elton's reasoning was grounded in a belief that pervaded the field of ecology in the 1950s: species diversity promotes stability, and a stable system should be harder for newcomers to break into. By the 1970s, the hypothesized link between diversity and stability had come under severe attack. Yet even today, you will find these notions commonplace in textbooks and in popular lore.[32]

Clearly, some communities and regions are now more heavily invaded than others. But does this mean those systems are inherently more invasible? Does it have anything to do with vacant niches, the number of native species, or low ecological resistance? Like the traits of successful invaders, the attributes of invaded communities have come under increasing scrutiny in recent years, but with little resolution. Some studies agree with Elton, others directly contradict him, and still others conclude that the species count may be irrelevant.

Here are two samples from a growing list of studies: Ecologist David Tilman of the University of Minnesota has found that in prairie grasslands, plots with higher plant species diversity do contain fewer invaders, just as Elton predicted.[33] Yet when ecologist Tom Stohlgren of the U.S. Geological Survey (USGS) and his team from Colorado State University counted plants in hundreds of plots, from the grasslands of Minnesota, Wyoming, South Dakota, and Colorado to forests and mountain meadows in the Colorado Rocky Mountains, they found the opposite: the more native species, the more exotics. Stohlgren believes that the same factors, such as fertile soil and abundant water, that may allow a large number of native plant species to crowd into a community may also make the site more inviting to newcomers.[34]

Some studies of animal invaders have also found little support for Elton's prediction. Ecologist Ted Case of the University of California, San Diego, looked at bird invasions on islands worldwide and found that a high diversity of native birds neither helps nor hinders invasions.[35] And for freshwater fish invasions in California, Peter Moyle of the University of California, Davis, concluded that physical factors in the stream were the key to invasion success, not resistance from the natives.[36]

"I think the take-home message is that species richness is not a good measure of invasibility," Simberloff says. Richardson agrees: "I think it's getting us nowhere fast. It's a nonissue." Besides, he points out, most studies count the diversity only in a single taxon—say, the number of plants, fish, or birds—instead of looking at the full range of life in a community. A newly arrived plant must interact not just with the native plants but also with an array of animals and microbes that have the potential to eat it or sicken it or give it a boost by pollinating it, dispersing its seeds, or forming a helpful (symbiotic) partnership with its roots. Furthermore, counting the surviving native species in some communities may tell us little about what native diversity was like when an invasion occurred, especially if an invader such as kudzu, rubbervine, mimosa, or lantana now smothers most of the landscape.[37]

Mack and many others, however, are not ready to give up on the idea that greater species diversity yields greater invasion resistance. Much of the work rejecting Elton's hypothesis is fatally flawed, he believes. For one thing, counting and comparing all alien species present in various plots provides only a snapshot of a dynamic process. Some of the aliens captured in the counts may be "waifs and transients" whose persistence and invasion success remain to be seen. More important, the studies seldom account for differences between sites in opportunities for invasion. "A site may appear to be invulnerable simply because it has received few immigrants," Mack points out.[38]

Williamson believes that although some communities clearly have suffered more invasions than others, it has yet to be proven that those communities are intrinsically more invasible. All communities, he suggests, "are more or less equally invasible." A community's fate rests on how many potential invaders arrive and how many of them have the right skills for colonizing the place.[39] If enough suitably prepared invaders assault any community, some will succeed.

Whether ecosystems such as islands are intrinsically more *invasible* than others, their native communities are clearly more *vulnerable* to the influence of invaders, especially those that add new levels to the food chain or bring other unprecedented roles or talents into a community.[40] The most vulnerable ecosystems are islands and other communities that have remained geographically or evolutionarily isolated, as well as agricultural systems, polluted or degraded systems, waterways, riparian areas, bays, and estuaries.[41]

Disturbance and Teamwork

It appears that invasibility, just like invasiveness, is not a fixed attribute but a continuing interaction between the character and number of would-be invaders and the character of the receiving community—the number and identities of its resident species, its climate, its soils, and myriad other ecological features. The character of both immigrants and communities can fluctuate over time, however, and so can the likelihood of invasion. Disturbance, for instance, whether human or natural, can greatly alter invasion potential.

Human activity always brings with it some level of disturbance. As the number of visitors to some of our most pristine and protected landscapes goes up, for example, so does the number of plant and animal invaders. The pattern has been documented for nature reserves in the Americas, Europe, Africa, Australia, Indonesia, and both tropical and temperate islands.[42] Human traffic can bring in more seeds, eggs, and other propagules even as it creates the sorts of disturbance that favor many invaders.[43]

Building roads, plowing the ground, changing grazing patterns, altering fire or flood frequencies, draining wetlands, and polluting streams and lakes—or, ironically, even cleaning up pollution—are all activities that can provide exotic species with the opportunity to establish and spread. Tiny wood-boring crustaceans known as gribbles, for example, probably arrived in the coastal waters off Los Angeles in the hulls of wooden ships during the nineteenth century, but they were apparently unable to colonize heavily polluted parts of the harbor until cleanup programs began in the late 1960s. As the water cleared, gribble populations exploded, even collapsing a wharf with their drilling.[44]

Roads—"daggers in the heart of nature," as ecologist Michael Soulé calls them—are particularly effective at opening new territories to invaders. Thistle-like plants known as cut-leaved teasels arrived in upstate New York before 1900 but apparently stayed put until the Interstate Highway System enabled them to move westward some seventy years later, forming dense monocultures along roads throughout the Midwest.[45] Likewise, logging roads have opened up the rain forests of Tasmania to invasion by root-rot fungus and the rain forests of Queensland to infiltration by exotic cane toads. In Yellowstone National Park, maps of incursions by noxious range weeds such as knapweed, St. John's wort, and leafy spurge form a figure-eight pattern that traces the major road network.[46]

Humans, of course, are not the only disruptive influences. Extreme natural disturbances can set off explosions of invaders. Cheatgrass seeds rained for decades onto the undisturbed tussock grasslands of Canyonlands National Park in southern Utah but never managed to take root through the intact biotic crust until an unusually rainy winter in 1994–1995 gave them an opening. By that spring, cheatgrass had blanketed the landscape.[47] Chinese mitten crabs hovered at the mouths of fast-flowing British rivers until severe droughts in the early 1990s reduced flows. Then hordes of the migratory crabs moved inland to establish themselves.[48] The blister rust fungus that has killed off whitebark pines throughout the wetter reaches of the American Pacific Northwest, including Glacier National Park, has been moving more slowly in the drier climate of Yellowstone. But USGS ecologist Kate Kendall expects that the fungus will eventually kill most of Yellowstone's whitebarks, too, just by continuing to advance in spurts during unusually wet years.[49]

Less visible global changes driven by human activities, such as rising levels of carbon dioxide in the atmosphere, may already be influencing the interactions between invaders and native species. Cheatgrass, kudzu, and Japanese honeysuckle exposed to elevated levels of carbon dioxide all respond with increased vigor. Some biologists suspect that rising carbon dioxide concentrations throughout the twentieth century may have played a role in the historic cheatgrass takeover of the American West, fertilizing the growth of the cheatgrass and thereby enhancing the buildup of dead grasses that fueled hotter and more frequent range fires.[50]

Sometimes, even where no humans tread, the burgeoning crowd of invaders we have introduced eases the way for yet more invasions. Ecologists now suspect that many of our lands and waters are becoming more invasible because potential invaders are just as likely to face a welcoming committee of earlier invaders as to meet the "ecological resistance" Elton foresaw. Although ecology has long focused on competition as the major factor in shaping natural communities, a number of researchers see the need for a "paradigm shift" in invasion biology that puts more focus on helpful exchanges, or mutualisms, such as animals pollinating plants or dispersing their seeds. In heavily invaded regions, newcomers often meet up with allies and partners that smooth their introduction into the community.[51]

Dan Simberloff calls the process by which exotics facilitate one another's advances "invasional meltdown."[52] The process may involve mutualisms, or it may involve animals and plants modifying the habitat in ways that favor other invaders. Some mutualisms are no-brainers, Simberloff and graduate student Betsy Von Holle point out, since they involve reunions between species that evolved together. Take figs, which require highly specialized wasps to pollinate them before they can produce viable seeds. The Indian laurel fig—one of more than sixty ornamental figs introduced into Florida—remained well behaved until its pollinating wasp accidentally caught up with it in the 1980s. Now it is spreading in Bermuda, Mexico, and Central America as well as Florida, helped first by its pollinator and now by non-native birds that spread its seeds. In southeastern Florida, one of those birds is a fellow import from India, the red-whiskered bulbul, which disperses the seeds of dozens of non-native trees and shrubs, many of them the same species it used for food and nesting materials back in India. Simberloff finds, in fact, that dispersal of exotic plants by exotic fruit-eating birds may be the largest category of helpful interactions between invaders.

What is harder to anticipate than the potential reunion of ancient partners is the development of new and unexpected relationships. Several pines with wind-dispersed seeds, for instance, have been introduced into Australia from North America and the Mediterranean Basin. In the new setting, two different species of native cockatoos in widely separated regions have "invented" a new seed-dispersal mechanism for these pines by plucking off their cones and flying into the native eucalyptus forests to pick them apart and scatter any seeds they fail to eat. The alliance between the American gray squirrel and the European stone pine in South Africa is a new relationship, as is the taste some South African baboons are developing for the seeds of Mediterranean cluster pines.[53]

Invading animals that create unprecedented disturbance in a community constitute Simberloff's second category of invasion promoters. Without the destructive aid of millions of imported cattle, for instance, cheatgrass and other range weeds might never have been able to dominate the grasslands of the American intermountain West.[54] In northern Australia, feral water buffalo are blamed for so damaging the floodplain forests and banks of the Adelaide River that they created ideal germination sites for the seeds of invasive mimosa, which has since swamped the region. Likewise, Japanese sika deer team up in a destructive synergy with rhododendrons to transform the woods of southwestern Ireland.[55] Indeed, sometimes invaders not only help one another invade but also magnify one another's adverse effects on the native community. This sort of synergy is what we saw in a previous chapter among exotic pigs, earthworms, plants, birds, mosquitoes, and avian malaria in the forests of Hawaii.

Well-traveled invaders much smaller than earthworms may also be changing the invasibility of communities worldwide. Dave Richardson and his colleagues note that many soil-dwelling mutualists, such as the nitrogen-fixing bacteria and mycorrhizal fungi that help plants draw nourishment from the soil, are becoming cosmopolitan as well. Many introduced plants cannot grow, let alone invade, without the right cocktail of such mutualists.

"We've made the invisible world much more diverse and enhanced the invasibility of a lot of sites," he pointed out. Indeed, he and others have concluded that because of the proliferation of potential partners, "ecosystems are becoming increasingly easy to invade."[56] And with greater opportunity for novel synergies among invaders, the devastating consequences of invasions are likely to escalate, too.

So far, then, we have neither a Holy Grail nor a set of powerful formulas for predicting invasion success. Developing more reliable predictions will require multiple approaches. And we will have to learn to scrutinize not only the potential fate of individual species but also their chances of succeeding in partnerships and teams. For now, the only predictor nearly everyone agrees on is history: a species that invades successfully in one place is likely to do it again elsewhere. That bit of knowledge, imperfect as it is, still offers a more powerful tool for preventing new invasions than most countries are using. Currently, even known troublemakers are seldom stopped at most borders; heeding the lessons of history would thus be a big step forward in most parts of the world.

The ability to make biological predictions about invasions is fundamental, but it is seldom paramount in our decisions to import new plants and animals. As I pointed out before, the making of a pest is a profoundly human rather than biological problem. Efforts to control the deliberate or even accidental movement of species will require putting limits on what goods can move, where they travel, and also how they travel. That's a prospect strongly opposed by many interest groups, from agricultural researchers, gourmet chefs, pet dealers, and garden centers to timber companies that feed their mills with imported logs. Every intentionally introduced exotic species, and certainly every trade route and transport method, has a constituency with a strong economic incentive to avoid restrictions. The questions we face are how to balance the benefits of such trade with the economic and ecological risks of invasions it creates, and how to ensure that the risks are borne by those who reap the benefits.

SEVEN Taking Risks with Strangers

"Individuals and their governments around the world observe a curious ambivalence about the potential spread of plant species. By no rational practice would the parasites of humans or their domesticated animals be freely dispersed. The lessons learned in one region are sufficient to cause erection of quarantine against these parasites' entry into a new range. . . . Curiously, this logic operates only intermittently concerning the deliberate spread of other taxa."

—Richard Mack and Mark Lonsdale, in the journal *BioScience,* 2001

"Modest gestures have already been made, such as special laws regarding ballast pumping and used tire inspection. But there is neither a general strategy for dealing with these invaders nor a widespread awareness of our vulnerability. We have made the globe a biological Cuisinart, and we will either have to deal with the consequences or use our scientific capacity to improve forecasting and monitoring."

—Donald Kennedy, in the journal *Science,* 2001

Let's say you would like to fly from Sydney to Perth carrying cuttings of a plant you fancied in a friend's garden, or ship hundreds of plant specimens or new crop cultivars from overseas to a nursery or research facility in Western Australia, or just order seed packets from a catalog to be mailed to Cousin Sarah in Broome. Your first visit should be to the World Wide Web site of Agriculture Western Australia.[1] There you will find an alphabetical listing of more than 12,000 plants and plant groups, from the popular flowering shrub *Abelia grandiflora* to *Zygostates* spp., a genus of orchids from the rain forests of Brazil. Scroll through the ever-growing list and you will see *OK* before most of the entries but also a fair scattering of *NOs*—10,310 OKs and 1,763 NOs by early 2001. If the plants you want to import are listed and labeled OK, you can proceed. If you find your plants on the list but with NOs beside them, don't bother—they have already been screened and found too risky. If they are not listed, they are among the 200,000 or so of the world's myriad plants that are banned from entry into Western Australia until plant profiler Rod Randall in Perth has scrutinized them for weedy potential. If you want to import them, you will have to request a weed risk assessment.

The computerized list may look to most of us like "red tape as usual," but it represents nothing short of a revolution for people concerned with slowing the movement of invasive alien species. Those OKs constitute what is known as a permitted list or "clean list," and they represent a guilty-until-proven-innocent strategy seldom applied to plant and animal introductions until the 1990s. That was when New Zealand, Australia, and the state of Western Australia, which has its own quarantine system, became the first to declare their borders closed to new plants that had not been screened for weediness. New Zealand, in fact, went even further, requiring preapproval for importation of any living thing new to the country, whether pet, zoo animal, or biocontrol agent. The United States, Europe, and most other nations that have any import controls at all still use mainly a prohibited or "dirty-list" strategy: every organism in the world is presumed innocent until it has proven itself guilty. Unfortunately, the proof usually comes not from the plant or creature failing muster in a risk assessment but from it invading and wreaking havoc in a country's fields, streams, or forests.

In the wake of a growing number of costly, high-profile pest and weed incursions, however, nations are under increasing pressure to adopt a more precautionary approach to the deliberate introduction of new organisms. Further, there is pressure to apply the same degree of scrutiny and precaution to all aspects of trade—the goods, the transport vessels, the routes, the packaging, and the pest, weed, and disease status of the exporting regions—to minimize unintentional introductions of damaging invasive organisms. Dirty lists are by nature reactive, and most nations react too slowly to keep even with the growing biological flotsam of global trade. What's more, most quarantine lists focus on pests of health and agriculture, to the neglect of invaders that threaten native species and natural areas.

Since the early 1990s, however, calls for greater precaution in the movement of people and goods have run head-on into the steamroller of free trade. In 1994, most of the world's trading nations committed to an unprecedented multilateral trade-liberalization package known as the Uruguay Round, which created a powerful new enforcement body, the World Trade Organization (WTO), to keep trade "fair" and settle disputes. One sweeping result of the agreement was the elimination of import quotas, tariffs, and other trade barriers that nations had long used to protect their agricultural sectors from overseas competition. As we saw in a previous chapter, with the relaxation of these and other barriers, the dollar value of goods traded globally soared to $6.2 trillion and the tonnage of merchandise in motion reached nearly 6 billion by the end of the decade. From the beginning, this rapid trade liberalization came under fire from many labor and environmental groups, and the clash reached a raucous crescendo in Seattle, Washington, in December 1999 when protestors caused the collapse of talks that were to open the WTO's Millennium Round of trade negotiations. The risk of invasive species introductions, however, remains the least publicized yet most certain environmental peril associated with unfettered free trade.

Every deliberate introduction of a new species and every new movement of goods around the world represents a trade-off between one group's expectations of benefit or profit and the usually unexamined potential for damage to public health, food production, the environment, and other vital elements of a nation's well-being. Too often, the sectors that stand to benefit from an import are not the ones at risk from unintended consequences. Indeed, the risks are usually borne by the general public in the form of higher food costs or tax-supported damage control.

Minimizing the risks inherent in the global movement of organisms and materials should be a fundamental part of any nation's efforts to protect public health, agriculture, industry, and the environment.

Yet efforts to restrict or scrutinize trade can bring powerful challenges, not only from governments in exporting nations but also from sectors within a country that want free access to overseas goods such as nursery stock, fruits and vegetables, cut flowers, aquarium fish, turtles, or timber. Nations must balance these and other competing interests and also deal with discrepancies between international commitments. Obligations under the WTO, for instance, may clash with commitments under a number of environmental treaties, including the Convention on Biological Diversity (Biodiversity Treaty). The WTO requires nations to use science-based risk assessment to demonstrate a risk and then mitigate that risk using the least trade-restrictive measures available. In the absence of scientific evidence, unrestricted trade is the default, although a nation is allowed to enact temporary measures while it searches for information to complete a rigorous assessment. In contrast, the Biodiversity Treaty and several other environmental treaties advocate the "precautionary principle," which means erring on the side of caution in the face of uncertainty. As the preamble to the treaty declares, "lack of full scientific certainty shall not be used as a reason to postpone measures to avoid or minimize a threat of significant reduction or loss of biodiversity." Given our limited powers to predict how a creature will behave in a new setting, the precautionary approach is especially relevant to the issue of invasive species.[2] A set of guiding principles developed for Biodiversity Treaty nations in 2001 recommends applying this approach both to prevention of unintended introductions and to decisions about species imports.

This philosophical face-off between treaties sometimes obscures the reality that free trade itself can persist only if trade pathways remain "clean." One has only to look at the recent conflicts among European nations and the United Kingdom over beef imports in the wake of "mad cow" disease and conflicts between the United States and Europe over the latter's rejection of hormone-treated beef and genetically modified grain to see that nothing throws up red flags to trade faster than real or perceived threats to human health. Likewise, nothing shuts down foreign markets so quickly as incursions of fruit flies or Asian long-horned beetles or outbreaks of foot-and-mouth disease. So it is not surprising that a series of pest and weed invasions in the 1990s, as well as increasing concern over the environmental threats posed by invaders, caused a number of nations to begin to rethink their quarantine strategies. As a result, both Australia and New Zealand have succeeded in applying precaution in ways that guard against species invasions without violating the rules of free trade.

Dirty List, Clean List

To understand how the state of Western Australia came to pioneer its clean-list strategy, let's start with another "what were they thinking?" tale. During the past century, farmers in Western Australia stripped away much of the native heath scrub and eucalyptus woodlands from that state's arid sand plains and converted them to wheat fields and sheep pastures. Unfortunately, the loss of the deep-rooted perennial native vegetation has led to rising subsurface water tables and salt intrusion. Rising salinity is now causing massive land degradation across rural Australia. In 1990, someone made the decision to import a shrubby Eurasian plant called kochia to rehabilitate salt-scalded lands. A brief look at the plant's track record might have made them think twice. In the United States, for instance, this same weedy variety of kochia long ago escaped cultivation as an ornamental to become a common crop and range weed in the West, where it sometimes poisons livestock. In 1990, however, Australia was operating with a dirty list, and kochia was not among the 250 species prohibited from importation. It was brought in and planted, and within only two years it had begun to run amok. Kochia seedlings were springing up as far as three kilometers away from parent plants. This shrub not only grows and spreads rapidly but also is highly competitive in pushing out desirable crop and pasture plants. It's also toxic to grazing animals at some stages in its life cycle. Researchers predicted that if kochia were allowed to keep spreading this way across southern Australia, production losses in croplands and pasturelands could reach $15 million to $20 million. So in 1993, only three years after the plant's introduction, the government launched an intensive kochia eradication program.[3]

By the time of the kochia fiasco, weed scientists in Western Australia were already lobbying for adoption of a clean-list policy and tinkering with flowcharts and decision trees to try to develop the needed screening tools. But it took one more incursion to convince state officials. That happened in 1996, when a batch of canola (oilseed rape) arrived from New Zealand contaminated with a number of weeds, such as catchweed bedstraw, that were new to the state and also not on the prohibited list. (Imported seed is inspected for contaminants, and a clean list allows authorities to reject shipments laced with unassessed or unapproved seeds.) Farmers had sowed much of the seed before the contamination was discovered and subsequently spent several years rotating crops and spraying with herbicides to knock out the weed infestation. Unlike Australia's other states, Western Australia has long had its own internal quarantine system to monitor people and goods arriving from eastern states and from abroad. Isolated by vast deserts, the state is effectively an island, free of many of the weeds and pests that infest the eastern half of the country. After the canola caper, the state government, buoyed by the demands of farmers for increased vigilance, finally embraced the clean-list strategy.

About the same time, Australia's federal quarantine system—the Australian Quarantine and Inspection Service (AQIS) in the Department of Agriculture, Fisheries and Forestry–Australia (AFFA)—was taking heavy criticism. A number of border breaches, including an infestation of chromolaena and an incursion of papaya fruit flies in Queensland, had prompted a major national inquiry into the effectiveness of AQIS. One issue highlighted by the resulting report was AQIS' failure to treat plant imports with the same stringency it used to oversee the importation of insects and other animals.[4] (Australia, plagued for more than a century by invasive exotic animals, from rabbits and foxes to water buffalo, has long made it extremely difficult to import animals into the country.) Legislative change took longer at commonwealth level than at state level, but within a year after Western Australia turned to a clean list for plant imports, the national government itself followed suit.

The Weed Risk Assessment (WRA) system put into action in 1997 at the state and federal levels was developed by Paul Pheloung, now at AFFA. It asks a series of forty-nine questions about a new plant proposed for import and yields a numerical score to indicate weedy potential. The questions probe such matters as a plant's past track record in new places, its climate preferences, its biological traits, and how it reproduces and disperses. The answers to some questions are given more weight in the scoring than others. The weightiest questions have to do with a plant's prior history of weediness—particularly whether it has invaded crop fields and conservation areas rather than infesting only roadsides and the like—and whether it grows well in arid and droughty regions. Aquatic plants also arouse particular suspicion, as do those with undesirable traits, including plants that are thorny, parasitic, able to chemically suppress other plants, unpalatable or toxic to grazing animals, allergenic, fire hazards, tolerant of shade or infertile soil, climbers, smotherers, or thicket formers, or plants that host insect pests or diseases that could attack crops. If a plant has been introduced into many new locales and has never taken hold in the wild, it is considered low-risk. If it has seldom been introduced and thus has had little chance to demonstrate its weedy potential, it is considered higher-risk. Scientists from more than a dozen organizations in Australia and New Zealand helped to calibrate the WRA's weighting scheme against known weeds and nonweedy plants to ensure that it picks up as many of the kochias and rubbervines and as few of the pansies as possible.[5]

To get the new screening system up and running without wreaking havoc in the nursery industry, Rod Randall and his colleagues needed to get a basic clean list compiled as quickly as possible. "Thousands of plants come into Western Australia every day from eastern states' nurseries, and all have to be checked for compliance with our regulations," he pointed out. "Not just the Permitted and Prohibited Lists but our disease and pest regulations, too." Randall started the OK list by identifying all exotic species that were already naturalized in the state and then adding all the grain, pasture, crop, and vegetable varieties in use. For the initial list of NOs, he included all the noxious weeds already on the dirty lists of AQIS and Western Australia. Then he began the assessments, and the lists grew rapidly.

"I asked the nurseries that regularly import into Western Australia to submit their catalogs to us so we could check them for weeds and add the nonweedy species to our growing Permitted List," Randall explained. "This took about four months, and we screened 7,500 taxa from 129 catalogs."

Randall and a single technician did the work, and the initial list was finished in December 1997. On average, only two to three species from each catalog failed to make the clean list, and most of these were rejected for insufficient information or lack of clear species identification.

Now that the basic lists are in place, Randall has only to assess requests to import new species. "But I still assess plants on a weekly basis," he said. "Over the past two and a bit years, I've assessed over 2,000 species." These assessments are provided for free as a public service. The extreme claims by some in the nursery industry that the system would shut off the supply of novel plants have not panned out: "The nursery industry, which is by far the biggest importer of new taxa, loses a few species along the way," Randall said, "but the vast majority are allowed in because they are just not likely to be weedy."

At AFFA headquarters in Canberra, where the national weed screening is done, Eve Steinke said she has been evaluating about 300 new species per year—fewer than in Western Australia because she deals only with new plants from abroad, not those moving interstate. (As in Western Australia, there is no fee for conducting an assessment.) About 56 percent of those are accepted for entry, 22 percent are rejected, and 22 percent are consigned to an "evaluate further" category. Usually, "evaluate further" indicates a lack of data. "Our biggest single problem is lack of published information," she said one day as I watched her search on-line databases in vain for information about an obscure plant from Madagascar that a nurseryman in Tasmania wanted to bring in. She finally assigned the request to the "limbo" category. It will remain there until enough information becomes available to screen it or someone considers it potentially valuable enough to pay for an evaluation in field or greenhouse trials.

By far, most of the plants in motion in Australia and around the world are nursery stock. But I was curious about what the new system means for all the exotic specimens now growing in botanical gardens throughout Australia and the massive and vital seed collections held by plant breeders and agricultural research organizations. Legally, I was told, these materials are now considered to be in quarantine.[6] Given Australia's troubling history with introductions of weedy pasture grasses, screening for invasive potential seems long overdue. At the time of my visit, Steinke had just screened 600 species of cereals, tropical legumes, and other plants with agricultural potential in the nation's germ plasm collection. Fully half of them would not be admitted into the country today, she found, including most of the legumes and even an ancestor of wheat with a shattering seed head. One-quarter of the material is permitted, and another quarter needs further evaluation.

"For now, we won't ask them to destroy the rejected material," Steinke said. "But they are to grow the plants only in glasshouses. We don't want them to plant these out in field trials or give them out to researchers."

As for Australia's botanical gardens, AQIS asked them as a first step to report anything they were growing that has weedy potential. "We think the list was skimpy," Steinke said, laughing. But for now, the gardens have only been asked not to sell any of that self-reported weedy material.

How effective will Australia's system be at reducing the number of ornamental and agricultural plants that become weeds in the future? Are there potentially invasive plants whose traits are not being captured by the WRA? Biologist and writer Tim Low believes the risk of weediness remains unacceptably high among the plants admitted by Canberra: "Every time a new gardening book appears overseas, nursery owners here are alerted to new prospects and Australia ends up with more sleepers [weeds in waiting]. The plants are coming in for no good reason except that a nursery somewhere wants something new to sell. Nurseries need not show that Australia will benefit in any way."[7]

Pheloung believes with the WRA in place, most of the risk now comes from illegal or undetected movement of plant material. Randall is pragmatic: "We can never be certain in this work, so we just do the best we can with the tools we have available. A realistic system only attempts to minimize risk." New Zealand, though, has created a system so strict it may be in danger of being self-defeating.

Like Australia, New Zealand became a role model in the 1990s for its unprecedented attempts to structure trade around the concept of "biosecurity," meaning the management of risks arising from pests, weeds, and diseases. Passage of the Biosecurity Act of 1993 gave the Ministry of

Agriculture and Forestry (MAF) the authority to prevent introduction of "unwanted organisms" not already established in New Zealand by developing standards for the import of "risk goods," which might be grapes from California, nursery stock from Australia, ballast water from Southeast Asia, or any other combination of material and source country that could deliver an unwanted creature or weed. In 1999, the Biosecurity Authority was created within MAF to carry out this mission. Despite its status as a role model, the New Zealand system—and indeed, that of Australia—is still very much a work in progress. Like quarantine agencies everywhere, MAF has traditionally focused on agricultural pests and pathogens, and the new biosecurity controls were motivated by fear of scourges such as foot-and-mouth disease or various pine diseases that could threaten the timber industry. A high-level review of the biosecurity system in 2000 called for greater attention to the threats invasive species pose to native biodiversity and also to marine ecosystems in this island country.[8]

A second law passed in 1996 set New Zealand on the path to developing a clean list for intentional imports of new plants and animals. That act created an independent agency, the Environmental Risk Management Authority (ERMA), to assess import requests and issue permits for genetically modified organisms as well as any new aquarium or aquaculture fishes, zoo animals, biocontrol agents, exotic livestock, ornamental plants, fruit trees, crops, microbes, or other living things that were not present in New Zealand on July 29, 1998, when the law took effect. ERMA first analyzes whether the organism is likely to pose a threat to the environment or public health and then weighs the costs and benefits of the organism to the environment, economy, and society. Even after an organism gains ERMA approval, it still must receive a biosecurity clearance from MAF like any other "risk good." The law bans outright importation of all snakes, venomous reptiles, beavers, gerbils, prairie dogs, moles, cane toads, mistletoes, Moreton Bay chestnuts, Indian swallowworts, and dozens of other organisms.[9]

As you might imagine, drawing up a big book of every living thing present in the country on a certain date—and investigating passionate claims by plant fanciers or commercial growers whose favorite hybrids, cultivars, and varieties of calla lilies, daffodils, or orchids were overlooked—has required a monumental ongoing effort. MAF maintains the official master list, the Biosecurity Index. It now includes approximately 20,000 exotic plants already in New Zealand, the legacy of lax controls on past imports, from acclimatization society days right up to the time of the law's passage. Some 2,000 of those plants have escaped cultivation and naturalized in the wild so far, and 160 of those are already being controlled at public expense.[10]

Another problem for ERMA was getting the word out that the law applies to pets and posies as well as to genetically modified salmon, sheep, pines, and canola, which have garnered most of the new agency's headlines. Indeed, it wasn't until May 2000 that ERMA officially approved the first two new organisms to be allowed into the country—two upmarket grass trees from subtropical Australia intended for interior decorating or use in protected landscapes. Importers have complained that the time and cost involved in ERMA approvals makes it hard to justify importing anything but fashionably upmarket new plants. Indeed, import applications have been few, and costs can run to thousands of New Zealand dollars. ERMA's chief executive, Bas Walker, agrees that the cost of gaining approval could sometimes exceed the value of the plant being imported; but, he cautions, "past experience shows us that if we fail to manage new organisms effectively, the whole country will face far greater costs in the long term."[11] The goal of making those who benefit from new introductions internalize, or bear the costs of, the potential risks is a step in the right direction. But the recent review of the biosecurity system expressed concern that ERMA's cumbersome process may be driving the plant trade underground and putting greater pressure on border controls to intercept smuggled or mislabeled plants and seeds. It recommended that ERMA, in consultation with conservation and biosecurity agencies, develop "a relatively straightforward process that will encourage importers of new plant species or seeds to have them assessed for their weediness, invasiveness, and other potential biosecurity impacts."[12]

Just as in New Zealand and Australia, the quarantine system in the United States came under critical scrutiny in the 1990s, a time when federal and state plant protection officials found themselves in "a constant state of emergency" thanks to multiple incursions of fruit flies, infestations of citrus canker and Karnal bunt fungus on wheat, invasions by the Asian long-horned beetle, introductions of Asian gypsy moths, and other breaches of border safeguards. The agency within the U.S. Department of Agriculture (USDA) responsible for keeping these pests, pathogens, and weeds at bay is the Animal and Plant Health Inspection Service, Plant Protection and Quarantine branch (APHIS-PPQ). The agency's major stakeholders launched a review of the situation, expressing concern that the onslaught of pests was increasing production costs on the farm and hurting the marketability of crops because of buyers' concerns about pest damage and pesticide residues as well as the pest infestations themselves. This review by agricultural interests concluded that APHIS-PPQ would not be able to enforce border quarantines effectively "without instituting profound changes."[13]

Among the recommendations was that APHIS-PPQ begin to apply the same clean-list approach to nursery stock it had long used on fruits and vegetables. Foreign produce is the only category of goods presumed to be a hazard and prohibited entry into the United States unless it has been assessed and found to be free of dirty-listed quarantine pests and diseases. Only preapproved fruits and vegetables from specified countries can be imported into the United States. In contrast, any plants from anywhere are presumed safe unless they consistently arrive at the border infested with quarantine pests. Any scrutiny focuses on the pests and diseases plants may carry, not on their own invasive potential. Only ninety-four taxa (one-quarter of which are species of mesquite) have made it onto the federal noxious weed list in the twenty-five years since the Federal Noxious Weed Act was passed and handed to APHIS to administer. For most of its history, APHIS-PPQ has focused its attention on crop pests, and agency culture has yet to fully embrace today's wider mission of excluding plant invaders and safeguarding forests, estuaries, and other natural areas from invaders. Environmental weeds such as melaleuca trees are sparse on the list, and not until 1999 did APHIS take steps to regulate interstate sales and movement of listed noxious weeds already present in the United States. Interest in a clean list for plant imports goes far beyond agriculture. The National

Invasive Species Council, an interagency group established in 1999 by President Bill Clinton, recommended in its first management plan the development of "a fair, feasible, risk-based comprehensive screening system for evaluating first-time intentionally introduced non-native species."[14]

How these recommendations will fare politically remains to be seen. The APHIS Weed Policy 2000–2002 proposes to "explore" weediness screening for nursery stock before importation.[15] The horticultural and seed trades, however, have long, and often vehemently, opposed such a move. "The biggest problem is turning off the faucet when it's been open for a hundred years or more," one APHIS scientist commented. The U.S. Fish and Wildlife Service, which regulates the import of "injurious" fish and wildlife, tried unsuccessfully three times in the 1970s to switch from a dirty list to a clean list or, at least, to substantially lengthen its short dirty list. The aquarium trade strongly opposed all those efforts, and the agency plan drew no major supporters at the time.[16] Recently, the agency has begun testing the waters again, seeking noxious candidates to beef up its dirty list. Since the 1980s, only a few critters, such as the brown tree snake, have been added to the list and denied entry. As mentioned earlier, many states such as Hawaii and California have enacted far more stringent prohibitions, although most have little money, personnel, or enforcement authority to block illegal interstate transport of pets and other animals. Throughout the world, in fact, the mechanisms for preventing movement of potentially invasive animals, as well as plants that threaten conservation areas, lag far behind those developed for agricultural weeds and pests.

Keeping Trade Fair and Clean

Whether a nation chooses a clean- or dirty-list strategy, any import restrictions must meet strict criteria under today's trade regime. Despite the assault on tariffs and quotas that got under way in the 1990s, the WTO agreement formally recognizes the continuing need for what it calls technical barriers to trade, including so-called sanitary and phytosanitary (SPS) measures designed to protect humans, animals, and plants. SPS measures include such things as limits on pesticide residues in foods, processing standards to reduce pathogens in meat products, and fumigation requirements to eliminate crop pests on grains and produce. However, out of concern that nations might misuse these measures as "green protectionism" to favor domestic producers, the WTO requires that nations be able to justify their SPS measures if challenged by showing that the underlying health and safety claims are based on scientific principles.

The easiest way for a nation to do that is to base its SPS measures on standards set by an acronym soup of international bodies specifically recognized by the WTO. The International Office of Epizootics (OIE), for instance, sets standards for safeguarding animal health (although not for dealing with animals as pests). Another, the Codex Alimentarius Commission, concerns itself with food safety and human health issues. A third, the International Plant Protection Convention (IPPC), sets international standards for dealing with weeds, insects, pathogens, and animals that threaten crops or natural flora (although the focus has been almost exclusively on crops).[17] There is no WTO-recognized source of international standards for protecting native plants and animals and the environment from invasive alien species that are not plant pests.[18]

The IPPC revised its standards in 1997 to mesh with the new trade agreements. The influence of those changes is evident in the new biosecurity regulations of Australia and New Zealand, which base their strict import requirements on scientific risk assessments and also comply with WTO provisions designed to keep trade fair. For example, because of the revised IPPC standards, not every plant on Western Australia's permitted list is actually "clean." Take the genus *Centaurea*, a group of knapweeds and thistles in the sunflower family. A dozen *Centaurea* species are permitted and another nine are prohibited. The rationale behind the prohibitions is easy to understand—all nine species are major range weeds, such as diffuse and spotted knapweed, that Western Australia does not have and justifiably does not want. But the permitted list also includes three noxious range weeds, including yellow star thistle. The difference is that these three are already widespread in Western Australia, so well established that the government is not even attempting to control them. Because of this, the IPPC standards do not allow the state to exclude further imports of these species, whether they show up as garden plants or seed contaminants. The goal is fairness. If a country's own farmers and ranchers are not required to spend money on weed control, current trade laws say that burden cannot be imposed on producers overseas.

This IPPC restriction has drawn heavy criticism from ecologists because, as we saw in an earlier chapter, continued arrivals of reinforcements can give a weed or pest population better odds of becoming established, spreading, and increasing its ecological influence. New introductions can bring different genetic strains with enhanced invasive potential. Also, every move improves a species' chances of moving again. Each new infestation provides opportunities for an invader to intercept other forms of transport and hitch a ride to yet more sites.[19]

Australia and New Zealand have taken a more conservative approach than the United States and most other nations to the risk of accidental pest introductions in trade as well as to deliberate imports of plants and animals. The IPPC standard allows this as long as a nation bases its restrictive quarantine measures on science and justifies them through formal risk assessments. Australia has, in effect, prohibited all new trade until its agencies can set import conditions under which it can proceed "in a manner consistent with Australia's very conservative approach to acceptance of pest and disease risk."[20] Thousands of risk analyses have been performed to set import conditions for everything from Chinese lychee nuts to African snow peas to U.S. grapes. New Zealand's Biosecurity Authority has taken a similar stance and uses risk analysis to develop a set of conditions known as Import Health Standards for each request to import new "risk goods" (a commodity from a specific source). In contrast, a report to the United States Congress in 1993 by the Office of Technology Assessment found that the USDA had apparently made "a policy choice to favor unburdened trade" and operated "under the presumption that unanalyzed imports will be admitted unless risks are proven." When the USDA did employ risk analysis, it consisted of little more than brief "decision sheets."[21] By the time of that report, however, burgeoning global trade in unprocessed wood was already beginning to force a shake-up in the way the United States approaches the risks of trade.

Risks from Uninvited Travelers

By the late 1980s, with timber harvests in decline on public lands in the northwestern United States, wood products companies were eying the bountiful supplies of softwood timber in Siberia. A handful of companies began negotiations to ship in whole larch logs from Siberia to keep their West Coast sawmills busy. At the time, no permits were required for importing timber. APHIS quarantine inspectors looked over shipments as they arrived in port to see whether they carried any dirty-listed quarantine pests. Prior to that time, most imported logs, lumber, and wood chips had come from Canada, whose forests abut those of the United States and share most of the same pests. But the prospect of raw wood shipments from Siberia alarmed both university scientists and officials of the USDA Forest Service, who warned of the potential for invasion by pests new to U.S. forests. In mid-1990, when two small test shipments of Siberian logs arrived in Eureka, California, the scientists' fears were confirmed. Two exotic pests were found on the logs, including the much-dreaded Asian gypsy moth. Unlike the European strain, which infests eastern forests and strips the leaves of oaks and other hardwoods, Asian gypsy moths feed on both hardwoods and conifers.

It was hardly surprising that pests were found on the log shipments. A long list of pests had previously been intercepted on Siberian logs arriving in Sweden, China, Japan, and Korea. Still, APHIS was prepared to allow full-scale log importation to proceed. Criticism from university scientists and intervention by a few members of Congress, however, pressured APHIS into halting shipments until a formal risk assessment could be completed. APHIS turned that task over to the Forest Service. The risk assessment team visited Siberia, looked at how larch trees were harvested and handled, and eventually identified 175 insects, nematodes, fungi, and other creatures on larches that could potentially hop a ride to the United States. The team had time and data to evaluate risks of only 6 of the 175 listed: the Asian gypsy moth, the related nun moth, the spruce bark beetle, several species of pine wood nematodes, larch canker, and annosus root disease.

The result, according to the Forest Service team: "This assessment clearly demonstrates that the risk of significant impacts to North American forests is great. The possible economic impacts range from a low of $24.9 million . . . because of introduced larch canker to a high of $58 billion because of introduced defoliators" such as the gypsy moth or the nun moth. The loss figures were for commercial timber stands only. The assessment described but did not attempt to quantify what extensive tree deaths would mean for biodiversity, recreation, erosion control, and other ecological values.[22] Indeed, risk analysis procedures worldwide take minimal account of risks to ecosystems or to native plants and animals.

In the end, the would-be importers were turned away, at least until they could propose effective treatment measures and handling procedures to mitigate the risk of pests traveling on Siberian timber. That experience—followed closely by requests to import Monterey pine logs from New Zealand and Chile (which APHIS approved)—prompted APHIS to develop a generic risk assessment process to evaluate pest risks for new imports of any commodity, from logs to cut flowers. First, a separate risk assessment is performed for each pest a commodity might harbor; then the results are combined to produce an overall risk rating.[23]

Looking over that painstaking risk assessment process helped me realize why hitchhiking pests, pathogens, and weeds present a different, and in many ways more formidable, challenge than imports of pets and nursery plants for countries trying to minimize the threat of bioinvasions. Uncensused and uninvited living cargo represents the corrosive underside of trade, a real but difficult-to-quantify threat that skulks along with every shipment. Yet in an era of free trade, it is not enough to know that certain goods from certain places are particularly rife with pests or weeds. Under WTO regulations, risk assessment must be specific to the pest, the commodity, and the region. Decisions made by the WTO dispute settlement body since 1995 indicate that an acceptable risk assessment must do the following:

- Identify the alien species the nation does not want.
- Spell out the potential ecological and economic consequences that could result if that species arrives, becomes established, or spreads.
- Evaluate the likelihood of the foregoing events occurring—it's not enough to conclude that they are possible.
- Evaluate how the import conditions the country proposes—say, visual inspection at the border, heat treatment, fumigation, or even a complete entry ban—affect the likelihood of the pest arriving, becoming

> established, or spreading. The chosen measure must be a scientifically rational choice, and it must also be the least trade-restrictive measure that will afford the country the level of risk it has chosen to accept.

If data are sparse or unavailable, a country can apply a temporary emergency restriction, but it must "actively" seek to fill in the information gap and do a more objective risk assessment. This allowance for a provisional restriction in the face of uncertainty is the only nod the WTO has made to the precautionary principle, despite the likely irreversibility of the threats posed by invasive alien species.[24]

Many ecologists believe, however, that the level of knowledge needed to create a scientifically unassailable case for certain trade restrictions will be a long time in coming. Millions of insects, fungi, and microbes that might be riding the global conveyor belts of trade remain unidentified and unnamed. The living diversity of vast expanses of the earth remains poorly studied. For many of the earth's named species, we know little more than a name. Inventorying what might be traveling on any given commodity shipped from any specific region, and gauging the probability of that occurring, along with the likelihood of damaging consequences, is a costly and time-consuming prospect at best. The risk assessment for Siberian larch logs cost U.S. taxpayers $500,000, although subsequent assessments using the generic process have cost much less. Few developing countries—including some of the most biodiversity-rich countries on the earth—possess either the money or the scientific and technical expertise to perform the level of risk analysis needed to fend off a challenge before the WTO, especially a challenge brought by more powerful trading partners.

Even when investigators can identify a list of candidate hitchhikers from a given region on a specific commodity, predicting which will be successful invaders is, as we've seen, far from an exact science. One of the biggest barriers to prediction is that many species are innocuous or minor players at home. In order to provide insight into the potential behavior of the six species it was evaluating, the Siberian larch risk assessment reviewed the histories of six of the most catastrophic forest pests and disease invasions in North America, including chestnut blight, Dutch elm disease, and European gypsy moth. "All but the gypsy moth were unknown as pests in their native habitats," the report pointed out.[25]

Not being able to name a creature presents another major problem. Unknowns obviously cannot be added to lists of excludable quarantine pests. Further, a hitchhiking bug that crawls out of a load of logs on the dock but cannot be identified as a listed pest cannot be sent packing. As entomologist Joe Cavey of APHIS put it: "You've got to be able to say 'This bug crawling on your $50,000 shipment is *X*, and *X* is on our prohibited list.'"

Fungi and microbes are even more problematic than insects, especially when they are riding aboard largely unregulated material such as nursery stock. Michael Wingfield of the University of Pretoria, South Africa, points out that mistakes in identifying and naming a fungus at the border can lead quarantine officials to ignore it. "The single tool we have to prevent pathogen movement is effective quarantine," he says. "But until we have more research and knowledge and a greater capacity to identify fungal pathogens, most of our rules are probably meaningless."[26] For instance, a number of fungal diseases of trees have apparently ridden into the United States undetected or unstopped on horticultural material in recent years, including dogwood anthracnose, which is devastating the East's beloved wild dogwoods, and pine pitch canker, which is endangering the last of California's native Monterey pines.[27] Likewise, new powdery mildews of poinsettias, petunias, and New Guinea impatiens probably entered recently on imported cuttings and propagation materials.[28]

Trade Packages and Pathways

Complexities such as these have convinced many people that species-by-species, country-by-country risk analysis is inadequate to address all the risks posed by the increasing movement of invasive alien species in global trade.[29] A concept I have heard voiced more and more often is, "Trade is a package." Think of it this way: a country does not simply take delivery of imported ceramic figurines, compact disc players, athletic shoes, bed linens, bananas, cement, steel pipes, or aircraft parts. What arrives at the dock is a ship, its ballast tanks sloshing with a witch's brew of water collected from unknown bays and oceans, its hull, holds, and superstructure infested with hitchhiking organisms. On board are crates and containers carrying not only goods but also wooden dunnage, straw, or other materials wedged into the slack spaces, plus whatever living things clung or crawled aboard as these items sat in weedy factory yards, docks, or rail sidings. This is the messy package we call trade.

Preventing the unintentional introduction of invasive alien species will require that we focus more of our attention on the whole package, the entire trade pathway. And we must do that proactively, starting before goods and vessels leave the exporting country and trying to make each part of the trade package as clean as possible. This approach involves costs, of course, but it has been clearly established that preventing invasions is far less costly and more effective than combating them after the fact. What's more, the costs will fall most heavily on those who benefit from the income generated by moving and selling a particular trade package rather than on the receiving society.

Some of the key management recommendations of the United States' National Invasive Species Council reflect just such a holistic approach to trade: to "identify the pathways by which invasive species move, rank them according to their potential for ecological and economic impacts, and develop mechanisms to reduce movement of invasive species," and to interdict pathways that serve as significant sources of hitchhiking invaders.[30] The International Maritime Organization (IMO) and a number of individual countries are taking a pathway approach to the ballast water problem. Rather than analyzing invasion probabilities and devising measures to prevent specific marine creatures from being transported, the search for solutions is targeted to finding appropriate ways to flush out or sterilize ballast water to minimize the number of all exotic organisms dumped into coastal waters.

The 1993 report by the Office of Technology Assessment mentioned earlier concluded that the USDA's tolerance "for unanalyzed risks is compounded by the low level of effort USDA devotes to researching where risky species are likely to come from and to proactively regulating so as to prevent problems before they arise."[31] That lack of attention to risky pathways hit home even as the USDA was bowing to pressure to scrutinize raw log imports. Asian gypsy moths showed up in the Pacific Northwest, not on logs but on grain ships. The female moths, attracted out of the forests and onto the docks by artificial lights, lay their eggs on anything from ship superstructures and cargo containers to automobiles and buildings. When grain ships carrying egg masses arrived in West Coast ports, larvae hatched and blew ashore in Seattle, Portland, and Vancouver, British Columbia. Eradicating that incursion took $17 million and four years and involved such unpopular measures as aerial pesticide spraying over urban neighborhoods. Another infestation appeared in North Carolina when Asian gypsy moths rode from Germany aboard a shipment of military vehicles. That infestation took three years and $6 million to eradicate.[32]

The USDA's APHIS now tracks vessels that call at Far East Russian ports during the gypsy moth egg-laying season and asks shippers not to charter those vessels for voyages that would bring them into U.S. or Canadian ports during the high-risk egg-hatching period. If such a vessel arrives and is found to be infested, it can be ordered to leave U.S. waters immediately.[33] (New Zealand requires ships that have visited those Russian ports to remain offshore for eight days and then be thoroughly searched upon docking, unless a ship has been certified by Russian authorities to be free of egg masses. The United States now accepts Russian certification, too.)

It took a headline-grabbing incursion by another forest pest, the Asian longhorned beetle, to bring U.S. and, indeed, world attention to a pathway more ubiquitous and problematic than traveling logs or moth-laden ships. That is the movement of wooden crates, pallets, and other low-grade wood packaging materials in trade. An enormous volume of wood is traveling the world at any given time in the form of pallets, crates, dunnage, packing blocks, spools, drums, skids, and other packaging. In the United States alone, an estimated 2 billion pallets are in use each day, from neighborhood grocery stores to bustling wharves. More than half of the $1.7 trillion in goods that moved in or out of the country in 1999 was encased or supported in some way by solid wood packing material (known in the trade as SWPM). Ten percent of U.S. lumber production goes into making more pallets.[34]

In 1996, infestations of Asian long-horned beetles appeared in trees in Brooklyn, New York. More followed at other sites around urban New York City, and in 1998 the beetles appeared in Chicago, apparently having crawled out of pallets or crates from China. The United States issued an emergency regulation banning imports of untreated SWPM from China—a costly blow that affected one-quarter to one-half of China's exports to the United States. China protested (although it was not then a member of the WTO), and APHIS began a risk assessment process to justify its requirements. In the meantime, the United States, Mexico, and Canada worked out a regional SPS standard that all three nations hoped to put in force in 2002.[35]

Both the beetle invasions and the new crate and pallet restrictions set off a global domino effect. Australia, New Zealand, Brazil, Argentina, the United Kingdom, and Finland, among others, all now require fumigation or heat treatment of SWPM from various countries. Not to be outdone, China in 2000 began requiring that pallets made from coniferous wood from the United States and Japan be treated to kill pine wood nematodes. The European Union planned to follow suit in 2001, requiring pine wood nematode treatment for coniferous wood pallets not just from the United States and Japan but also from Canada and China.[36]

Clearly, an international standard for treatment is needed to ensure that the wood packaging pathway is free of all forest and other pests. Pest-by-pest, country-by-country standards are ultimately inadequate because wooden pallets and dunnage, visible in piles near any dock or warehouse, are reused and recirculated the world over and their origins obscured. Multilateral negotiations on a standard were well along in the IPPC by late 2001.[37]

Packaging and raw logs, however, are not the end of the story of traveling wood. Since January 2000, for instance, U.S. officials have intercepted two other long-horned beetle species crawling out of shipments of bamboo garden stakes from China.[38] From 1997 to 1999, Australian officials also intercepted exotic pests arriving on bamboo garden stakes as well as on wooden cricket bats, coffee grinders, mousetraps, mirror frames, rainsticks, cane chairs, camphor wood chests, wooden toys, wine racks, carved masks, and other items. (A large proportion of these timber products are fumigated before export, but Australia has had major problems with overseas fumigation providers. Causes of treatment failures have ranged from fraud to fumigation of goods after they had been shrink-wrapped.)[39] The world has a long way to go to ensure that all trade pathways involving wood are free of invasive species.

A Push for Invader-Free Trade

The sweeping round of global trade liberalization agreements that began with creation of the WTO is not yet a decade old, and the drive to level remaining trade barriers is far from over. The 142 member nations of the WTO agreed at a November 2001 meeting in Qatar to launch another multiyear round of trade talks in 2003. Among other issues to be addressed, nations will be seeking ways to balance their international commitments under multiple treaties and protect their own industries and natural resources while pursuing a share of the promised benefits of globalization. That process was under way before the Qatar meeting. Six months after protestors disrupted the proposed Millennium Round of trade talks in Seattle, the WTO Committee on Trade and Environment (CTE) met with representatives of what are known as MEAs—multilateral environmental agreements, such as the Biodiversity Treaty—to open a channel for discussion of tensions between WTO rules and MEA obligations to halt biodiversity loss, climate change, ozone depletion, deforestation, overfishing, and problems of hazardous wastes. The United Nations Environment Programme is encouraging further integration of trade and environmental policies on the grounds that "preventing environmental damage is cheaper than fixing it later, and some environmental damage is irreversible."[40]

Although the Biodiversity Treaty calls on its 180 member nations "as far as possible and as appropriate" to "prevent the introduction of, control or eradicate those alien species which threaten ecosystems, habitats or species," the treaty as currently framed has no enforcement mechanism. (Initial versions of the treaty drafted by the World Conservation Union [IUCN] provided for creation of a scientific authority modeled on the Convention on International Trade in Endangered Species of Wild Fauna and

Flora [CITES] to single out high-priority invaders.[41]) Treaty nations in 2000 adopted the Cartagena Protocol on Biosafety to guide international trade in genetically modified organisms. That protocol turns the burden of risk analysis around, putting the responsibility on the country that seeks to export the organisms. So far, however, treaty nations have shown no support for a similar protocol or annex that could add teeth to the provision on invasive alien species.

Stemming the tide of invasive species moving in trade will quite likely require a combination of many approaches at a number of levels. In the case of deliberate movements of exotic animals and plants, a proactive clean-list approach provides the best hope for countries to exclude known and suspected invaders. The IUCN guidelines on invasive species prevention released in 2000 recommend requiring "the intending importer to provide the burden of proof that a proposed introduction will not adversely affect biological diversity."[42] The Global Invasive Species Programme (GISP), in its 2001 *Global Strategy on Invasive Alien Species,* recommended that more rigorous environmental risk analysis procedures be developed that shift "the burden of proof to those individuals proposing the intentional introduction of a potentially invasive species."[43]

To minimize unintentional introductions of invasive species in trade, nations and international standard-setting bodies must begin to look more broadly at pathways by which unwanted organisms move. The current patchwork of regulations and standards contains glaring gaps created by the traditional focus on crop pests and human and livestock pathogens and by the neglect of plants and animals that invade our protected lands and threaten biodiversity. These gaps must be filled, either by broadening the coverage and outlook of current standard-setting bodies or by designating new sources of international standards for protecting the environment and native biodiversity from invasive alien species.

WTO representative Erik Wijkstrom, after listening to participants at a GISP workshop in Cape Town, South Africa, speak of the need to "put invasive species on the WTO radar," noted that the organization is essentially reactive. "WTO is member driven," he told the group. "Governments must bring up the issue. The staff can't. Are your government representatives to SPS, IPPC, OIE, etc., sufficiently aware of the invasives issue? Are there gaps in their coverage of issues?" It seems clear that until concerns about invasive alien species loom larger on the radar screens of nations and their quarantine and trade bureaucracies, they will not move up on the WTO agenda.

One way to supplement import protocols is to encourage various industries and trade organizations to develop voluntary codes of conduct that minimize the risk of introducing invasive alien species, just as the IMO has done with its voluntary guidelines for managing ballast water. Key sectors for such codes should include the pet trades, horticultural industries, container shipping, and aid and development organizations. In many cases, self-regulation by such sectors could avert the need for new government or international regulations, taxes, or penalties to enforce safe practices.

Another neglected approach is the development of economic tools that shift the risk burden of invasive species to those who benefit from international trade and travel. This is equivalent to a "polluter pays" strategy, which requires those involved in trade to internalize the environmental costs created by their activities. This can take many forms, including civil fines, criminal penalties, special taxes, mandatory insurance, fees, and bonds. It is usually difficult, and often impossible, to trace the exact source of an invasion—say, which ship dumped the first zebra mussel larvae from its ballast tanks into the Great Lakes or which crate harbored the first Asian long-horned beetle to set foot in Chicago. Thus, fines and penalties may be difficult to apply, especially in cases in which there is a time lag of years to decades between the arrival of an invader and the appearance of harmful consequences.[44]

Attorney Peter Jenkins of the International Center for Technology Assessment in Washington, D.C., who has been involved in the issue of trade and invasive species since the early 1990s, believes instead that economic tools need to operate proactively: "The key to doing this is to apply the economic instruments at the very time the potentially damaging activity occurs and at the industry sector level." The appropriate tools for this would be fees and taxes, including fees levied on those who import live animals, plants, or seeds; on passengers who travel between continents by air or sea; or on cargo ships or planes as they arrive to deliver goods. The goal, Jenkins noted, "is to be proactive in assigning prevention responsibility to these intercontinental sectors, these beneficiaries from globalization, rather than trying to assign culpability after the fact."[45] The money raised could fund surveillance, early detection, and rapid response to new invasions. Australia took just such an approach in the wake of the foot-and-mouth disease outbreak in the United Kingdom and Europe, increasing levies on overseas travelers to help raise an extra $250 million to beef up its border controls and extend its disease surveillance and response capabilities.

Economist Charles Perrings of the University of York envisions empowering the precautionary principle in the form of a "precautionary economic instrument" such as an environmental assurance bond. Such bonds would require that importers of potentially invasive species accept financial responsibility for the risks and uncertainties involved. "What we should be looking for is a regime that allows the social benefits of new introductions, whilst protecting society from the associated risks," Perrings points out. "The difficulty with new introductions is that the associated risks are generally uninsurable commercially for the reason that they are fundamentally uncertain and potentially very large."[46]

We will need not only multiple approaches such as these but also the political will to apply them, at the borders and beyond. Unless we succeed in cleaning up global trade pathways, we can expect increasing threats to native biodiversity and increasing economic losses resulting from invasive weeds, pests, and diseases. Already, severe economic and conservation problems caused by invasive alien species have helped scientists and policy makers involved in GISP to spotlight the threats created by liberalized trade and bring the issue to the forefront of the international agenda. So far, however, only a few countries have acted zealously to guard against the arrival of new invaders, and the difference in vigilance is obvious as soon as you arrive at the border.

EIGHT Stemming the Tide

"[T]he most striking feature of biosecurity for New Zealand is that it is every bit as important as national security. The invaders that pose the greatest risk to our unique ecology and biotic economy will not be two-legged warriors in twenty-first-century wakas or spitfires. They are likely to have six or more legs, be microscopic, green, hard to spot on any radar screen, and great infiltrators if they slip through our defenses."

—J. Morgan Williams, *New Zealand under Siege,* 2000

"We are a weed species, and wherever we go we crowd out natives and carry with us domesticated species that may become weeds in the new environment. What we destroy we often do not intend to harm. What we import, we import with the best intentions. But I find myself wondering . . . why we should have had to repeat so much dreary history."

—Wallace Stegner, *American Places,* 1981

Some 63 million parcels and letters arrive in New Zealand from overseas each year, and their first stop is the International Mail Centre, a low, warehouse-style building near the Auckland International Airport. One fall day, Kerry McGuire led visitors into a small holding room lined with metal shelves. The shelves were heaped high with parcels, all of them opened and each sporting a white tag taped to one flap. McGuire picked up a small box and read the attached form. A woman in the town of Matamata must have received a duplicate of that form by mail a few days earlier, and as I listened to McGuire, I could easily imagine her surprise, and probably annoyance, at getting this notice from the New Zealand quarantine service instead of her package.

The box contained a small aromatherapy pillow filled with lavender seeds sent from Australia.

Under New Zealand's Biosecurity Act of 1993, seeds are "risk goods," meaning they might be or might harbor an invasive pest, pathogen, or weed. Some are prohibited entry altogether. Others, like these lavender seeds, must be fumigated or heat-sterilized before release.

"We've sent her a notice offering her three choices," he explained. "She can return it to sender for NZ$4.80 (US$2.00). We can heat-treat it for NZ$6.65 (US$2.80). Or she can have us destroy it at no cost." Officially, the woman had twenty-eight days to respond before the pillow would be destroyed. "But normally we'll let it sit here for a year. You destroy something and it suddenly turns out to be a family heirloom," he said, only half joking.

Since 1999, every parcel and letter mailed to this island country has been run through a gauntlet of X-ray scanners and detector dogs on the lookout for suspicious organic materials, from this seemingly trivial pillow to bug-infested fruit lurking unseen behind cardboard and brown paper wrappings. McGuire is group leader of one of the Ministry of Agriculture and Forestry (MAF) Quarantine Service inspection teams at the center, and the little package with the lavender pillow in it was just one of dozens that fail to pass muster each day. "At Christmas, these shelves are overflowing," he said.

The adjoining shelves were stacked with prohibited items for which there were only two options: return to sender or destroy. Parsing through these, McGuire found citrus fruit, walnuts, a boxthorn plant, dried meat floss from Thailand, honey, flower bulbs, and milk powder. At the end of the room stood a refrigerator and freezer, ready to receive items, such as fresh fish or oysters, that start to smell after they are unwrapped.

"We get everything from dead seagulls to live turtles," McGuire commented. Some items are sent by unsuspecting friends or relatives, others ordered over the Internet by people ignorant or careless of the law, and still others mailed by flagrant smugglers.

New Zealand, as we saw in the previous chapter, has enacted some of the most restrictive entry requirements in the world, both on deliberate imports of plants and animals and on commodities that might be infested with unwanted organisms. But regulations are only as rigorous as their enforcement. Thus, since the mid-1990s, New Zealand has pioneered a level of quarantine and border controls that no country except the vastly larger island continent of Australia comes close to matching. From the mail that arrives here to the used cars that roll off ships from Japan to the tents and boot treads of hikers flying in on holiday, everything that enters New Zealand is subject to examination by humans, dogs, and X-rays in an attempt to stem the tide of new pests, pathogens, and invasive plants. It's a system that serves as a model and inspiration for other regions, from South Africa to Hawaii to Ecuador's remote Galpágos Islands.

Some might look at these efforts as locking the barn door after the horse has escaped, but even this heavily invaded country has a long list of unwanted organisms that would pose a threat to human health, agriculture, forestry, native plant and animal communities, and the "clean and green" image that underlies the tourism industry. Mediterranean fruit flies (Medflies), for one. Analysts fear that if the Medfly were to become established, it could destroy the nation's NZ$600 million (US$250 million) industry of Chinese kiwifruit cultivation. And in fact, it was a Medfly outbreak in 1996 that helped to spur creation of the detector dog and X-ray programs, first at the airport and then at the mail center. Others on the most unwanted list include Asian tiger mosquitoes and pathogens such as the dengue fever and Ross River viruses they carry; foot-and-mouth disease of livestock; all snakes, whether they menace birds or humans; forest pests such as the Asian gypsy moth and white-spotted tussock moth; pests of stored products such as the khapra beetle; giant African snails; and the voracious northern Pacific seastar, which has already invaded Tasmanian coastal waters to the west, having arrived either in ballast water or clinging to the hulls of ships from Japan.

Even though New Zealand has arguably the tightest border biosecurity system in the world, no country that annually admits 63 million pieces of mail, nearly 3 million air travelers, 400,000 shipping containers, and millions of tons of other goods and commodities can remain impregnable. The costly quarantine effort is calculated instead to hold the risk of new pest incursions to a level the nation finds acceptable. The quarantine operation, from mail room to wharf, seizes more than 80,000 items each year and detects approximately 90 percent of the biosecurity risk material entering the country.[1] Yet as the flow of people and goods increases—and as quarantine agencies such as MAF are pushed to broaden their focus to include environmental as well as agricultural and human health threats—this level of biosecurity gets ever costlier to maintain. It is, as we've seen, a work in progress, with questions remaining about goals, agency roles, who should pay for what, and how to engage the public in helping to ease rather than exacerbate the problem.[2]

In most of the world, both quarantine laws and their enforcement are considerably more relaxed, and admittedly the level of scrutiny New Zealand maintains would be vastly more expensive to sustain in larger countries with land borders. (The United States, for instance, has 83 million people arriving at its borders each year, compared with New Zealand's 3 million.)[3] Yet there is increasing pressure on other countries to tighten their border controls, and I thought the front lines of New Zealand's biosecurity effort would be a good place to observe what many consider the state of the art in quarantine technique and a model for the rest of the world. That fall day, I joined a handful of people from the World Conservation Union's (IUCN's) Invasive Species Specialist Group based at the University of Auckland for a behind-the-scenes look at what MAF's Peter Barnes, our tour leader and national manager of international operations, called "the fingers of the border." The "border" on an island is not a fixed line but any point where people step ashore, planes land, vessels dock, or mail, parcels, or cargo touch down. And so we set out to sample the activity at the mail center, airport passenger terminal, air cargo depot, freight stations, and wharf.

Turtles, Bulbs, and Tramping Boots

From the holding room, McGuire led the way through a receiving room, where staffers were emptying mailbags from a recently landed plane, piling parcels into large gray laundry carts, and stacking letter mail into plastic trays. We followed the carts and trays through double doors into a large room and found ourselves alongside three parallel conveyor belts. Two of the belts were running, filling the room with a loud rumble. Immediately, the incarnation of global trade slid past us in the form of dozens of boxes marked "Amazon.com" moving toward the X-ray units as the operators stared intently at the passing images on their screens.

These X-ray machines, and the ones we would see later in the airport arrival hall, produce images equivalent to or better than those from medical computed tomography (CT) scanners, McGuire told us. Seeds, honey, meat—how can you identify these on an X-ray? I wondered. McGuire led us over to the unused belt in the center of the room and fed a training videotape into the monitor. As X-ray images scrolled from right to left across the screen, McGuire kept up a running commentary.

"This works on the atomic weight of the goods, so something like this lead candleholder shows up very dark, and then you can see a kitchen whisk there, with the metal showing up a greenish color. This is some form of bulbs." The image was brownish, but the small bulbs with their clipped stalks were clearly discernible. Smuggling of seeds, bulbs, and plant cuttings has been an increasing concern since tight restrictions on new plant and animal imports took effect in 1998.[4]

"There you can see wooden eggs, and next to them, in the same box, proper hen's eggs. With hen's eggs, the calcium in the shell comes up blue, whereas the wood is that orangey brown color. Next, dahlia bulbs. A lot of the organic stuff will show that orangey color."

Images of apples scrolled by, then a picture frame with an arrangement of wheat seed heads mounted inside. "That's got fungal rusts and smuts. They need heat treatment."

A small apple tree with soil clinging to its roots appeared. "Other soil will look darker than that because of greater clay content," McGuire said.

Next, palm seeds. "These could be worth $200 each because they're on the CITES list. People try to smuggle them all the time." (CITES is the Convention on International Trade in Endangered Species of Wild Fauna and Flora.) Next came bear paws and bear gall, both restricted CITES items, too.

Honey. Dried dates. A whole pressed duck, its skeleton clearly imaged in greenish blue, thanks to its calcium content. Then a melon from Japan, books, cake, a shirt, chestnuts in a tin, meat, boots.

"Tramping boots. We're keen on those. We clean those before we release them. Flat-soled shoes we don't bother with."

Plant tissue culture vials. Coconut seeds.

"Very small quantities of seeds are difficult to pick up, which is why we need the sniffer dogs, too. These dogs can pick up a blade of grass in a single envelope in a tray with hundreds of letters."

A wooden chair from Malawai appeared on the screen, its carved design visible as a road map of white lines.

"See those little white lines? That's actually *Bostrychid* beetle damage. When we opened this parcel, it was just a mass of shavings, as if somebody had run a sander on it. The beetle had chewed its way through."

Not everything, of course, is so recognizable. "Sometimes you can't identify what you're seeing, but you know it's different from the item listed in the customs declaration on the box. You think, 'That's not quite right.' You get suspicious. That's how a lot of people get caught trying to smuggle things. Of course, people with a better knowledge of the technology can get quite good at declaring something that will look right on the screen."

A staffer was ready to run the day's express parcels, so we yielded the X-ray station and moved aside to watch her. Now and then, she turned from the screen to pull a parcel off the belt, slap on a green sticker, and toss it into a cart beside her. This cart and those from the other belts would go to a handful of quarantine officers in another room, who opened packages and made decisions about the fate of risk goods such as the aromatherapy pillow now sitting in the back room. If operators send out too many nonrisk items, the quarantine officers can note and report back on the types of goods that are tripping them up.

One small table out in the inspection area is reserved for the excruciating task of seed inspection. A telephone book–sized catalog on the table lists 40,000 seeds by their scientific names and gives the import specifications for each. Many are prohibited entirely, and others must be fumigated. Any seed shipment of five kilograms or less requires 100 percent inspection for contaminants such as fungi, weed seeds, or insects. A single seed inspector has been handling the job for two decades now, coming in two days each week to pore over the incoming seed lots. Yet the task is far more vital than the part-time hours imply. Seeds accounted for nearly one-quarter of the risk items seized from New Zealand's incoming mail in 1999–2000.[5]

The quarantine officers now open less than 1 percent of the total incoming mail, McGuire said, but because of the X-ray and detector dog screening, what they do open represents a substantial percentage of the risk goods that arrive. The Biosecurity Act of 1993 authorizes MAF officers to open and inspect any mail suspected of containing a risk good. (In contrast, the U.S. Postal Service requires quarantine inspectors to obtain federal search warrants for inspection of first-class mail.)

I picked up an express parcel from Japan that the operator had just tossed into the cart. What might be in it? Possibly meat products, McGuire suggested. "We just had a big foot-and-mouth disease scare in Japan and Korea, so all of a sudden we've had to stop a lot more things from Japan than before." (It was the same virulent Asian strain that would find its way to South Africa in human food scraps from a ship six months later and sweep the United Kingdom six months after that, borne by contaminated feed from Asia.)

I was curious about how long nonexpress parcels and mail are delayed for this inspection process. "If the normal travel time from the States is five to seven days, your letter will probably spend a day or two of that here," McGuire says. "And if we detain an item, it will be here even longer."

Now that anyone can follow a parcel's tracking number and whereabouts on the Internet, a smuggler can tell quickly whether a parcel has been delayed in the inspection center, McGuire pointed out. That has made smugglers harder to catch. "They don't put return addresses on when they're smuggling, and if a package gets held up too long now in the mail inspection, the addressed recipient will just refuse it. We had that happen with four live turtles last year. They came in a parcel from China, packed in a wet sponge. One died on the way over, and we had to put the other three down."

When the express parcel inspection was finished, McGuire pointed out the small ramps running up onto the end of each conveyor belt, ramps the detector dogs use to access the belts. In the short time since the X-ray machines and dogs arrived, the rate of interception of risk items in the mail has risen from 55 percent to 85 or 95 percent. Indeed, the number of seizures more than doubled in the first year.[6]

That day, however, no dog was on duty—an unusual circumstance. Later we learned, however, that all the detector dogs that could be freed up were making a show of force on the waterfront. The day before, a front-page story in the *New Zealand Herald* had sounded the alarm that an Australian tiger snake had been captured on the wharf, the third snake to appear in the country in the past three weeks. Invading snakes grab the attention of government ministers in this snake-free land far more readily than weed seeds or even gypsy moths.

Passive Beagles, Active Mutts

New Zealand's Quarantine Detector Dog Program is housed in a small metal building not far from the mail center on the grounds of Auckland International Airport. A low chorus of barking hangs in the air from the kennels outside. All the country's detector dogs are kenneled here, with the exception of those that work in Wellington and Christchurch. Inside, we found ourselves in a large room arranged like an airport arrival hall. Clusters of cardboard boxes and well-traveled luggage of various types and sizes were scattered around the concrete floor. Instead of a baggage claim belt running down the center, however, we saw a conveyor belt like those at the mail center.

"Dogs can be trained to detect anything," program manager Rene Gloor told us as we took ringside seats in a scatter of folding chairs. "Paper, plastic, termites in houses, currency, gas leaks, smuggled cigarettes." Detector dogs fall into two categories, based not on what they are trained to find but on how they respond. Active dogs that work the mail belts and other behind-the-scenes locations are trained to bite and scratch at items they detect. In contrast, international convention calls for beagles to be trained as passive response dogs to work among the traveling public—"civilians," as the quarantine people call us. When a trained beagle detects a targeted item—an apple, say, or flower bulbs—it sits down and is rewarded with food. "Beagles are used at airports because of their cute, nonthreatening looks," Gloor said. "They don't frighten people, and they're good for public education."

At that point, senior dog handler Pete Crocker brought his partner Benny into the room on a leash and the two began walking the fake concourse, moving quickly from one cluster of bags to the next. It's a sight most international air travelers now take for granted as they wait with their carry-on luggage for their checked baggage to appear on the carousels.

Airport beagles such as Benny are trained to detect all sorts of restricted items, including fruit, meat, plants, seeds, foods, animals, and eggs. "Last week, we had a beagle pick out a lady carrying frog meat strapped to her waist," Gloor said. "The beagles don't bite and scratch at what they detect like the active dogs do, but it kept jumping up and sitting down next to her, ignoring her bags. Finally, the agent touched her waist and got suspicious. We found forty or fifty packets of frog meat from Asia."

Beagles are actually latecomers to the detector dog world, Gloor pointed out. In the 1960s, some governments began to use dogs such as Labrador retrievers and German shepherds to detect narcotics, explosives, and various contraband items. In the 1970s, Mexico and then the United States began to use dogs to search incoming mail and baggage for fruit and other agricultural quarantine goods. But these dogs did their work out of sight of the public. Then, in 1984, the U.S. Department of Agriculture (USDA) pioneered the first of the so-called Beagle Brigades for the express purpose of training gentle-natured, appealing dogs that could work among the public in noisy, crowded airports. Canada and Australia also developed Beagle Brigades, and in 1995 New Zealand followed suit.

Passive response dogs have come to play a vital role, and Gloor expects that other countries will develop Beagle Brigades as the threat of new bioinvasions grows along with increases in airline travel. But Beagle Brigades cannot replace the role of active dogs.

Unlike the beagles, which are bred for the detector dog program, active dogs are usually what Gloor calls "purebred bit-o'-this, bit-o'-that" beasts chosen from the pound for their insatiable play drives. What beagles do for a snack these dogs will do for a game of catch with a rolled-up towel.

Gloor introduced us to Basil the mail dog, a large, brindled hound with four white feet. Crocker turned on the conveyor belt and began to toss packages from a wheeled cart onto it. As we watched, Gloor unleashed Basil, and the dog charged up the ramp and began to scamper back and forth over the length of the belt, sniffing and whimpering excitedly as he dodged his way between boxes. Within seconds, he had chomped onto a small box as eagerly as though it were a slab of bacon. Gloor took the box and brought it over to show us. Inside was a handful of leaves wrapped in a rag. He turned back to Basil, pulled a rolled towel from his back pocket, and, with a toss, sent the dog happily charging off to fetch it. After a few rounds of toss and fetch, Gloor sent Basil back up the ramp, and the dog resumed his frenetic whimpering and sniffing. This time, he gingerly pulled a single envelope from a large tray of letter mail. Inside the envelope we found two seed packets.

While we were watching, a small Styrofoam container was delivered to the building. Gloor finds it hard to train detector dogs to sniff out snakes with only an old snakeskin to offer them, so he had put in a bid for possession of the snake killed on the wharf a few days earlier, to provide his dogs with more realistic lessons. All of us gathered around to watch as Gloor and Crocker opened the container and looked at their prize. What we saw was, well, a very frozen snake.

It was hard for me, as a native of the American West, to muster the sense of novelty and apprehension New Zealanders must feel about any snake. Instead, seeing the snake left me musing about why no one in the nineteenth-century acclimatization societies had come up with the brilliant idea of introducing snakes into New Zealand—perhaps a python intended to swallow up the introduced ferrets, stoats, and weasels that were supposed to have gobbled up the introduced rabbits. It is sometimes comforting to know that not every foolish thing that could have been done actually was done. That is also why there is still great value in quarantine today, even in a place as hammered by invasive species as New Zealand.

Crocker took the snake into a back room to defrost it.

Dogs have proven themselves invaluable in mail rooms and airports, but whether dogs are the best means of detecting snakes and other unwanted creatures—spiders and scorpions, for instance—that stow away in tightly packed, six-meter-long sea containers was still under study a year later. As a result of the snake finds, MAF funded new projects to examine the risks posed by containers, ways of detecting those risks, and ways of decontaminating containers. With the vast bulk of the world's cargo traveling by ship, and an ever-increasing percentage of that volume packed into sea containers, all countries will eventually have to pay more attention to the biosecurity risks posed by this trade pathway.

Containerships, RO-ROs, and Yachts

Thousands of containers each week, nearly 400,000 per year, are offloaded at Auckland's commercial seaport, which occupies a fenced ribbon of concrete on the city's downtown waterfront. As our little group drove through a checkpoint and onto the camera-monitored wharf, two containerships sat docked alongside a pier beneath towering cranes. One ship still hovered low in the water as two quayside cranes worked to relieve it of containers stacked as many as six high on its open deck.

As forklifts moved the containers into long queues on the wharf, a couple of MAF personnel checked each one on all six sides for any giant African snails that might have slithered aboard in Asia or on any of the Pacific islands where the predator has been introduced. The voracious snail, which can weigh as much as a kilogram and attack hundreds of species of plants, is another great-idea-gone-bad that New Zealand has somehow dodged. One landward area of the wharf contained dense rows of "reefers," refrigerated containers holding fresh produce. Electrical cords snaked from each reefer and collected by the hundreds at a row of power posts, giving the row of reefers a surrealistic appearance, like windowless trailers in an abandoned recreational vehicle park.

Quarantine officers on the wharf open and inspect at least 5 percent of arriving containers, including those carrying declared risk goods (such as produce), as well as a fraction of the others, chosen at random. If officers opening a container door see live insects heading for the light, they close it up and send it to be fumigated. Container shipments do seem to be getting "dirtier." The percentage of contaminated containers increased by 18 percent from 1993 to 2000.[7]

Most containers are not unloaded on the wharf, however, but at some 500 freight stations around Auckland run by importers and exporters. These stations are some of the fingers of the border Barnes had talked about. "We send inspectors out to these sites to clear low-risk goods," he had said. "We have to operate this way because otherwise, with the amount of cargo there is, we would completely choke up the waterfront."

He had taken us to one of these, Tappers Freight Station, before we arrived at the wharf. A MAF quarantine inspector is stationed there daily. (Altogether, MAF inspects more than 25 percent of sea containers that arrive, compared with 1 to 2 percent inspected at U.S. ports.) Whereas most of Auckland's freight facilities are specific to a company or product, the Tappers warehouse receives containers loaded with mixed goods. Tappers is a place for "devanning," that is, breaking down those big, tidy-looking metal boxes; sorting out the jumbles of crates, pallets, and wooden filler material inside; and storing or forwarding the individual shipments. Quarantine officer Zoran Sinovcic was on duty when our group arrived, and he led us down one of dozens of rows of tall metal shelves. The newly arrived shipments, each on its own pallet, included machinery, drums of oil, cases of wine, sacks of coffee beans, and a consignment of personal goods, from golf clubs to well-used suitcases.

Shouting over the din of forklifts and the trucks idling along the loading docks, Sinovcic took us to MAF's own corner of the warehouse. The first thing that caught our attention was a pallet with a unique antique strapped to it—an old motorcycle sidecar, dark red and dirty, unacceptably full of leaves, bugs, and whatever else had fallen into it during the decades it undoubtedly sat abandoned in some field or barn in Australia. Some collector had a bit of cleaning and fumigation to pay for before he would be allowed to claim this prize. Also in the corner were several wooden crates with visible insect damage. One of the main duties of the quarantine officers at the freight stations is to check all the wood packaging and filler material that comes out of each container for excessive bark or insect damage or signs of fungus. The first two can be treated with fumigation, but fungus-infected wood must be burned as a precaution against the spread of blights, cankers, rusts, root rots, and other forest scourges.

Clearly, the goods and packaging that arrive by sea freight are vastly different from the parcels and cargo that come by air. Air cargo into New Zealand can be anything from cut flowers, produce, and seed shipments to live goldfish and expensive racehorses. For weight reasons, air cargo includes much less wood crating and filler material than does sea freight.

Wood packing materials, lumber, and raw logs are high-risk pathways for moving forest pests such as the Asian gypsy moth, but they are not the only ones. The previous chapter pointed out that gypsy moths have succeeded in reaching the United States on multiple occasions as egg masses attached to ships or cargo. New Zealand, like the United States, restricts the entry of ships that have docked in Russian Far East ports during the gypsy moth egg-laying period. Furthermore, any containers that have ridden the rails through moth-infested forests aboard Russian trains bound for Far East seaports are banned altogether from entry into New Zealand. There is another equally troublesome pathway for these moths, however, that New Zealand cannot close. That pathway is used cars, New Zealand's number two import by value.[8]

By the mid-1990s, thanks to the withdrawal of tariff support and protectionist measures, the last of New Zealand's domestic car assembly plants had closed down. Now all cars must be imported, and each year those imports include 120,000 to 140,000 used cars, 95 percent of them from Japan. Japan's forests, like those of the Russian Far East, are home to Asian gypsy moths, and the moths not infrequently lay their putty-blob egg masses on the undersides of vehicles. Spotting and destroying these eggs, either before they leave Japan or as soon as they reach New Zealand, is a priority for MAF, which has five people stationed permanently in Japan just to preinspect used cars before they are shipped. As many as 80,000 used cars, however, arrive at the Auckland wharf each year uninspected and needing immediate attention.

We could see beyond the rows of containers to a section of the wharf that looked like a used-car lot, with hundreds of vehicles of every make, color, and age parked in haphazard rows. At times, more than 2,500 vehicles arrive here in a single week aboard ugly, blunt-ended roll-on, roll-off vessels, or RO-ROs. Each car, as it is driven off the ship, must be run up onto an arched metal ramp so two MAF inspectors can check the undercarriage with spotlights, peering up at the dark vertical surfaces around the wheels, where moths are most likely to lay eggs. It takes a trained team only thirty seconds to inspect each car.

The egg check, however, is just the beginning. Each car that has not been preinspected must pass a thorough boot-to-bonnet search for other contaminants. One in four vehicles fails and requires cleaning or fumigation. Besides gypsy moths, at least four mosquito species with disease-carrying potential, including the Asian tiger mosquito, have been intercepted at New Zealand ports around imported vehicles, tires, and machinery. And that is not all.

We walked along the wharf beyond the inspection ramp to look at a cluster of cars with red tags attached to their windshield wipers. One was an older white Nissan pickup truck with an insulated topper installed over the bed. An inspector opened the back door of the topper and pointed out tiny bugs crawling around inside. They were khapra beetles, a stored-product pest that has earned a spot among the world's worst invaders. The khapra beetle is a MAF target pest that shows up all too frequently in these vehicles. This truck would soon have a heavy tent thrown over it and be fumigated right on the wharf.

Another failure, a black Toyota Celica, sat nearby. A quick look showed it was filthy, with leaves and dirt in the trunk, under the hood, and jammed beneath the windshield wipers. As we watched, a "closed transporter"—a tractor-trailer rig—arrived to take the Celica and some of the other red-tagged cars away for steam cleaning. If such cars are carried in an open transporter, they must first be shrink-wrapped.

New Zealand imports other used vehicles and heavy machinery from Japan that are even more likely to be contaminated than passenger cars. Used bulldozers, excavators, concrete mixers, drilling rigs, cranes, log haulers, rubbish trucks, farm tractors, and buses regularly arrive at the

Auckland wharf. From the looks of that day's arrivals, much of the mud-caked equipment had been driven straight from its last job in forest, farm field, or construction site right onto the ship. About 70 percent of the machinery that arrives requires decontamination to avoid introducing whatever pests, weeds, or pathogens it picked up on the last job.

We drove from the car inspection site to a nearby section of the wharf where a steam-cleaning facility had been built for decontaminating equipment too heavy to be carried to facilities off the dock. A huge yellow excavator was sitting atop the paired concrete ramps of the cleaning station, its tracks dripping muddy water into a well that funneled the overseas dirt, debris, and water into a chlorinated tank. The excavator had just been sprayed with high-pressure hot water. I followed an inspector down into the well between the ramps and watched as he checked the underside of the machine and the recesses in the track links and rollers for any remaining soil and contaminants.

In addition to the containerships and RO-ROs, hundreds of vessels arrive in Auckland with holds full of bulk grain or fertilizer, sawn timber, palletized boxes of bananas or pineapples, and myriad other bulk commodities. More than 1,500 consignments of imported fruit and vegetables arrive each year, and inspectors must examine some fraction of each—say, 600 oranges chosen at random from a shipment. In one-third of these produce shipments, they find quarantine pests such as fruit flies. For shipments of poles or lumber, inspectors must unwrap, unstack, and examine 10 percent of it for wood-boring insects and other forest pests.

And then there are all the noncargo vessels that arrive from overseas ports, from naval vessels, cruise liners, and yachts to traditional Polynesian sailing ships. MAF inspectors must clear their stored foods, pets, hulls, baggage, passengers, and crew for entry. What's more, on behalf of the Ministry of Fisheries, MAF monitors all vessels' handling of ballast water, an immense invasion pathway that we'll come back to shortly.

Despite all this effort, invaders inevitably breach the borders. MAF estimates, in fact, that about fifty new organisms enter the country each year, and many are not as readily seen and stopped as the occasional snake. In

1999, two separate infestations of painted apple moth were discovered by members of the public in the suburbs of Auckland, provoking an expensive, multiyear eradication effort. Painted apple moths from Australia, which belong to the same family as gypsy moths and white-spotted tussock moths, probably rode into the country on shipping containers. The voracious moths are considered a major threat to New Zealand's orchards and forests.[9]

An outbreak of white-spotted tussock moths in 1996, around the same time as the Medfly outbreak, cost New Zealanders about NZ$12 million (US$5 million) to eradicate. It also created a public relations nightmare because the solution relied on both aerial and ground spraying of pesticide in one of Auckland's wealthiest neighborhoods.[10] "I don't know if we'd ever get away with it again," Peter Barnes said. "The public doesn't understand moth danger like they do snakes and spiders."

Weeds are an even more difficult threat than moths to rouse the public about, and they clearly rank lower than pest insects among MAF's priority targets. Illegal plants are undoubtedly getting past the border, and some of them will turn invasive in the future. Since the 1960s, about eight new weed species per year have been establishing themselves in New Zealand. Other pests and pathogens besides moths are getting through the defenses, too. In mid-1997, someone—most likely a farmer—smuggled a rabbit-killing virus (rabbit calicivirus, the agent of rabbit hemorrhagic disease) into New Zealand from Australia, where it had been released as a biocontrol agent against introduced rabbits.[11] And even as the Australian tiger snake captured on the wharf was making front-page news that week, a little one-paragraph item inside the *Herald* caught my eye. It reported the discovery of bee-killing *Varroa* mites in a hive in South Auckland. Within weeks, that story would burgeon to headline news as the extent of the mite infection and the threat to New Zealand's bee and honey industry was revealed. By July, three months later, a commission would have decided that the mites had spread too far and into wild country, making eradication appear impossible.[12] The most likely source of the mites was not some hitchhiking bee in a sea container or used car but a person, probably a beekeeper, smuggling a foreign queen bee in his luggage through the Auckland airport.[13]

Flying with Forbidden Fruit

A cluster of us stood with MAF group leader Howard Hamilton in the international arrivals hall at Auckland International Airport as a beagle and his handler skirted the oval pods of the baggage carousels. Newly arrived passengers waited for the belts to rumble into motion and watched without alarm as the beagle sniffed quickly past the carry-ons at their feet. Beagles are the first line of defense against risk goods being carried into the country by airline passengers, but they cannot work every flight.

"In the case of this flight from Hong Kong, there will be a lot of dried foodstuffs and quite a few tour groups," Hamilton said. "The dogs will run around the hand luggage on all these flights."

But there are certain flights, he said, for which there is no reason to use the dogs because "we know the whole flight's got food on it." Particularly on flights from Samoa, Tonga, and Fiji, island people traditionally carry food gifts from home. "More Pacific Islanders live here in New Zealand than in those countries, and they all have their traditional foods and gifts that they bring," Hamilton said. "That's something you learn to understand better on this job, what these different cultures are likely to be carrying, and why. But it's still our job to find it."

New Zealands' passenger arrival form, which all of us had filled out on the plane before landing, is very explicit about what passengers must declare to the quarantine officers. That includes all meat and meat products, all fish, all dairy products, eggs, honey, pollen, honeycombs, beeswax, feathers, bones, tusks, furs, skins, hunting trophies, pets, reptiles, birds, insects, fruits and vegetables (fresh, dried, frozen, or cooked), raw nuts, herbs and spices, live plants and cuttings, dried plants, cut flowers, bulbs, seeds, fresh or dried mushrooms, pinecones and potpourri, noodles and rice, bamboo, cane, rattan, straw, baskets, woven mats, wood carvings, camping gear and boots, riding tack or any equipment and clothing used around horses or farm animals, animal foods and medicines, and "soil and water in any form."

Some of these items are prohibited; others are allowed entry after inspection and, if needed, treatment. In the austral summer, as many as twenty to thirty tents arrive per flight, often covered with dirt and full of leaves and insects from overseas campouts. Tents and boots are taken into a back room, where quarantine workers wash and dry them while their owners wait to claim them at a counter in the airport lobby.

Nowhere in the world is the quarantine declaration form more detailed or explicit. Yet not everyone reads it the same way. The concept of agricultural, much less environmental, quarantines is also foreign to many people. "People come down here from many countries and see the red and green lines and initially think it's for customs goods to declare," Hamilton said. "People from California understand. They're used to reading about Medflies. But other people don't understand. We try to explain to them our economy and give them a whole spiel and then send them on their way." Without their fruit, meat, and honey, of course.

We heard the baggage belts spring to life. As people began gathering up their checked and carry-on bags and pushing laden trolleys toward the exits, we followed along. New Zealand is a long flight from almost anywhere, and most of these travelers were undoubtedly hoping to do what the vast majority of airline passengers arriving anywhere in the world do: walk quickly out the green "nothing to declare" line. But that doesn't happen here.

On average, about forty overseas flights per day land in Auckland. In the first three months of 2000, those planes brought 573,000 passengers. "We managed to get about 46 percent of them through the X-rays," Hamilton said.

We watched as passengers queued up to hand their declaration forms to a MAF officer at one of the exit lines. The officer asked a few questions before motioning some to the exits, others to one of the X-ray checkpoints, and a few to a row of long stainless steel search counters beyond. "We look at various things: nationality, occupation, where they've been, what they've checked on the form, what their luggage appears to be," Hamilton explained. Non-English speakers, for instance, are more likely to have misunderstood the forms. Tourists returning from Bali are almost certain to be bringing home wood carvings.

I asked whether they find risk items on every flight.

"Between the dogs and hand-searching and X-rays, oh yes."

Is most of it declared?

Asian kudzu vines blanket trees along a Florida highway. Kudzu invades another 50,000 hectares in the southeastern United States each year, overtopping and obliterating trees, telephone lines, barns, derelict vehicles, farmland, and stream banks. (Photograph by Ann Murray/University of Florida)

Rubbervine, imported to Australia from Madagascar in the 1860s as a garden plant, smothers vegetation along the Gilbert River in North Queensland. The exotic vine now infests 350,000 square kilometers of Queensland, an area more than half the size of Texas. (Photograph by Mike Nicholas)

Many of Tahiti's mountain forests have been replaced by dense stands like this one of invading miconia trees. The shallow-rooted miconia increase the risk of landslides. The size of a single miconia leaf suggests how the invading trees are able to shade out native vegetation. Miconia infestations now blanket two-thirds of Tahiti, directly endangering forty to fifty endemic plants by overtopping, shading, and crowding them from forested slopes. (Mountainside photograph by Paul Holthus; leaf photograph by Robert Hobdy)

Before

After

A swimming area at Wakulla Springs, Florida before and after an invasion by the Asian waterweed hydrilla. First introduced as an ornamental pond plant, hydrilla now causes severe problems by blocking recreational swimming and boating, impeding commercial navigation, clogging irrigation canals, shading out native plants, degrading water quality, slowing water flow, and interfering with water supply systems. (Photographs by Victor Ramey/University of Florida)

Large mats of invading water hyacinth choke an important wetland in the Solomon Islands. A wayward ornamental from South America, water hyacinth has become a scourge of waterways worldwide. In Africa, for instance, the weed has exploded across the continent since the 1950s, threatening rice cultivation, fisheries, navigation, hydroelectric power generation, tourism, and even human health (by providing habitat for snails and mosquitoes that vector schistosomiasis, malaria, and other diseases). (Photograph by Peter Solness)

A diver looks down on a cascade of the invasive seaweed *Caulerpa taxifolia* on the Mediterranean seabed. The lush, fern-like tropical seaweed was dumped into the sea from a public aquarium in Monaco in the mid-1980s and within fifteen years had carpeted vast areas of the shallow seabed off France, Spain, Italy, and Croatia, replacing much of the native life of the ocean floor. (Photograph by Alexandre Meinesz)

Workers use a high-pressure hose to blast away zebra mussels encrusting the walls of a Detroit Edison power plant in Monroe, Michigan. These mollusks, native to the seas of Eastern Europe and Western Asia, showed up in the North American Great Lakes in 1988, apparently discharged from the ballast tanks of a freighter. They spread quickly and aggressively, invading twenty states and reaching the mouth of the Mississippi within a decade. Zebra mussels encrust and foul anything solid in the water, from rocks, boat hulls, and pilings to water-intake pipes and filtration equipment. A cross-section shows a water pipe completely blocked by a dense infestation of mussels. (Photographs by Peter Yates)

The Nile perch, introduced several decades ago into Africa's Lake Victoria, has dramatically expanded the tonnage of fish caught from the lake even as it helped eliminate hundreds of species of small native cichlid fish that supported traditional village economies. Perch are shown here being unloaded at an export processing plant at Jinja, Uganda. (Photograph by Yvonne Baskin)

Village women who used to dry and sell small cichlid fish caught by local fishermen now fry and sell perch scraps from the processing plants along the road outside the city of Kisumu, Kenya. (Photograph by Yvonne Baskin)

A feral dog kills a Galápagos land iguana. Dogs, pigs, and human predators had exterminated so many of the unique reptiles by the late 1970s that a captive breeding program was launched at Charles Darwin Research Station to rescue the species. (Photograph by Tui De Roy)

Invading goats compete with giant tortoises for food on Isabela Island in the Galápagos. Goats also strip bare the water-capturing trees and other vegetation, creating drier conditions and eliminating vital waterholes and wallows that sustain tortoises and other native wildlife. (Photograph by Godfrey Merlen)

The exotic southern house mosquito shown on the eye of an apapane, or Hawaiian honeycreeper, transmits an avian malaria parasite brought to Hawaii by imported birds. (Photograph by Jack Jeffrey)

North American bullfrogs weighing up to nearly half a kilogram have escaped from frog-leg "ranches" worldwide to become major predators of other frogs as well as fish, salamanders, and even mice and birds. (Photograph by W. Stephen Price)

A feral cat kills a bird in Weste Australia. In the United States, feral cats and outdoor pet cats kill half a billion birds each yea (Photograph by Evan Collis/ Dept. of Conservation and La Management, Western Austral

"On average, we find between 350 and 700 undeclared items a week at this airport," he said. "Of all the things that are undeclared, 50 percent of it is actually permitted but may need to be inspected for the presence of insects or soil or seeds—things like wooden items, didgeridoos, tents, boots, packaging."

Detailed information about quarantine activities at the Auckland airport—the number of arriving passengers checked by dogs, X-ray scanners, or hand searches; the number who exit without scrutiny; the number and types of declared and undeclared items found; the types of luggage in which various items were found—ends up in a database that allows MAF to evaluate how well the different parts of the system are working and calculate the level of risk New Zealand faces—its "risk exposure." All this is the province of Carolyn Whyte, a biometrician, who applies mathematics and statistics to biological problems.

At the time of the 1996 Medfly outbreak, before dogs and X-rays were stationed at the airports, quarantine inspectors were hand-searching the baggage of only 16 percent of arriving passengers (even then, a far larger percentage than in most of the world), and 84 percent "were going straight out the doors," Whyte told us. Surveys determined that only about 55 percent of fruit carried by passengers was being detected. On the basis of the proportion of seized fruit that was infested, Whyte calculated that the risk of an outbreak was 18 percent per year. That meant the country could expect one outbreak every five to six years. (As in other quarantine operations worldwide, the system is calibrated against the arrival of agricultural pests, not the arrival of species that threaten biodiversity.) Then came the X-rays and dogs.

Eighteen months later, only 38 percent of arrivals were going out the door with their baggage unscreened by X-rays or hand searches, and many among that 38 percent had been on flights met by detector dogs. Whyte had calculated that the risk of pest outbreaks would drop to 5 percent at that point, but in actual practice, the risk dropped even lower.[14]

However, passenger arrivals have continued to increase—by nearly 14 percent in 2000 alone—and the X-ray and search lines are at capacity more often. That means more people go out the door unchecked. By January 2000, the proportion going out the exits without physical searches or X-rays had risen to 44 percent.

Even if MAF can keep up a 95 percent detection rate, if the number of passengers grows, so does the volume of risk material in that 5 percent that goes undetected. Consequently, the actual risk of a pest outbreak also rises. Whyte predicts that by 2020, the risk of a pest outbreak could increase to one every three to five years, even if MAF continues to detect 95 percent of arriving fruit.

Setting the acceptable level of risk is up to policy makers. But Whyte's data and models help them gauge where to target resources to achieve an acceptable risk. The law of diminishing returns says it will be increasingly expensive to lower risk by continuing to add more technology. Queuing more people up longer to go through the current inspection system has its limits. International guidelines call for all arriving passengers to be out of an airport within forty-five minutes of landing, and most of us would find a holdup of even forty-five minutes maddening. Public awareness and compliance with the system are critical in holding down the risks.

What New Zealand is turning to next to jolt public awareness is an instant fine of NZ$200 (US$84) for careless passengers caught with undeclared fruit, plant, or animal products. Liberal use of the fines is intended to encourage more people to present questionable items voluntarily for inspection.[15]

Stretched Thin at Continental Borders

Larger countries such as the United States are under increasing pressure to modernize their quarantine operations, too. In the previous chapter, I reported the deep concern expressed by stakeholders in agriculture and state plant protection agencies that the border safeguarding system run by the USDA's Animal and Plant Health Inspection Service, Plant Protection and Quarantine branch (APHIS) was ineffective and unable to cope with the increasing frequency of exotic species introductions. The stakeholders' alarm was provoked largely by a bevy of high-profile outbreaks of crop and forest pests. That report was not the first criticism. In 1997, the

United States' General Accounting Office found that APHIS inspection capabilities had been unable to keep up with the increasing flow of passengers and cargo. Furthermore, scientists and citizens concerned about threats that invasive exotic species pose to native species and ecosystems have been even more critical, both of permissive import policies and of quarantine efforts themselves.[16]

Of course, the United States has a much bigger job to do than New Zealand, with its handful of international airports and seaports and no land borders. The United States operated 300 ports of entry for goods, parcels, and people as of 1999, and more were emerging each year. Many have no quarantine presence at all. Along the Canadian border, for instance, only five of the twenty-five crossings were monitored by quarantine staff, despite the fact that Canada provides a "back door" for entry of exotic fruit flies. (Because fruit flies cannot become established in Canada's climate, that country has traditionally maintained less restrictive entry requirements for exotic fruit.)[17]

During the 1990s, imports and exports through U.S. ports of entry increased by more than 30 percent and passenger traffic doubled. Air cargo volume is doubling every five to six years, and an increasing proportion of this cargo consists of cut flowers, fresh fruit and vegetables, and other perishable commodities that require rapid quarantine clearance. Following the worldwide trend, more and more sea freight is traveling in containers. From 1993 to 1997, containerized cargo arriving at the Port of Long Beach, California, more than doubled.[18] APHIS has some 2,000 inspectors, along with X-ray machines and beagle teams (fifty-nine teams at twenty-one airports and two post offices), staffing 100 ports of entry, and it intercepts 1.5 million prohibited items and 50,000 quarantine pests each year. APHIS personnel also inspected 71,000 ships, 838,000 cargo shipments, and 83 million passengers (aboard planes, ships, cars, and railways) and pedestrians arriving from foreign countries or flying from Puerto Rico or Hawaii to the continental United States in 1998–1999.[19] (Unfortunately, despite Hawaii's vulnerability to alien invaders, that state does not receive similar protection through prescreening of goods and passengers traveling from the mainland to Hawaii.) According to APHIS entomologist Joe

Cavey, inspectors must use "educated triage" to judge which cargo is most likely to be "dirty" because they can inspect only 1 to 2 percent of the items that come through.

Smuggling presents the greatest danger for pest introductions, according to the stakeholder report. But in the ever-increasing bustle of cargo arrivals, quarantine inspectors can miss even properly labeled materials—just as U.S. Fish and Wildlife Service inspectors overlooked brown tree snakes on a manifest in 1998. A shipment of an officially listed noxious waterweed known as bur reed arrived from Holland in 1999 and was distributed to Home Depot stores in thirty-five states for sale as a pond plant before the error was discovered.[20]

The stakeholders concluded that even with more resources, APHIS cannot hope to keep pace with continuing increases in cargo movement unless it broadens its focus beyond the ports of entry. That means putting more effort into both mitigating pest risks at the point of origin and identifying and reducing risks from new pest pathways. Likewise in Australia: a 1996 review of that country's quarantine system recommended that the focus be changed from primarily border operations to "a continuum of quarantine" that deals with risks preborder, at the border, and postborder.[21]

Offshore inspection, treatment, and certification of the sort New Zealand encourages for used vehicle imports are already a burgeoning trend worldwide. The United States conducts offshore inspections in about thirty countries as a way of both reducing the risk of pests arriving on U.S. shores and speeding the clearance of perishable goods, such as fruit, at ports of entry. Two other major pest introduction pathways that have been closely monitored in New Zealand are already receiving attention from most of the world's trading nations: solid wood packaging and ballast water.

Ballast Water Roulette

Tens of thousands of ships travel the world's oceans each day, some laden with containers, vehicles, grain, or ore and others empty of cargo between ports. But a truly empty ship would float high like a cork, rolling in heavy seas or high winds, its bow lifting and slamming the surface, its propeller lurching in and out of the water. So ships empty of paying cargo need something heavy, cheap, and easily accessible to help them ride low in the water for stability and better handling on the high seas. That something is ballast, a word that stems from the Middle Dutch for "useless load." Until the 1880s, rocks, sand, or other heavy materials served the purpose. Since then, however, ships have converted to seawater ballast. Even a ship fully loaded with cargo may burn so much fuel on a long voyage that it must take on ballast to replace the weight. Many vessels also ballast or deballast in port to maintain stability while cargo is being unloaded or loaded. Sediment in the ballast water settles out and builds up in the bottom of the tanks, and ships must be put in dry dock periodically to remove it.[22]

Jim Carlton of Williams College–Mystic Seaport in Connecticut has often heard shippers argue that ballast discharges "can't be bad anymore because they've been going on for so long and everything that can be moved already has moved." But the nature of ballast water risks is changing, just like the risks from other sectors of global trade and travel. Bigger and faster ships take in more organisms and allow more of them to survive the journey. New trade routes connect parts of the world that never before exchanged goods, much less marine organisms. And the more invasions that occur, the more species any given region can contribute to the global ecological roulette: Tasmania can now export northern Pacific seastars; the Great Lakes, European zebra mussels; Long Island Sound, Japanese crabs; Peru, Asian cholera bacteria. The result of all these factors is that ballast-related invasions are accelerating.[23]

One finger of New Zealand's border that I was unable to observe—and that even MAF inspectors have little way to monitor—is the 4 million to 6 million metric tons of ballast water discharged into the country's waters each year by some 3,000 visiting ships. Ballast water is a "risk good" under New Zealand's Biosecurity Act of 1993, and in 1998 the country became the first to enforce mandatory controls on ballast discharges along its entire coastline. Ships must gain permission from MAF before discharging any ballast. Exceptions can be granted if storms or heavy seas make it dangerous for a ship to exchange its stabilizing ballast in midocean, but even that exemption is not available for ships that have taken on water in certain high-risk ports. Those include Tasmania and Port Phillip Bay, Australia, where the northern Pacific seastar has invaded. However, although MAF can determine an inbound ship's previous ports of call, the agency has little way to verify whether a ship's master has actually filled or flushed the tanks when and where he reports.[24]

The idea behind ballast exchange is that organisms collected from harbors, bays, lagoons, or river mouths are not likely to survive release in the deep ocean, and organisms picked up in the at-sea exchange are unlikely to survive when released along a coastline. Everyone involved in the issue, however, will tell you that ballast exchange is a stopgap measure, neither completely safe for ships to perform at sea nor fully effective biologically. The biological ineffectiveness stems from the fact that ballast exchange never fully empties a tank or flushes out all the organisms. Ecologist Greg Ruiz and a team from the Smithsonian Environmental Research Center (SERC) on Chesapeake Bay, for instance, conducted experiments aboard oil tankers sailing up the West Coast to Alaska. During flow-through procedures, in which water is pumped into the bottom of tanks and allowed to flow out the top, dye tracers showed that "dead zones" can form in the tanks where eddies keep the water from mixing. Organisms may be trapped in these zones, or creatures that can swim may deliberately move away from the zones of turbulence and avoid being flushed. Procedures wherein tanks are emptied and then refilled are more effective at exchanging the water. Thus, various practices may replace 70 to 90 percent of the ballast water and purge some organisms more thoroughly than others.

Because flushing is imperfect, the contents of a ballast tank reflect a ship's travels and its ballasting history, and every release or addition changes the living brew inside. Just as with quarantine practices on land, ballast exchange cannot reduce to zero the risk of new species introductions. Research is under way around the world on a variety of options for filtering ballast water as it is loaded or sterilizing it with biocides, heat from the engine, or other means.[25]

The 156 member nations of the International Maritime Organization adopted voluntary guidelines in 1997 for managing ballast exchanges to minimize marine invasions, and the organization has a committee working to develop a legally binding instrument. Until a mandatory global standard is in place, however, nations are on their own.

In response to invasion by the zebra mussel—a "motivational species," as Carlton calls it—the United States in the early 1990s began requiring overseas ships to change ballast water before entering the Great Lakes. At other ports around the country, the U.S. Coast Guard requests but does not require the 71,000 commercial ships that call each year to dump or exchange ballast in the open ocean before entering. The master of a ship is required, however, to file a detailed report about where water was taken up and where it was released. The National Invasive Species Act of 1996 gives the Coast Guard authority to make exchanges mandatory in 2002 if voluntary compliance proves ineffective.[26] To help determine whether this is the case, SERC scientists are tracking patterns of ship arrival, ballast discharge, and reporting to determine how the shipping industry is behaving, monitoring which organisms are being introduced into key U.S. waters, and watching whether the rate of invasion changes along with ballast patterns.

Australia, too, prompted by an outbreak of toxic red tide organisms from Japan in the late 1980s, began requesting that ships voluntarily exchange ballast at sea before entering Australian ports. In July 2001, Australia tightened its ballast controls, declaring that ships would not be allowed to discharge ballast within Australian waters without written permission from a quarantine officer.[27] The officers' decisions will be based on a new risk management system that profiles water just as other efforts profile risky airline passengers.

Risk analyst Keith Hayes at Australia's CSIRO Marine Research described four key factors in the assessment: the likelihood that a specific "donor port" is infected with an unwanted organism, the likelihood that a vessel ballasting there will become infested (for instance, whether the harbor is deep enough that propeller wash is unlikely to stir bottom sediments and organisms into the water column, where they can be sucked up in ballast), the likelihood that an unwanted species will survive the journey, and the likelihood that the species will be able to survive conditions in the receiving port. These factors have been incorporated into the new Australian Ballast Water Decision Support System (BWDSS), which allows quarantine officers to assess on a tank-by-tank basis whether to allow a ship to discharge ballast in specific ports. By using the Inmarsat automated satellite system to gain access to information about a vessel's ballast while it is en route, an officer can notify the ship's master of the results in time for him to decide how to handle any high-risk ballast. His choices include exchanging those tanks at sea, retaining the water, or transferring water internally between tanks for trim and stability without discharging it. At some point in the future, the options may include newly devised ways of sterilizing or filtering the water before release. A ship's agent could also gain access to the BWDSS by means of the Internet, Hayes explained, to run hypothetical routings, find out how the system rates the risks, and plan ballast management well before a ship sails.[28]

How well the system will work in practice remains to be seen, as does the effectiveness of other new risk-based border controls pioneered in recent years. "It's like turning on the lights one by one as you go down the corridor, with no mental map of what's ahead," Hayes said.

At best—and we are far from that in most of the world—strictly enforced border controls can slow the tide of new invaders to a trickle. Especially for continental countries with extensive land borders such as the United States, it may be physically or economically impractical to achieve the level of vigilance that is possible—although still seldom achieved—on islands. Vigilance by exporting countries, tighter controls on entire trade pathways, and cooperation by the traveling public can further reduce the threat of new introductions. As long as trade and travel continue to burgeon, however, our best efforts cannot stop all harmful species at the border. To prevent invasions, we also need to develop strong surveillance and monitoring efforts to detect incursions and the capacity to act rapidly to prevent potential invaders from gaining a beachhead.

NINE Beachheads and Sleepers

"There must have been plenty of them about, growing up quietly and inoffensively, with nobody taking any particular notice of them. . . . And so the one in our garden continued its growth peacefully, as did thousands like it in neglected spots all over the world. . . . It was some little time later that the first one picked up its roots and walked."

—John Wyndham, *The Day of the Triffids,* 1951

"Who would have thought when the first triffid weed, *Chromolaena odorata,* was collected in 1961 on a hill in Hluhluwe Game Reserve, that by 1984 it would be smothering vast areas of the reserve and threatening to engulf it completely? Who would have predicted in 1940 when it was first discovered near Durban that it would become the major challenge to all the coastal reserves of Natal?"

—Ian A. W. Macdonald, in *African Wildlife* magazine, 1988

Fly north along the Great Barrier Reef until you have cleared the tip of mainland Australia and you will find yourself looking down on one of the sea's most treacherous and storied crossroads. The Torres Strait, barely 150 kilometers across, separates New Guinea from mainland Australia. The separation is both geologically and politically recent, however, and not nearly as effective as a quarantine officer might wish. Geologically, Australia's Great Dividing Range sweeps north through Queensland and up the Cape York Peninsula and then plunges into this shallow strait, finally petering out as it approaches the swampy shore of New Guinea. During the great ice ages, the range formed the spine of an exposed land bridge that connected the two landmasses and allowed humans and other creatures to walk into Australia. As the glaciers melted about 15,000 years ago, rising sea levels flooded the low headland, sparing only the isolated mountain crests that now form a line of stepping-stone islands across the turquoise waters of the strait. By that time, however, even the engulfing sea could not sever the human connections between the two lands. Ever since, Aboriginal peoples have plied the strait in canoes and sailing vessels, populating the land bridge islands and coral cays and maintaining a bustling commerce in fish, dugong, and turtle meat as well as taro, bananas, and yams amid the maze of reefs and shoals. European sailors discovered the strait four centuries ago, but it remained for Captain James Cook in 1770 to pioneer the routine use of the strait as a convenient, if perilous, passage between the Pacific and Indian Oceans. Later, Captain William Bligh and his loyal followers, set adrift in a small boat by mutineers on the *Bounty,* also found their way through the strait to safety in the East Indies.

Politically, the Dutch, British, and German colonial empires carved up the region, with the Dutch claiming the western half of the island of New Guinea—now Irian Jaya, a part of Indonesia—and the British and Germans splitting the eastern half, now united as Papua New Guinea (PNG). The southern part of PNG, along with the Torres Strait Islands, was administered by Queensland as part of British Australia and remained part of the state of Queensland even after Australian independence (German occupation of northern PNG ended in 1914). White settlement of Australia began about the same time as the mutiny on the *Bounty,* but Europeans at first found little to exploit in the islands besides sea slugs, which were dried and sold to the Chinese as bêche-de-mer. Then, in the 1860s, the lure of profits to be made from gathering pearl shells set off the equivalent of a gold rush that brought adventurers, traders, and missionaries flocking to the Torres Strait from throughout the Pacific islands. As a consequence, the Melanesian heritage of the islands is now mingled with that of Polynesians, Japanese, Malays, and Europeans.[1]

PNG gained its independence in 1975, but the strait and its islands, right up to the New Guinea shoreline, remain Australian. Shifting political boundaries, like the rising seas of an earlier age, however, cannot automatically sever the human ties or stem the movement of goods, people, and pests. Coastal Papuans retain treaty rights to travel freely among their relatives on most islands of the strait, trading pandan mats, goanna-skin drums, wongai wood spears, seed beads, seashells, yams and sago, and the meat of sea turtles and marine mammals known as dugongs. Although trade in live plants and animals, fresh fruits and vegetables, and meat, milk, and hides of land animals is forbidden, the ease of travel in the strait creates a border control nightmare. What's more, bêche-de-mer and pearl shell trafficking in the strait have been replaced by "New Guinea Gold" cannabis and boatloads of illegal immigrants from western Asia and the Middle East.

The strait thus creates for Australia the equivalent of a land border, one that separates a poor and strife-ridden region from another that is relatively wealthy and secure. North of that border, minimal internal quarantines restrict movement of people or goods among the scattered islands of Indonesia, from Sumatra to Java to Borneo to Irian Jaya. In fact, the "transmigration" policies of the Indonesian government have uprooted whole villages in densely populated areas and moved them to sparsely settled Irian Jaya. Little but a line on a map keeps whatever insects, weeds, and pathogens that also reach Irian Jaya from crossing into PNG.

Instead of relying on border quarantines alone to prevent pest outbreaks in the islands of the strait or on the "Top End" of Australia, however, the country has developed a unique frontline early warning system known as the Northern Australia Quarantine Strategy (NAQS). Since the 1990s, NAQS personnel have been actively surveying for potential new pests, diseases, and weeds arriving in PNG and even Irian Jaya, helping those countries to combat them before they establish a beachhead for invasion, monitoring their movement toward the islands and the northern coastline of Australia from Broome to Cape York, and identifying rapid response measures that would be needed in the event of an incursion. The target list for this Top Watch effort includes papaya fruit flies, screwworm flies, red-banded mango caterpillars, grapevine leaf rust, Panama and blood diseases of banana, citrus canker, sugar cane stem borers, Asian honeybees and the *Varroa* mites they carry, Japanese encephalitis, bluetongue and other cattle diseases, chromolaena, and forty other noxious plants.

Even the strictest border controls can never be 100 percent effective at preventing pest invasions, but thankfully there is often a window of opportunity between the arrival of a pest and the time it needs to establish a beachhead and begin reproducing in a new locale. That window offers the most practical and cost-effective opportunity for eradicating a pest before its population explodes and disperses. First, however, the pest incursion must be detected and a rapid and effective response mounted. The time window may be longer for plant invaders, and effective surveillance must include keeping watch not just for new arrivals but also for "sleeper" weeds, potential invaders that have secured a beachhead but not yet roused from "slumber." Although small-scale surveillance efforts are increasing around the world—from patrols for brown tree snakes in Hawaii to tracking of prickly pear cactus and Indian myna bird infestations in South Africa's premier national park, Kruger—the NAQS effort remains unique in its scope and longevity. Most countries still have no systematic surveillance and rapid response strategies to deal with new incursions or explosions of sleepers. The NAQS effort, then, offers a glimpse of how other countries might go about filling a glaring gap in their efforts to manage species invasions.

Surveying the Strait

The Coastwatch chopper was carrying us northeast under a heavy cloud layer. The sea stretching before us was calm but wrinkled, an expanse of aqua elephant hide. To give me a firsthand look at how their surveillance system operates, NAQS scientists were letting me join them during a periodic check of two small coral islands and two long-abandoned volcanic nubs, Gabba and Nagir Islands.

Six of us had lifted off that April morning from the small heliport on Thursday Island, the administrative center of the Torres Strait. In front with the pilot was a customs officer, who was keeping an eye out for suspicious boat traffic. In back, NAQS botanist and team leader Barbara Waterhouse, plant pathologist Richard Davis, entomology technician Kylie Anderson, and I sat knee to knee for the hour-long flight to Yorke Island (also known as Masig Island), in the central island group. (Every island has multiple names bestowed by Europeans and traditional inhabitants.) The sliding windows on the rear chopper doors remained slightly ajar to capture a cool but muggy breeze. I was grateful both for the air and for a seat by the door, facing forward on my first helicopter ride.

We were flying over numerous coral cays, some mere patches of white sand, others fringed with mangroves, still others sprouting graceful casuarina trees. Although casuarinas are destructive invaders in many places where they've been introduced, from Hawaii to Florida, here they are native pioneers. Their closed cones float out from shore, land on newly emerged cays, and sprout. As they grow, the trees provide perching sites for fruit bats and Torresian imperial pigeons migrating to PNG. The bats and birds drop the seeds of fleshy-fruited trees such as the wongai, a revered tree whose wood the island people use to make traditional spears for hunting dugongs. Soon, bird- and bat-planted forests spread in circular patterns around the casuarinas.

There are 133 rocks, cays, land bridge islands, and volcanic remnants that breach the surface of the strait (and thousands more barely submerged hazards to navigation), but only 17 islands are permanently inhabited. Pointing off to the left, Waterhouse wrote on a notepad for me, "All mangroves and crocodiles—Sassie Island." Talking to the pilot over the radio, she recounted how, during the first papaya fruit fly outbreak on the islands in 1993, the chopper had landed on Sassie just long enough for them to jump out and check fly traps while the pilot kept the rotor turning to fan away the clouds of mosquitoes.

That season, Waterhouse and entomologist Judy Grimshaw had just arrived in Torres Strait for a routine survey when word came from a laboratory in Brisbane that a fly found earlier in one of their island traps had been identified as a papaya fruit fly. Despite their common name, these housefly-sized creatures infest almost all edible fruit and some vegetables, laying their eggs just under the skin. When the larval maggots hatch and feed, they cause the fruits to decay and drop before ripening. "We dropped the scheduled survey and immediately went into response mode," Waterhouse told me later. That included setting out more than 200 traps on sixty-plus islands in the first two weeks and then flying from island to island for the next five weeks checking their catch. Their efforts showed that the fly was confined to the five northernmost islands, and the outbreak was successfully eradicated. Later incursions, including one into Cairns in 1995, were also eradicated. "However, since then, due to the buildup of fly populations along the PNG coastline, where no active management is practiced, each wet season when the northwesterly winds blow, we receive a new crop," she said. "For the last three years, there has been preemptive treatment of each of the highest-risk islands to reduce fruit fly populations, and no establishment has occurred."

Bill Roberts, whose Plant Protection Office oversees funding for eradication efforts, told me later that the direct cost to the government for the eradication program had been A$34 million (US$17.5 million). The nation's trading partners had also responded to the outbreak by temporarily shutting down imports of produce that might host the fruit flies—not just papayas but also bananas, mangos, coffee, and cashews—a move that cost the industry A$100 million (US$51 million). The costs would have been far higher, however, if the pest had been able to establish a base in the rain forests of Queensland, where it could also damage a wide array of native plants.

Our chopper passed quickly through a rainsquall, and as we emerged we saw a fleet of five prawn trawlers anchored in the lee of a tiny cay, their nets draped on long booms to dry. Prawns provide a major source of export revenue for the Torres Strait region. The trawling is done at night, and the season is strictly regulated. The pilot dropped low over the sleeping fleet so the customs officer could take surveillance photographs.

The clouds were closing in again as we dipped toward the grassy landing pad on Yorke, an island bisected beach to beach by an airstrip, this one paved. The white sand road leading to the community center was pocked with deep puddles. We could see Hilda Mosby weaving through them on her Yamaha all-terrain vehicle as she drove out to greet us. Mosby, dressed in a uniform of khaki shorts and shirt, is the Australian Quarantine and Inspection Service (AQIS) officer for Yorke Island. Waterhouse and Richard Davis climbed on the fat-tired vehicle and rode off with Mosby to check in with the island council, which represents the 300 or so residents on this two-and-one-half-square-kilometer island. (The Torres Strait Islanders were the first of Australia's Aboriginal people to demand and win control over their ancestral lands. Since the 1980s, island councils like the one on Yorke have had jurisdiction over their own communities.)

Half an hour later, all four of us were together again, clustered underneath a house with Mosby and the home owner, waiting out a torrential downpour. Many island homes, although built by the councils from modern wood siding, are raised on stilts in the Queenslander tradition, a practical design in this monsoon climate. For most of the year, the prevailing wind sweeps into the strait from the southeast, "but in the wet season, the wind comes from the northwest or the north, and that's straight out of New Guinea and Indonesia," Waterhouse pointed out. "Some of our pest incursions, like the papaya fruit fly, come associated with the wet season. That's also the time when things grow most actively, so you get your annual species of plants sprouting, and the trees put on their fruits. And it's a time the pathogens start developing as well."

NAQS scientists survey various sites one to four times per year, depending on their risk level. When they cannot make their periodic visits here, however, Mosby is alert for the arrival of suspicious organisms. Since 1989, starting here on Yorke and on Saibai, the island closest to the PNG coast, AQIS has been training resident quarantine officers such as Mosby. Now fourteen islands have full-time quarantine officers. Back on Thursday Island, Torres Shire's mayor, Pedro Stephen, at that time coordinator of quarantine operations in the region, would explain with understandable pride that AQIS was the first national government agency to provide career-level employment in the islands. Islanders trained as quarantine officers are fully qualified to request transfer to any AQIS post in Australia. "So when we employ them, we do not employ them as 'black trackers,' so to speak," he told me. The officers' presence, status, and public education efforts in their communities have made these islands some of the most quarantine-aware places in the world.

The local officers' duties include helping the NAQS staff monitor fruit fly and screwworm fly traps and "sentinel" cattle, pigs, and beehives that provide early warning of the arrival of targeted diseases such as Japanese encephalitis, which is now endemic in PNG. Sentinel pig herds were established on several strategically located islands and on Cape York to be blood-tested regularly for encephalitis after a fatal outbreak on one island in 1995 and the first recorded case on the Australian mainland in 1998.[2]

Now that Yorke has a paved airstrip, at least six commercial airline flights each week arrive directly from Cairns, and Mosby must also inspect all departing baggage. The numerous international cruising yachts that ply the strait are not allowed to land until they have cleared Australian customs and quarantine on Thursday Island, but the temptation to drop anchor and visit what look like postcard paradises is often irresistible. It is Mosby and the other AQIS officers who must send them on their way, for there is a real fear that dogs, cats, or pet birds aboard such vessels could be carrying diseases such as rabies, avian influenza, or Newcastle disease of poultry.

Most of the boats that arrive, however, are visitors from coastal PNG, now traveling in motorized dinghies instead of the traditional vessels envisioned under the treaty. That treaty between Australia and PNG gives certain coastal villagers the right to travel and barter a limited set of goods—although not sell them for cash—within what is known as the Torres Strait Protected Zone, which encompasses the islands north of the Thursday Island group. Only 8,000 to 9,000 people live on the islands, but 26,000 people in the coastal villages of PNG retain treaty rights to visit. "If you live on Saibai in a nice council-built house, there are people in a village only one nautical mile away living in grass huts," Mayor Stephen later explained. The coastal people see islanders getting Australian social welfare benefits and shopping in council supermarkets, whereas their own villages are among the poorest and most remote in PNG. What's more, fishing jurisdictions do not jibe with the border or treaty lines, so foreigners can fish quite near islands where they cannot land or, even if they have treaty rights to land, are not allowed sell their catch for cash. The temptation to break the rules must be powerful. A series of random checks in 1999 found that one of every seven dinghies intercepted in the strait was carrying prohibited items, including mangoes, bananas, and other fruit.

Island families usually invite relatives from PNG for major feasts and celebrations such as weddings and "tombstone openings," ceremonies at which grave markers are unveiled. Family members traditionally bring food gifts for feasting, including allowed sea turtles and dugongs but also forbidden items such as bananas—not so different, I imagined, from visiting relatives anywhere in the world. What would it hurt, Aunt Mary thinks, if I slipped a bit of fruit or a cutting from the backyard through customs to give the family a little gift from home? Thanks to Mosby's public outreach efforts in the school, at community meetings, and on Yorke Island radio broadcasts, however, she doesn't have to detect every violation like this single-handedly.

As the rain slackened to a light sprinkle, we moved on to another house across the road. Mosby chatted with the householder, who welcomed us, smiling, into her backyard. Waterhouse came over to me with a stalk of tall grass she had picked from the edge of the road. "This is *Imperata cylindrica,*" she said. That was one name I knew, Cogon grass or alang-alang, a major weed in many parts of the world, including the Gulf Coast of the United States. "It's native here, and it's just not a problem. It used to be used for thatching here, and it still is in New Guinea and parts of Indonesia." Now the council-built houses here, painted in a serene tropical palette of pinks and aquas, have metal roofs.

She pointed out a climbing glory lily with a beautiful orange flower. "It's considered an environmental weed near Brisbane," Waterhouse said. "I've noticed it here a lot, but it never seemed to be spreading before. I wonder if it's going to start to misbehave." She pressed a sample of this potential "sleeper" weed in newspaper between the plywood covers of her collecting case to take back to the NAQS laboratory in Mareeba, near Cairns.

When she joined with Judy Grimshaw and a mycologist to pioneer the NAQS surveys in 1990, "everyone was keen to have an entomologist and a plant pathologist," Waterhouse had told me, "but when they heard that I looked for weeds, they thought that was a joke." Weeds, particularly environmental weeds, were not considered a significant quarantine issue then. "One of the reasons they decided to begin to look at weeds was the threat of chromolaena coming in from Indonesia," she recounted. This extremely aggressive plant has the potential to take over forest plantations, native forests, and grazing land, and its seeds can travel undetected in soil, on clothing or machinery, or on the wind. In 1992, Waterhouse and the NAQS team found chromolaena on the New Guinea mainland for the first time, and the PNG government has subsequently been able to take advantage of a biocontrol program developed in Australia by Rachel McFadyen to keep it in check.

"Rachel was quite vocal that we must look for it in northern Australia and adamant that it would arrive here on cyclone winds if it wasn't already here," Waterhouse said. But it turned out that the weed was already in Queensland. Waterhouse first spotted it on an overgrown hillside during a visit to the Tully region, 160 kilometers south of Cairns and outside the NAQS survey area. It had arrived not via the strait but apparently in pasture seeds imported in the late 1960s, a time when AQIS was not set up to look for contaminants in seeds and the pest potential of chromolaena was unrecognized.

"The aim of NAQS is either to give early warning and try to implement measures to stop introductions or to find new pests early enough in the stages of incursion to be able to do something about them," she said. "Although chromolaena was well established in Tully District, it had not yet become more widespread, and eradication was thought to be possible. As long as the momentum is maintained, we're pretty well close to getting rid of it." The discovery of that infestation helped to prompt a review of quarantine that eventually led to expansion of the AQIS mission to include the integrity of the environment as well as agriculture.

Davis had been across the yard picking sweet potato leaves stunted by sweet potato little leaf disease. "I think little leaf is spreading on this island," he called out to us. "I don't recall seeing a backyard like this before. There's hardly a leaf that's not infected."

Waterhouse called back, "I don't think we've been in this yard for a while, and there's a grapevine growing here." Turning to me again, she noted that the grape industry is increasing in commercial importance in Australia, with some commercial enterprises now in the Tropics. "There's quite a nasty disease of grapes, a rust that occurs just to our north, in Indonesia, so we try to inspect any grapevines that we come across for symptoms of the rust disease because the spores readily spread on the wind."

Davis came over to inspect the undersides of the grape leaves, peering at them closely with a small magnifying glass. I asked about the yellowing on many of the grape leaves, but he told me it was not rust.

"No, it's a very distinctive orange rust on the underside of the leaves, literally rusty, with powdery masses of spores coming off. It's in northern Irian Jaya. We've found it in two places there." Fortunately, he found none on this grapevine.

Waterhouse explained: "We work from target lists, which help focus our search efforts. We try to report anything that is new or unusual, but certain pests have been prioritized according to important commercial hosts that we're especially concerned about. That's why we check sugarcane, grapes, and bananas. There's any number of other things that could come in that we would hope to pick up incidentally. But otherwise we would literally be looking for a needle in a haystack."

Davis added: "We've all got our own private target lists, if you like. There are a number of plant diseases on New Guinea that Australia doesn't have, but they don't make the official target list because they're diseases of crops that are not commercially important. The most obvious one is taro. There are a couple of really important diseases of taro in PNG that I'm always looking for. But whether anything would be done about it if we found it is a completely different story." In fact, failure to respond has in the past allowed some detected pests to reach Australia.

Later, as we walked back toward the waiting chopper, Waterhouse said that one such pest is a relatively recent arrival on Yorke, the spiraling whitefly, named for the way it lays its eggs in a spiral pattern on the undersides of leaves. She and Judy Grimshaw originally found the whitefly along the New Guinea coast in 1990. Despite its potential to devastate a wide range of food plants, the pest was not on the initial NAQS target list drawn up by consultants, and managers took no action. "We said, 'But this is going to arrive in Torres Strait very soon.' And it did. It was here within six months, and it gradually spread. But they were still saying, 'It's not on the list, so it can't be that much of a problem.' These days, with the pest rating system we have now, it would have been on the list. And eventually, through Judy jumping up and down a lot, a biological control program was implemented here in Torres Strait, which reduced its impact locally."

The best indicator that the biocontrol agent, a parasitoid wasp, is working can be found on Boigu, the island where the spiraling whitefly first arrived, Waterhouse said. Like Saibai, Boigu sits within sight of the PNG coast. "For a year they couldn't grow chilies or tomatoes or aubergines. Once the parasitoid was up and running, keeping the whitefly population to reasonable levels, people could resume growing these crops. Unfortunately, that pest has subsequently spread to the mainland from Torres Strait because the proper action wasn't implemented soon enough."

A Gateway for Banana and Bee Scourges

After a twenty-minute flight from Yorke Island, we came upon another coral island, Warraber, flat and green as a lily pad in the aqua sea. The helicopter put down on the grassy airstrip in bright sunshine, and Ted Billy, the new AQIS inspector and chairman of the Warraber Island Community Council, came out from a small building nearby to greet us. We climbed into his van for a journey of a few blocks—Warraber is but a kilometer square—and assembled again behind a house where half a dozen banana trees were growing behind a low fence of corrugated sheet metal. Stepping gingerly over the fence, Davis began inspecting the black edging on the older leaves. The concern of everyone was that it might be black Sigatoka, a devastating fungal disease of bananas.

Black Sigatoka is present in PNG and on several of the outer islands of the Torres Strait, Davis explained. "On mainland Australia, the Queensland Department of Primary Industries has maintained a buffer zone, whereby throughout the peninsula north of Cooktown only resistant bananas can be grown." The most susceptible bananas, unfortunately, are the important commercial varieties, the "supermarket" bananas such as Cavendish and Lady Finger. The government has removed all those plants in the Cape buffer zone and provided backyard growers with resistant varieties. "The idea is, if the spores of the fungus blow in or get in some other way, they won't establish because there are only resistant bananas."

Most of the bananas traditionally grown on the islands have some resistance, and "for that reason, its presence in the outer islands has been tolerated and nothing has really been done about it," Davis said. But the disease appeared on the mainland last year, and the eradication campaign had been expanded for the first time to some of the islands in the Thursday Island group. "They basically ripped out the diseased bananas and removed all the susceptible Cavendish and Lady Finger varieties, too," he said. "But that program never came as far as this island." Indeed, one of the banana plants he was examining looked to him like a Lady Finger.

"But, even having said all that, I'd put even money on this being something else, another leaf spot disease, and not black Sigatoka," he added. With a pocketknife, he began cutting strips off the blackened leaf to take back to the lab. "All we can do is take it back and test it." (His diagnosis would turn out to be right, I learned later.)

The sky was beginning to cloud up again as we left that garden and walked up and down the island roads, poking into bushes and turning up the undersides of leaves. This is one of only two islands without the spiraling whitefly, and Waterhouse was showing Anderson some potential host plants to watch.

As the rain began, we spotted a tin-roofed shelter attached to a house, an open-sided pool and pinball emporium, perhaps defunct, certainly dilapidated. No one seemed to be about. We gathered underneath and Davis explained NAQS' newest banana disease worries to me as we sidled around the perimeter of the sagging pool table, avoiding the drips and rivulets that began to pour through the leaky tin sheet above us.

On a NAQS survey in Irian Jaya and PNG recently, the team had spotted two banana diseases for the first time: Panama disease, a fungal wilt, and blood disease, a bacterial wilt. Panama disease is now a major scourge of bananas worldwide, and it presents a serious threat to the plantains, or cooking bananas, that are a staple food in PNG. "The greatest risk is that the disease can be transported on planting material," Davis said. "New banana suckers can be absolutely loaded with spores of the fungus and still look absolutely healthy." AQIS has produced a poster and informational flyers in three languages for the NAQS teams and others to distribute, urging people not to move planting material between villages or across the border. Blood disease, also a scourge of cooking bananas, appears to have originated on the Indonesian island of Sulawesi about eighty years ago. "It has really ripped quickly through these cooking bananas and eliminated them from the diet of the people of Sulawesi," Davis explained. "And now it's threatening to do the same thing on the island of New Guinea, where we found it last year. Whenever it enters a new area, its impact has been positively explosive." Again, the team has been working to get information out to farmers on how to keep the disease from spreading.

Waterhouse pointed out that the government of PNG has welcomed Australia's assistance because it has little money for either surveillance or extension work among subsistence farmers. When NAQS started in 1990, bananas along the PNG coast had been devastated by a moth called the banana skipper. Australia helped introduce two biocontrol agents that have greatly reduced the damage suffered by PNG farmers and also made it unlikely the skipper will get into the strait and onto the Australian mainland.

Banana diseases represent much more than an economic threat in the strait, Mayor Stephen later pointed out. "Not only is the banana an important fruit in our culture, but we use the leaves for cooking as well, and the actual stem we use for traditional costumes." Even now, traditional food for feasting is wrapped in banana leaves and set atop chopped stems laid on hot coals in underground earthen ovens to steam. The ability to continue growing fresh fruits and vegetables is vital not only for tradition but also for health, Stephen said. Islanders, like many other indigenous peoples from the Pacific to the American Southwest, suffer high rates of diabetes, which is aggravated by modern snack food and supermarket diets.

"You find quarantine so noticeable around the community now because people have actually taken ownership," he told me. "It's not difficult for us to deliver programs here because everyone knows that if there is an introduction of some very nasty pest or disease, we have nowhere to move. This is our home."

A slow rain greeted us back on Thursday Island and kept up all night. At dinner, Waterhouse filled me in on the next day's target: Asian honeybees and their parasitic *Varroa* mites, which attack both adult and larval bees and quickly debilitate whole colonies. The mites, which have helped depress bee populations from the United States to southern Africa and recently invaded New Zealand, threaten Australia's bee and honey enterprises as well as the pollination services provided to both crops and native plants. We had gathered to eat in the tin-roofed courtyard of the Federal Hotel—fondly called the Feral Hotel for its boisterous frontier ambience and a poolroom crowd that often looks wild enough to take on the creatures of the Mos Isley cantina in *Star Wars*.

Asian honeybees were spotted on the northern side of PNG in 1989, Waterhouse recounted, but it was assumed they would take a long time to move south because of the high mountain barrier in between. Instead, they appeared near the southern coast three years later. Fortunately, NAQS had prepared quarantine officers on Boigu and Saibai Islands to recognize bee swarms. The following year, 1993, the inspector on Boigu reported the first incursion. Two weeks later, bees turned up on Saibai. Examination of the bees' nests showed they were carrying *Varroa* mites. The northwestern winds of the wet season, it turns out, coincide with the bees' swarming season and carry them to the nearby islands. But how far over the open water could the bees fly? From Saibai to uninhabited Gabba Island is thirty-five kilometers. If bees could make it to Gabba, they could easily island-hop to the mainland. In the absence of more sophisticated monitoring techniques, the NAQS team tries to survey Gabba every three months. Also, Gabba and smaller Nagir Island to the south both have remnant mango populations, a few bananas, wild cassavas, yams, and a bit of sweet potato left by former inhabitants, and Waterhouse wanted to examine some of these plants to see what pests might have found refuge there.

The rain was still falling in the morning when we arrived at the heliport—an unpromising day for bees. For Gabba, too, as it turned out. Heading north, we soon found ourselves facing the iron-gray wall of an advancing storm front. After a few attempts at finding a way around the storm, the pilot gave up on Gabba and turned back south for a few minutes' flight to Nagir. As we passed over the slopes of Nagir's main hill, Waterhouse spotted the bright green of mangos amid the dense forest on the hill slope. Long ago, villagers had planted their gardens there to take advantage of runoff water. We circled back toward the beach, where old tin sheds and a small graveyard with white crosses also testified to Nagir's human past. At the base of the slope, we set down in a rotor-whipped frenzy of shoulder-high speargrass. Davis headed toward the beach to look for diseased leaves while I followed Waterhouse and Anderson upslope. Within minutes, a drenching monsoon rain had overtaken us.

As we fought our way upward through a sea of tall, wet grass, we could see fire scars on the eucalyptus trees, evidence that people still came here periodically to set fires to improve the growth of thatching grasses. Waterhouse warned us to watch for feral pigs, which islanders sometimes put ashore for future hunting. As we advanced up the rocky slope, we moved into a rain forest, the trees festooned with tangles of wild and feral yams and other clinging vines. Gray rocks appeared to be clustered around the bases of tall trees, another artifact of human cultivation, much like the overgrown rock walls one comes across in the reforested old fields of New England. A huge kapok tree had fallen over, its roots tipping up a gray boulder adorned with a large dendrobium orchid. "In a mainland forest, that would be in somebody's garden by now," Waterhouse commented. Unfortunately, the mangos that had been so clearly visible as we banked over this slope were all but invisible to us from the forest floor. Besides, it was hard to stare upward into a cascade.

We came to an impenetrable wall of bamboo and detoured to one side, looking for an opening. At a steep, slippery gully we were able to push upward again. Waterhouse pointed to a vine, its large leaves yellowed and spotted with disease, and asked whether I recognized it. I did not, for it looked sickly and bedeviled here in its native range, not at all like the robust, dark green kudzu that smothers so much of the southern United States.

After barely two hours on the ground, with no letup in the rain, it became apparent that even if honeybees or other pests such as mango leafhoppers were here, they were not going to be out and about for us to find. The three of us made our way back downhill to the chopper and found the others ready to go.

Five months later, on "a day that was brilliant for bee-hunting," the NAQS team did land on Gabba. They found no bees, but the first mango Waterhouse picked from a long-neglected tree and sliced open contained another target pest. It was a red-banded mango caterpillar, a potential threat to Australia's A$80 million mango industry should it ever reach the mainland. Gabba became the sixth site in Torres Strait where the caterpillar had been detected in the past decade.

A Blind Spot in Managing Invasions

Beyond the NAQS region, detection of most exotic pest incursions in Australia has relied on "someone noticing a problem and reporting it to authorities," according to Chief Plant Protection Officer Roberts. In 2001, however, Roberts' agency was preparing to launch an ongoing program of exotic pest detection surveys to be carried out within a five-kilometer radius of all ports and "hazard sites" such as container depots and devanning stations, and also in all forests within a fifty-kilometer radius of such sites. Surveys would be conducted one to four times each year at each of the country's fifteen seaports and eight international airports, with the most frequent surveillance at the busiest ports. Already Roberts' office manages a trapping program for specific pests such as Asian gypsy moths and exotic fruit flies at some of these ports. The new surveys would actively seek out not only forest pests but also pests of agricultural and horticultural crops. The proposal is based on a similar program in New Zealand, which has for decades conducted periodic surveys for forest pests and pathogens in all parks, reserves, and open spaces within a five-kilometer radius of ports and hazard sites.[3]

In most of the world, however, the lack of systematic surveillance, early detection, and rapid response to invasions constitutes a large blind spot in invasive species management. There are a number of single-purpose surveillance programs, of course, such as the interagency effort to intercept brown tree snakes arriving on Hawaii in airplane wheel wells or cargo from Guam, and numerous fruit fly trapping programs around the world. A country that experiences an incursion of Mediterranean fruit flies (Medflies) or other quarantine pests, for example, can reestablish its pest-free status—and thus get others to lift embargos against its exports—only by using scientifically valid, active surveillance methods to confirm that the pest has been eliminated. This kind of highly targeted detection effort by its very nature cannot spot any of the myriad other pests, weeds, or pathogens that might have slipped past a country's border.

Most exotic pest incursions in most parts of the world, in fact, are spotted first by members of the public or people out doing their jobs in ports, marinas, fields, forests, or conservation areas. The discoverers may be quarantine officers, agricultural extension agents, park rangers, foresters, gardeners, scientists, tour guides, photographers, hikers, or amateur naturalists. In the United States, for instance, a visiting marine ecologist chanced upon the first European green crab discovered in Washington State while looking at exotic cordgrass along the coast.[4] In New Zealand, it was a householder in suburban Auckland who found the first invading white-spotted tussock moth in a peach tree. Likewise, members of the public are credited with discovering the painted apple moth and various mosquitoes, frogs, snakes, and termites that eluded New Zealand border controls.[5] In Fremantle, Western Australia, it was marine builders who noticed what turned out to be dreaded Formosan termites—only the second known arrival in that country—in an imported motor launch they were repairing.[6] In Brisbane, a supplier of catering equipment called quarantine officers to report what turned out to be exotic auger beetles and long-horned beetles that had bored into dunnage used to pack imported refrigerators and vending machines.[7]

Most countries today rely almost solely on such hit-and-miss reporting by the public and yet make little effort to educate the public about what to be on the lookout for and why. That's unfortunate because most of us cannot tell native insects, leaf blights, or plants from new arrivals, much less gauge their potential for wreaking havoc. Good public education campaigns, such as the miconia alerts in Hawaii and French Polynesia, have resulted in numerous reports from the public about formerly undetected infestations of that invasive tree. Australia is now funding a pilot program to educate and enlist fishermen, boaters, divers, surfers, beachcombers, schoolchildren, naturalists, and people in the marine industry as "active watchdogs" to help detect new marine invaders before they become established.[8]

Once a pest or pathogen is detected, by whatever means, rapid containment and eradication is the most effective response. But timely action requires advance efforts such as preparation of contingency plans and training and equipping of staff to be available for emergency efforts—something few countries do well and none do well enough. The longer a pest, pathogen, or weed incursion goes unnoticed or ignored, the more likely it is that the window of opportunity for eradication will close—just as happened with *Varroa* mites in New Zealand and West Nile encephalitis virus in the eastern United States.

Taking a cue from NAQS' successes, countries can deal more effectively with incursions if they first gather "foreign intelligence" about pests on the horizon, help to combat those pests in exporting countries, and develop emergency plans to deal with the pests' arrival. Quarantine agencies can build up watch lists of potentially harmful species that are most likely to arrive as they assess the risk of new trade pathways, such as timber imports from specific countries. The Global Invasive Species Database developed for the Global Invasive Species Programme (GISP) and other on-line databases and information networks now under construction can also help countries target their surveillance and response preparations.[9]

Some have suggested that teams of experts be trained, funded, and kept available to react to pest emergencies in the same way the World Health Organization and the Centers for Disease Control in the United States respond to outbreaks of infectious disease. Attorney Peter Jenkins of the International Center for Technology Assessment also points to the industry-funded oil spill response plan as a model. That plan is designed to make oil booms, skimming equipment, and personnel available quickly anywhere in the world to battle spills, to avoid delays while money for cleanup is collected from the owner of a wrecked ship.[10]

As of 1999, the U.S. Department of Agriculture (USDA) had prepared emergency response guidelines to deal with incursions of fewer than two dozen pests, including fruit flies. In most other outbreaks, the agency's response is delayed while a pest risk assessment is conducted. Even when an assessment shows that an emergency response is warranted, bureaucratic problems delay the freeing up of personnel and money.[11] The National Invasive Species Council has recommended that new legislation be drafted to create a well-coordinated and well-funded response system that integrates federal and state efforts.[12]

Some observers doubt, however, that responses to invasive plants, especially weeds of concern for conservation but not to farmers and ranchers, can ever be addressed adequately under the leadership of the USDA. One firm doubter is Randy Westbrooks, formerly noxious weed coordinator for the USDA's Animal and Plant Health Inspection Service (APHIS) and now invasive plant coordinator for the U.S. Geological Survey's Biological Resources Division (USGS-BRD) in the Department of the Interior. He points out that the crop protection laws under which APHIS operates, unlike environmental protection laws, enable but do not require the agency to act. Congressional funding and agency priorities "are driven fundamentally by stakeholder interest and need," he says. But wetlands and wildlife habitat have no stakeholders. That is one reason so few invaders of natural areas, such as melaleuca, have ever been added to the federal noxious weed list. It is also why the APHIS budget for battling witchweed, a parasitic plant that threatens corn production, was three times higher than his APHIS budget for controlling all other noxious weeds throughout the country, and why the agency "barely yawned" while the aquatic invader salvinia fern spread to nine states.

"What we need is a third, parallel system, a biological protection system, probably somewhere in the Interior Department, to complement the environmental and crop protection systems," he says. That system would coordinate early detection, rapid assessment, and response to invasive species that do not have an organized constituency, such as salvinia fern, garlic mustard, melaleuca, and purple loosestrife. "Weeds know no borders, but the people who care about them do," he adds. "People are going to work only within their areas of responsibility. We just shouldn't try to rely on the crop protection system for everything."

Catching Sleepers as They Wake

Although rapid responses to the arrival of aggressive weeds such as salvinia fern and chromolaena are vital, the border is generally not the place to look for most of tomorrow's invasive plants. Species that long ago secured a beachhead on new shores may "sleep" through years or even multiple human generations before they explode into damaging invasions. "The next generation of weeds is already here," Australian plant ecologist Mark Lonsdale says. I heard the same sentiment when I visited a knapweed-infested hillside near my home with Montana State University rangeland ecologist Roger Sheley: "If nothing else ever arrives, we have enough weeds now to destroy the West." Many of these are already advancing across the landscape, but others are undoubtedly in a "lag phase," growing unobtrusively in gardens, forests, pastures, and roadsides. These are the ones ecologists call sleepers, time bombs, or incipient or emergent invaders. Detection and surveillance efforts for sleeper weeds, previously almost nonexistent, have recently been implemented from South Africa to Micronesia, Hawaii, Australia, and New Zealand.

The difficulty, of course, is telling which seemingly well-behaved plants are permanently benign pansies and which are sleepers that will awaken to overwhelm the habitat. It is worth repeating an observation by conservation biologist Michael Soulé: "Time is on the side of the invader." Because of that, he would like to see the precautionary principle supplemented with the "principle of promptness."[13] The longer a species is established in a place, the greater are its chances of exploding out of control. That is especially true if the species has already proven invasive elsewhere. Yet, like Randy Westbrooks in the United States and Barbara Waterhouse in Australia, "weed warriors" everywhere too often find it difficult to whip up a sense of urgency or a commitment of resources, even for control of known invaders. Sometimes the wait-and-see attitude means no action is taken until weed populations have advanced beyond the point at which eradication is practical or even possible. Weeds are a problem we too often leave for future generations.

In the long run, both New Zealand and Australia hope their new guilty-until-proven-innocent import policies for plants will reduce the flood of potential sleepers to a trickle. For decades to come, however, both countries can expect new invaders to arise from gardens, pastures, roadsides, and even abandoned agricultural research plots. In Auckland, more than 600 exotic plant species have naturalized outside cultivation—a figure no other city in the world, except perhaps Honolulu, is likely to beat—and four new species escape into the wild each year, usually from gardens.[14] The pool of potential escapees includes 18,000 alien plants already being cultivated in New Zealand.[15]

In Australia, too, shutting off the supply of new imports will take a long time to slow the rate of new weed invasions. Plant ecologist Richard Groves of the Commonwealth Scientific and Industrial Research Organisation (CSIRO) has documented that the number of introduced plants becoming established in the wild in Australia has increased in a straight-line fashion over time and is continuing at a rate of four to six species per year. Quarantine efforts have yet to reduce the rate of increase.[16] To try to reduce the adverse effects of sleeper weeds already in the country on both production and conservation lands, the country has developed the comprehensive National Weeds Strategy.[17] One of the first actions under that strategy has been the selection of twenty "weeds of national significance," a list of priority targets ranging from salvinia to mesquite, lantana, gorse, rubbervine, and mimosa.[18] Also, to discourage gardeners from using weedy plants already in the country, the nursery industry teamed up with research and government agencies to draw up a list of fifty-two "garden thugs" to target, including tree of heaven, giant reed, asparagus fern, pampas grass, cotoneaster, Cape ivy, English ivy, St. John's wort, Monterey pine, blue morning glory, and wandering jew.[19]

Finally, under the auspices of the new Cooperative Research Centre for Australian Weed Management (CRC Weeds), CSIRO "wants to explore the idea that we might put in place a system for detecting sleeper weeds that are just starting to emerge from their slumber," Lonsdale told me. "We want to have a network of sites around Australia that are regularly monitored to look at which of the species are on the upward part of the exponential curve. And having detected them at an early stage, we'll be looking to eradicate them."

In a growing number of countries, individuals, land managers, and agencies are taking similar steps, developing ways to monitor the behavior of naturalized exotic plants, just as Waterhouse keeps an eye out for changes in climbing glory lilies and other potential invaders in the Torres Strait. Keeping an eye out means inventorying what plants are present, investigating their weedy reputations elsewhere, prioritizing the potential sleepers to watch, and periodically assessing their status. Assessment methods may include aerial surveys, satellite monitoring, and other technologies that can help detect and map changes in the extent and density of weeds. For now, however, assessment is mostly done—when it is done at all—by eye and on foot.[20]

The Hawaiian island of Maui is a good example. USGS-BRD ecologist Lloyd Loope and two research assistants have begun scouring the island, from suburban yards and gardens to roadsides, nurseries, and parks, for signs of exotic plants that may be starting to spread. Their ultimate aim: to devise a surveillance and response system that can protect Haleakala National Park and surrounding slopes that support many of the island's 100 endangered species. The system they develop may also provide a prototype for use around other parks and protected areas. The difficulty of cataloging sleepers on Maui is that there are thousands more exotic plants than natives present. When I first heard of the project, I pictured the plight of frenzied airport beagles set loose in a fruit market. How would they decide where to begin?

I met Loope and research assistants Kim Martz and Forest Starr one day near Makawao, a historic cowboy town now booming with "lifestyle" settlers, art galleries, and boutiques. Makawao is in up-country Maui, on the slopes of the dormant 3,000-meter-high Haleakala volcano. This is a land of cattle ranches, vineyards, fields of sweet Maui onions, botanical gardens, and commercial flower farms growing South African proteas and other exotics for shipment worldwide. My tourist guidebook told me that I would "notice plenty of natural vegetation, as clumps of cacti mingle with purple jacaranda, wild hibiscus, and towering eucalyptus trees." All natural somewhere, I suppose, but not here.

As the four of us drove around the two-lane roads at a pace of less than ten kilometers per hour, Starr explained that he and Martz had planned to start by getting a list of all the plants on Maui and cross-checking it with global weed databases. "Well, we got to step one and said to Lloyd, 'We're ready for the list now.' But there was no list." So Starr and Martz, both self-taught botanists, have spent much of their time documenting what is growing on Maui, exploring the suburbs and waste places of this small island with the passion other plant hunters reserve for wild new frontiers.

"If you look up the mountain, you can see big pastures of kikuyu grass and eucalyptus stands," Starr pointed out. "It's all alien but relatively monotypic." What we could see from below was all there is in those single-species landscapes. "The real diversity of sleepers is down here in people's yards," he added. "This is where things are going to start, and this is where the winnable battles are, so that's where we've been focusing."

We parked alongside a residential road and began to walk. Starr stopped to show me a vine in the morning glory family that was draping a rock fence. Using a global positioning device, he took coordinates for the vine to add to their baseline map showing the distributions of potential sleeper weeds on the island.

"This vine—we'd never seen it before on the island, and it's already showing invasive tendencies. Look, it's going up in the trees and suckering out over there. We sent it in to the Bishop Museum (in Honolulu), and it's a new state record, known only from this small patch right here. Now, that is a winnable battle. We could search databases from around the globe and ask, 'What has this species done elsewhere? Has it just sat benignly for centuries, or has it exploded as soon as it's been introduced?'" If the latter, then the Maui Invasive Species Committee (MISC), a voluntary partnership of private, government, and nonprofit agencies, could send its weed crew to clear the vine—if the landowner agreed. Most owners do cooperate, Starr said.

When Starr and Martz encounter an infestation of any of MISC's six target species—miconia, pampas grass, fountain grass, ivy gourd, giant reed, or rubbervine—they report Global Positioning System points to the group directly.

"We can't ignore the really aggressive, widespread weeds," Starr said.

"But we also have to focus on the ones that are brand-new, that haven't gotten to the park at all yet but show invasive tendencies and merit a close look while they're still potentially eradicable."

A year after my visit, Loope's team had assembled a target list of more than 100 species, and the numbers were continuing to edge up slowly. They had also logged more than 1,600 kilometers of painfully slow survey drives and recorded and mapped 13,067 occurrences of target plants.

Unlike Australia and New Zealand, of course, Hawaii and the rest of the United States still offer an open door to plant imports, so the team's task is constantly compounded. That day on Maui, we had stopped at a house with a vine-draped cinder-block wall fronting the road. "What I want to show you is this Japanese honeysuckle, which in eastern North America is considered to be one of the worst weeds," Starr pointed out. "And yet it has recently been promoted as a landscape plant for ground covers on steep slopes in Maui. So here's this highly invasive weed, and yet it's being promoted here by the landscape industry. To me, that says, 'If we're letting Japanese honeysuckle in, what wouldn't we let in?'"

Loope and his team are continually finding troubling new answers to that question. During the following year they would discover, among other things, a few individuals of a species with such a bad reputation, from Florida, California, and the West Indies to Australia, that it would become number seven on MISC's priority eradication list: the Jerusalem thorn tree.

In a number of other places in the world, scientists and land managers are trying to nip emerging plant invasions in the bud even as new candidates continue to arrive. On present and former U.S. territories in Micronesia, for instance, a project called Pacific Island Ecosystems at Risk (PIER) was launched in 1998 to survey and assess exotic plant species that threaten pastures, gardens, or native plant communities. PIER, a project of the USDA Forest Service, supplies land managers and quarantine organizations with information on both the status and risks posed by plants already present and known invasive plants that have not yet reached specific islands.[21]

South Africa's Kruger National Park, an island of wildland amid a sea of human homes, farms, and game ranches, began a set of surveys along its major rivers in 2001 that officials hope to develop into a formal, ongoing monitoring program to get control of an escalating weed invasion problem. Only 6 alien plants were found in the park in the 1930s; that number rose to 43 in the 1950s and 160 in the late 1980s.[22] There are more than 300 today. The new surveys are designed to spot recent infestations that are limited enough to eradicate and to track increases in the density or rate of spread of established invaders. The targets include not just plants such as chromolaena, lantana, and prickly pear cactus but also Indian myna birds, Nile tilapia fish, and *Varroa* mites in bees.

"We accept that we have hundreds of alien species that we'll never eradicate," said alien biota manager Llewellyn Foxcroft. "But we hope to maintain acceptable limits of the priority species."

Like Kruger, every region has invasive alien species that are now too firmly entrenched to eradicate. In too many cases, the window of opportunity closes even before an incursion is detected. Once an invader is established, land managers are faced with the decision of whether to try to control it and, if so, how much effort to spend to minimize its influence. At that point, the focus should shift from a one-on-one battle with an uninvited species to a larger management vision of what we need and expect from the land and the living communities it harbors.

TEN Taking Control

"Effective prevention and control of biotic invasions require a long-term, large-scale strategy rather than a tactical approach focused on battling individual invaders. An underlying philosophy of such a strategy should be . . . to address the underlying causes rather than simply destroying the currently most oppressive invaders. System management, rather than species management, ought to be the focus."

—Richard N. Mack et al., in the journal *Ecological Applications,* 2000

"The control and extirpation of damaging aliens, especially species like fire-conducting grasses, will remain among the most urgent of all management activities. But a policy of blanket opposition to exotics will become more expensive, more irrational, and finally counterproductive as the trickle becomes a flood. Only the most offensive exotics will be eliminated in the future, using such criteria as degree of potential hazard to humans and to indigenous species and the probable cost of extirpation."

—Michael E. Soulé, in the journal *Conservation Biology,* 1990

"I listened carefully for clues whether the West has accepted cheat[grass] as a necessary evil, to be lived with until kingdom come, or whether it regards cheat as a challenge to rectify its past errors in land use. I found the hopeless attitude almost universal. There is, as yet, no sense of pride in the husbandry of wild plants and animals, no sense of shame in the proprietorship of a sick landscape."

—Aldo Leopold, *Sketches Here and There,* 1949

On a narrow, wind-whipped plateau 1,500 meters above the South African Cape, I joined a small cluster of people sheltering in the lee of a massive boulder pile. The helicopter that had just delivered us lifted off and then headed back down the rugged face of the mountain. A storm was building, and there were several more loads of mountaineering gear, food, and camping supplies still to be ferried up from a rugby field in the town of Ceres, below us. And then there were the chain saws and slasher tools, for this was to be no ecotourism adventure. Instead, the mission seemed to a visitor as unlikely as the team that would stay on this mountain to carry it out. Take the team supervisor, Mariska Farmer, twenty-nine, mother of two young boys, a small, vibrant, confident figure now checking through a pile of gear. She was born in Ceres, an orchard town that tourism promoters call the "Switzerland of South Africa" because of the rugged scarps that flank it. Yet until three years ago, Farmer had never ventured into these mountains. Besides her, there was Evelyn Tokwe, forty-four, whose friends had told her she was too old to learn to ride in a helicopter, much less winch out of one onto a mountaintop or rappel down a cliff face while cradling a chain saw. But that is exactly what she and the others on this crew—four women, three men, all previously unemployed—had learned to do. The target of their slashers and saws this trip was to be the errant pine seedlings that dotted the ridge in front of us and sprang from the crevices and ledges on the mountain face. Each invading alien pine, if allowed to grow to maturity, would rain seeds into the already tree-choked ravines and water catchments in these mountains, usurping more water needed by people and pear trees, native wildflowers, and springbok antelope. The job of clearing invading plants is most effectively done from the top down, and it is Farmer's team and a few others like it in the Cape that take care of seed sources at the top. These high-altitude teams form the elite forces of a program called Working for Water, a bold marriage of poverty relief, invasive plant control, water conservation, and biodiversity preservation that was launched by South Africa's first democratic government almost as soon as it took office in 1995.

Although the optimal strategy for avoiding damage caused by invaders is to prevent their arrival or establishment, prevention is not always an option. Centuries of commerce in plants and animals have left South Africa and most other countries with a troubling legacy of well-entrenched invaders. Fortunately, there is a growing tool kit of techniques that can be used to minimize their adverse effects, from physically clearing them out and judiciously employing pesticides to altering land-use practices and introducing natural enemies as biocontrol agents. I begin here with the Working for Water story because its success in integrating invasive species control into the larger fabric of a nation's life and hopes for the future has made it a model for weed and pest management elsewhere in the world. Working for Water also illustrates two important points about invasive species management: the most effective efforts make use of a range of techniques, and they focus not on demonizing and defeating particular invaders but on restoring the economic or ecological conditions we value in our lands and waters. For Farmer and Tokwe and the 20,000-plus other weed warriors of Working for Water, clearing invasive alien trees and shrubs from the nation's stream banks and water catchments is a means, not an end. To understand why they are targeting weedy trees and what they hope to achieve, however, we need to leave the team to their work on the mountain and return to earlier times, when this arid land was largely treeless.

South Africa is one of the world's "megadiversity" countries. In the Cape region alone, the "Cape floral kingdom," with its 8,600 species of flowering plants and ferns, is recognized as one of the earth's six distinct floral kingdoms. Much of the Cape's plant life is found in shrubby, heath-like communities known as fynbos, and 70 percent of its native plants are found nowhere else on the earth.[1] Or, at least, they were not until the great plant-hunting expeditions of the eighteenth and nineteenth centuries. Now gardeners the world over grow proteas, bird-of-paradise, ice plants, gladioli, amaryllis, geraniums, arum lilies, and many other treasures from the fynbos.

But floral wealth was not what first brought Europeans to the rugged Cape of Good Hope. In 1652, the powerful Dutch East India Company set up an outpost there to provision its ships with fresh meat and produce midway on the grueling sea voyage from Europe to the East Indies. By the nineteenth century, overgrazing by livestock had left the sandy flats of the Cape Colony open to severe wind erosion, and the colonists began importing many species of acacias and hakea shrubs from Australia to stabilize the dunes.[2] Thousands of other trees, shrubs, and flowers were imported for windbreaks, for fuelwood or timber, or to grace the gardens of the colony, even as botanical explorers were collecting trophies from the native fynbos to take back to Europe. Altogether, more than 750 species of trees and 8,000 shrubs and flowering plants from Europe, Asia, Australia, and the Americas have been introduced into the Cape and other regions of South Africa. So far, 161 are regarded as invaders, and some of the others are likely to be sleepers that will eventually explode from slumber.[3]

Ironically, the most aggressive and problematic invaders have been the trees, large, thirsty life-forms that are now diminishing scarce water resources, displacing native fynbos, and fueling hot, destructive fires that open the way for subsequent flooding and erosion. I call this ironic because a century ago, learned people expected the imported pines, eucalyptus, acacias, and other trees to increase water supplies, not decrease them. In the 1860s, for instance, the Colonial Botanist recommended planting trees, in the belief that they would bring more rain and alleviate a severe drought. In the 1880s and 1890s, foresters promoted the planting of European cluster pines on Cape Town's imposing Table Mountain to increase water supplies (and also because the mountain's "bare and stony slopes" were "an offense and eyesore to Cape Town"). Within a few decades, seeds from these and myriad other plantings throughout the region had begun to spread far beyond where the trees were initially cultivated.[4]

The centuries-old belief that trees make rain persisted in popular lore until well into the twentieth century (and it does have some validity in certain climatic regions, such as the Amazon rain forest and oceanic islands).[5] But scientists in South Africa had already begun to realize the truth: replacing the drought-resistant, low-stature fynbos with large thickets of trees was reducing stream flow. The trees, with their greater mass of wood and greenery and deeper roots, drew up larger quantities of water from the soil and exhaled (transpired) greater amounts of water vapor from their leaf pores into the air. Instead of condensing to fall again as rain, however, the moisture is lost from the region, blown away to fall elsewhere.

Today, 110 species of alien trees, including willows, poplars, pines, mesquites, black wattles, and thicket-forming acacias, along with aggressive shrubs such as hakea, lantana, chromolaena, bugweed, and prickly pear cactus, have invaded more than 8 percent of the country's land surface, and the invaded area is expanding by 5 percent per year.[6] These trees and shrubs are transformer species, invaders that alter the processes of water, earth, and fire and change the terms of life for everything in the community.

By the 1960s, the South African government had acknowledged the problem by requiring that areas beyond the boundaries of forest plantations be actively cleared of pines, black wattles, and other escaped commercial species. But the law was only sporadically enforced, and clearing was a low priority for most plantation owners. Besides, some government agencies continued to promote the planting of invasive trees and shrubs for windbreaks and erosion control.

It took scientists decades of tedious experiments to quantify the water lost to these spreading infestations of trees, and it was their hard numbers that finally won support for the all-out invasive tree-clearing efforts of The Working for Water Programme. Some of the most convincing numbers came from work that began in the 1930s, when the Jonkershoek Forestry Research Centre was established in the Eerste River basin, near Stellenbosch, forty kilometers east of Cape Town.

Ecologist Brian van Wilgen of the Council for Scientific and Industrial Research (CSIR) Division of Water, Environment, and Forestry Technology drove me out to Jonkershoek a few days after I visited the Ceres team. A rare spring snowstorm had just dusted the mountains of the Cape a glorious white. At Jonkershoek, van Wilgen turned onto a series of rutted switchbacks that took us high up the southern slope of the river basin. We could see a series of shallow ravines slashing down the opposite slope, a few of them still bordered by native fynbos vegetation, a fuzzy gray carpet from this distance, but most of the catchments were now hemmed between dark green walls of pine planted in neat blocks. In the 1930s, forest hydrologist

C. L. Wicht had installed concrete weirs at the bottom of each ravine. He monitored the flows from each catchment, through wet times and dry, and then he planted some with pines and began to track how flows changed as the trees grew over the years. By 1977, Wicht's successors at the station were able to report that replacing the low fynbos shrublands with tall pine woodlands had reduced the flow in a stream by almost 60 percent.[7]

Studies in other parts of South Africa where small watersheds were heavily afforested—the equivalent of a dense invasion—have yielded similar dramatic results. In KwaZulu-Natal Province, filling a small grassland catchment with pines resulted two decades later in an 82 percent reduction in stream flow. In Mpumalanga (formerly Eastern Transvaal), grassland streams dried up completely six to twelve years after dense plantings of pine and eucalyptus.[8]

The results would be less dramatic, of course, in larger catchments with sparser plantings (or invasions). But van Wilgen pointed out that 80 percent of South Africa's water comes from just 20 percent of the land area, mostly from the higher-elevation, higher-rainfall areas that are also at most risk of invasion. Another disturbing finding is that the amount of water a pine or wattle or eucalyptus tree uses depends on how much is available. A tree growing on a riverbank or in a streambed uses twice the water taken up by a tree growing away from the river. Escaped trees that now clog watercourses on public and private land alike use up almost 7 percent of the water that flows into the country's rivers, more than twice the water commercial forestry plantations use, and they also occupy a much greater share of the country's land area.[9]

Long before the numbers on water loss had been tallied, Wicht had also recognized invasive alien plants as "one of the greatest, if not the greatest, threats to Cape vegetation."[10] Today, 40 percent of the land area of the Cape Peninsula has been claimed by agriculture and urban development. More than 10 percent of the rest is densely invaded, and another 30 percent lightly invaded. Without intervention, the prognosis for native plant diversity looked grim.[11] As conservation awareness rose in the 1970s, many private groups and government agencies began actively hacking and applying herbicide to alien trees and shrubs in fynbos reserves and parklands. It was soon clear that a much larger effort was needed, but at the time there was little hope of mobilizing government support purely on conservation grounds.

Why, then, had something like Working for Water not been created decades earlier? "Because the scientists talked only to each other," van Wilgen said. "We were saying as scientists, 'There's an impact of trees on water resources.' But Water Affairs thought like engineers. They didn't think about manipulating vegetation. They build dams and tunnels and purification systems."

Then, in the late 1980s and early 1990s, two threads began to come together. One came from the scientists. Van Wilgen and David Le Maitre of CSIR, along with a number of colleagues from universities and conservation agencies, began to pull together the results of more than half a century of stream-flow measurements at Jonkershoek and roll them into models that projected what the advancing tree invasion in the Cape, if left unchecked, would mean for the region's water supply. The results were alarming: if the invasion were allowed to proceed, the projected loss would eventually total more than 30 percent of the water supply to Cape Town.[12] Next came a detailed analysis of whether the benefits of stopping the invasion outweighed the admittedly high costs of clearing trees and rehabilitating native vegetation. The cost-benefit analysis showed that the financial benefits of clearing the catchments significantly outweighed the costs because improved stream flows would reduce the need for more dams, waterworks, and other costly or questionable schemes that were being considered, from desalinization of seawater to cloud seeding. Even with the ongoing cost of clearing regrowth factored in, the unit cost of water from a dam built in a treeless fynbos watershed would be 14 percent lower than that of water from a dam in a forested catchment because the latter would yield less water. And the analysis did not attempt to put numbers on other values that clearing trees would yield, from preserving biodiversity in the intact fynbos to abating fire and erosion.[13]

The second thread that converged to make Working for Water possible was a political and social watershed for South Africa: the 1994 election, which brought Nelson Mandela's Government of National Unity to power.

At that point, in a nation of 43 million people, 12 million lacked adequate access to drinking water, 20 million lacked adequate sanitation, and more than half the rural African population was unemployed.[14] With the change in government, hundreds of thousands of those rural Africans flocked to the cities, where they erected the expanses of "informal settlements" that now ring the outskirts of Cape Town and other urban areas, creating overwhelming new demand for water and sanitation services.

With water supplies already tight, previous national governments had been considering new engineering schemes in the dam and tunnel system to the east of Cape Town. Guy Preston, national leader of The Working for Water Programme, was then an academic scientist arguing passionately, but with little success, for water conservation instead of more dams. In late 1993, Preston and others concerned with water conservation created an informal alliance with ecologists and environmental managers working in the fynbos. The groups decided to prepare a "road show" on their proposals and bide their time until a post-apartheid government took power.

In July 1995, they presented their findings to the new minister of water affairs and forestry, Kader Asmal, stressing the theme of job creation. Asmal ran with the idea. Within three months, thousands of previously unemployed and unskilled people, "the poorest of the poor," more than half of them women, were out chopping and pulling and spraying the stumps of invasive trees throughout the country. By March 1998, employment in the government's Working for Water Programme had reached a dizzying high of 35,000. In the program's first five years, private sources and international donors joined with the South African government to invest $R800 million (US$90 million) in it.[15]

"We needed to move fast, put people to work, and get money into workers' pockets," recounts Preston, who took over the reins of the new program with characteristic intensity. Working for Water has set up crèches, or day-care facilities, for the children of workers, established clinics, and built partnerships with social service agencies, which provide workers with information on family planning, AIDS, and financial management. Workers also receive job training and certification in skills such as chainsaw operation and fire fighting. Five years into the program, employment had stabilized at more than 20,000 people and the battle against weedy trees had been firmly institutionalized.

The number of people employed, along with the number of trees chopped and expanse of land cleared, are significant measures of success, but they do not indicate whether the goals of enhanced water supplies and thriving biodiversity are being met. In truth, there are few "before" studies with which to compare the flows in any given stream after clearing. But recent studies do show that just as growing trees reduces water runoff, clearing them can increase it.[16] When Monterey pines were cleared from the stream banks in one of those Jonkershoek catchments, for instance, stream flow was boosted by as much as 44 percent for each 10 percent of the catchment that was cut.[17]

Particularly heartening are the observations and enthusiasm of people who live in the now weed-free catchments. "We've heard so many good stories," Derek Malan, a Working for Water area coordinator, told me on another outing. "A farmer will say, 'I've been on this farm for fifty years and I've never seen water in that stream, and now it's flowing.' And they tell their neighbors." An engineer by profession, Malan had spent several decades working for the water agency on irrigation and drilling projects in the Western Cape before joining Working for Water.

We were driving across the Overberg, the region across the mountains east of Cape Town, toward a village called Elim at the southernmost tip of Africa. Moravian missionaries created Elim and other settlements like it throughout the Cape in the early nineteenth century to serve the mixed-race people the government designated as "colored." As we entered the main road of the village, driving between rows of neatly whitewashed Cape Dutch houses with dark thatch roofs, it was like stepping back a century or more into a vibrant remnant of history. The land and the buildings are still owned and managed by the community, and many families' roots go back to the founding of the settlement, in 1824. The modern cash economy, however, had generated few job opportunities out here in recent decades, except for migrant workers.

In 1998, Working for Water was able to secure American and Norwegian aid as well as national poverty relief funds to build economic opportunities based on the rich human and natural capital of Elim. Besides its tradition and its people, Elim boasts a unique botanical wealth—a rare and distinct floral community known as Elim asteraceous (daisy-like) fynbos interspersed with patches of acid sand fynbos and marsh fynbos. The community's 3,000 hectares of native vegetation showcase at least 500 species, including rare heaths, gladioli, an endemic protea known as the "bashful sugarbush," a dwarf purple protea, and an endemic leucodendron whose yellow flowers often dominate the hill that rolls up next to the village—thus named Geelkop, yellow hill.

At the Working for Water office in the back of the mission store, we met Karl Richter, community member and fynbos guide. One of the first targets of the clearing project, Richter said, had been a line of pines and hakeas that had commandeered the top of the Geelkop. Now the community runs the restored Geelkop as a private nature reserve with a six-kilometer hiking trail developed to showcase its endemic wonders. Had I arrived in Elim a day earlier, in fact, I would have encountered forty-four busloads of botanically minded tourists enjoying the community's sixth biennial flower show, hiking the Geelkop, and admiring a magnificent display of thousands of cut flowers in the village's new metal flower barn. Fortunately for me, the displays of flowers—more than 100 species—were still massed in the barn, and a quick tour through the building reminded me why plant lovers find it so hard to resist stuffing exotic seeds and cuttings into their luggage.

Besides allowing Elim to develop nature tourism, restoration of the fynbos has helped the community double its exportation of wild-picked flowers (not the rare endemics, of course). Now there are plans to begin cultivating proteas and other fynbos products for the South African and European markets on some of the cleared agricultural land. Even the aspirations of tiny Elim, however, are buffeted by the whims of the global marketplace.

"Europe had a good summer last year, with lots of poppies and other flowers, so there was less of a market for imports," Richter said. Besides, thanks to the plant hunters of old, New Zealand, Australia, and Hawaii outstrip South Africa in protea exports. South Africa, however, now does a booming business cultivating Western Australian banksias for the cut-flower market. The sting in this plant swapping is that some of these banksias, like the Australian hakeas before them, are expected to become future invaders of the fynbos.[18]

In Elim, Working for Water crews also cleared trees from the watercourses, including the permanent springs that prompted the missionaries to choose and name this place (Elim was the biblical oasis where the Israelites rested on their way to the Promised Land). After the pines were cleared from the springs, Richter told me, water flow increased enough that the community had recently built a second reservoir.

Sustaining the Gains, for People and the Land

The trends in both water and biodiversity conservation are positive, but The Working for Water Programme is not free of controversy or criticism, even from some of the scientists who helped create it. The emphasis on hiring the chronically unemployed, for instance, though necessary to keep government enthusiasm and funding high and meet the social goals, has led to problems with absenteeism and low productivity. The less land cleared, of course, the smaller the water gains for the money. Also, project jobs have been distributed around the country rather than focused on biodiversity hot spots such as the Cape fynbos, which hosts only one-quarter of the efforts. Still, there are other teams doing vital conservation work by clearing mesquite trees from the Kalahari Desert and prickly pear cactus and chromolaena infestations from the rivers of Kruger National Park.

"I've tried not to be a purist," says Preston. "Making mistakes is character-building, and our characters have really been built over the last five years."

Working for Water is intended to work itself out of a job after twenty years, and that means it must prepare its trainees to compete in the private sector. Under a new labor strategy, the program is training and equipping workers and helping them organize into teams under a single contractor or as cooperatives. Landowners and government agencies such as Working for Water then contract with these teams to clear blocks of land. Enabling these crews to go after private clearing, wildland fire fighting, and restoration contracts is one of the "exit strategies" Working for Water is using to graduate its trainees and bring in a new round of workers every two years. Christo Marais, manager of program development and planning, has also set up pilot projects for secondary industries that make use of much of the debris generated by clearing: firewood and pulpwood sales, charcoal making, and even furniture making and crafts, from mesquite deck chairs to black wattle walking sticks. His challenge is to generate long-term economic opportunities such as these without creating an ongoing demand for invasive plants, and that means actively seeking native plant substitutes that can sustain such secondary industries in the long run.

Some exit strategies will not rely on harvesting the land at all. In places such as Elim and Ceres, there are new jobs as tourist guides. Back on the rugby field at Ceres, I learned that the high-altitude team had just incorporated as the Ceres Mountain Guides. They had leased and begun to rehabilitate the formerly derelict Ceres Mountain Fynbos Reserve, which includes the mountain they were clearing during my visit, to run as a hiking and climbing center.

"Over time, our hope is that the percentage of a team's time in clearing alien plants will go down and the percentage of these other activities will go up," Marais explained. "If this comes true, this will be the best conservation program ever in the world, I believe. We will have changed a negative environmental impact into a positive economic sector."

The enormous cost of Working for Water is justified now because it fills an immediate need for human development and for rapid clearing of major water catchments. But intensive, costly hand labor cannot sustain either goal over the long term. Physical clearing provides fast, dramatic results, but trees and shrubs can grow back rapidly from huge numbers of seeds that remain viable in the soil. To provide a sustainable, long-term means of controlling invasive trees and ensure that the gains from twenty years of labor are not lost, Working for Water is making a substantial investment in research on biocontrol.[19]

Historically, 103 biocontrol agents have already been released in South Africa against forty-six weed species, and twenty-two of these weeds are considered substantially or completely controlled. Two of the most striking successes were quite visible as I traveled about the Cape. Dead and dying stands of a thicket-forming acacia known as Port Jackson willow covered the flats, their branches deformed by knobby, gall-like growths caused by a rust fungus released in the mid-1980s. An analysis by Le Maitre indicates that the rust fungus has saved enough in clearing costs alone to pay back the amount invested in researching and releasing it by 800 to one. A second acacia species known as the longleafed wattle has been reduced "essentially to non-invasive status" by the effects of a seed-feeding weevil and a gall-forming wasp that were also released in the mid-1980s.[20] Thanks to the wasps, only green, marble-sized galls can be seen on the branch tips, where the wattles would ordinarily be producing flowers and seedpods. More recently, a seed-feeding weevil was released in an attempt to limit the spread of a third acacia species, the black wattle. Scientists are also exploring the consequences of introducing biocontrol agents that can kill mature pine and acacia trees—an idea that troubles the forestry industry, which might have to use pesticides to protect its plantations from these agents.

In another strategy to help bring weed invasions under long-term control, the Department of Agriculture has drawn up new weed legislation expanding to nearly 200 the list of invasive plant species that are prohibited or restricted. Certain plants classified as Category One—chromolaena, lantana, and water hyacinth, for example—will have to be eradicated wherever they occur and can no longer be sold or traded. Category Two plants include invaders that are in commercial use, such as eucalyptus, pines, and wattles. Category Three includes invaders that are used as ornamentals, such as guava, cotoneaster, giant reed, and the widely used and cherished but non-native jacaranda tree. Plants in the latter two categories may be grown only in demarcated areas, and the noxious ornamentals in Category Three are banned from future commerce. More significant still, landowners are now required to obtain weed-free certification from the Department of Agriculture before they can sell or subdivide land, a requirement that should provide plenty of job opportunities for the graduates of Working for Water as well as for biocontrol agents. Next up on the policy agenda: import controls on new plants with invasive potential.

Choosing Our Battles

Just as in South Africa, public and private landowners in the United States and elsewhere are faced with a rising tide of aggressive alien invaders, plants and animals that slowly and progressively degrade vital natural processes and native communities and interfere with productive uses of the land and water or with human well-being. The task of eliminating invaders or reducing their adverse effects is taking an ever-increasing share of the budgets of farmers, foresters, watershed managers, park administrators, public health agencies, and international development organizations. Yet it can often be money well spent: well-thought-out control programs can often limit the spread of an invader or hold its numbers to sufficiently low levels that it ceases to cause significant economic or ecological damage.

The techniques available for control, as we've seen in The Working for Water Programme, fall into several categories: physical or mechanical methods, from use of bulldozers and chain saws to hunting, trapping, or fishing of animal invaders; chemicals such as herbicides, insecticides, or poisons; biocontrols, including strategies such as fertility control in animals as well as the release of exotic natural enemies; and land management techniques such as burning, rotating crops, or changing grazing patterns. Integrated pest management (IPM) makes use of a combination of these techniques to control an invader while minimizing the collateral damage to the environment and to other values.[21]

At times, these techniques can be used to eliminate an established invader completely, although often at great cost and through the use of military-style campaigns involving poisons, pesticides, disease, or shooting.[22] Eradication is much more likely to succeed early in an invasion, before a species is widespread, but catching an invasion at this stage requires that a country put in place early detection and rapid response capabilities. The prospect of eradication has a decisive ring to it, an air of finality that appeals to politicians and the public and, in the case of quarantine pests, comforts trading partners. Where eradication seems feasible, then, it is usually the best course to take.[23]

A wide range of invading animals have been successfully eliminated, both on land and in watery habitats, from Great Britain and Central America to islands the world over. Fruit fly and screwworm fly infestations, for instance, have been eradicated from various parts of the world through the tactic of rearing and releasing sterilized male flies that mate with wild females, a union blessed with no little maggots. Destructive larger animals such as feral pigs, goats, and cats, as well as mice, rats, and rabbits, have been eradicated from a growing number of islands worldwide, from islets around New Zealand to the Galápagos Islands, by hunting, trapping, and poisoning.[24] In the east of England, a fifty-year trapping campaign eventually succeeded in eliminating South American nutrias.[25] Fisheries managers in the United States, faced with the challenge of restoring native fish populations, have poisoned non-native fish such as pike, large-mouth bass, and brown trout out of numerous stream reaches and small lakes—a "scorched earth" process often vehemently opposed by sportsmen and neighboring communities and carried out by the same agencies that once deliberately planted the exotic fish. Australia pioneered the use of a similar tactic in marine waters in 1999, eliminating an infestation of black-striped mussels, marine relatives of zebra mussels, from Darwin Harbor by killing off every living thing with bleach and copper solution. Native marine life quickly recolonized the harbor from nearby waters.[26]

Other types of organisms have defied most eradication efforts. The United States has had little luck, for instance, in eradicating or even slowing the advance of established forest pathogens, from chestnut blight and Dutch elm disease to white pine blister rust. The same can be said of most widely established insect populations, such as the gypsy moth in the eastern forests of the United States. The Asian long-horned beetle, too, is quite likely in the country to stay, although concerted action may limit it to areas around New York and Chicago.

Likewise, it would take a massive commitment of money and years of effort to attempt to eliminate even recently arrived human disease vectors such as the Asian tiger mosquito, which is now established and spreading in the United States, Australia, New Zealand, and parts of Europe, Africa, and South America. Most human pathogens, such as the West Nile encephalitis virus, are also unlikely candidates for eradication, although there are stunning exceptions—namely, the decade-long global vaccination and quarantine campaign that succeeded in the late 1970s in eradicating smallpox from the earth (except for stocks of the virus retained in the laboratories of several countries).

Plants, too, are seldom feasible to eradicate once they have spread beyond a small infestation. To quantify just what that means, ecologists Marcel Rejmánek and Mike Pitcairn scrutinized efforts to eradicate invasive plants by the California Department of Food and Agriculture. They found that professionals almost always succeed at eradicating exotic weed infestations covering less than 1 hectare. Where invaders cover 1 to 100 hectares, the success rate for eradication drops to about one-third of the efforts. Beyond 100 hectares, the scientists found it "very unlikely" that eradication will succeed "with a realistic amount of resources."[27] There are successful exceptions, of course, particularly where resources and political support make extensive and sustained eradication efforts possible. In an eight-year effort, Australians have almost eliminated kochia from more than 3,200 hectares (see chapter 7 for how it got there). Sometimes, if the measures needed to control an invader are the same ones that would be used in an eradication attempt, an all-out effort against a widespread noxious weed can be worth the cost. In Hawaii, crews on Maui and the Big Island are attempting to find and eradicate widely dispersed infestations of miconia to prevent the aggressive trees from spreading seeds into the mountain watersheds and remnant native forests, where most of the islands' endangered birds have taken refuge.

Because eradication has more political appeal than does control—a word that implies a never-ending need for money—some programs that might better be termed area-wide suppression continue for decades under the label of eradication. The eradication campaign against the cotton boll weevil in the southeastern United States, for instance, has gone on for nearly four decades, at a current cost of $100 million per year. Another example is recurring eradication efforts in California to quell outbreaks of the Mediterranean fruit fly (Medfly), which have occurred sporadically since 1975 and annually since 1986, and which many scientists believe represent eruptions of an established population rather than new incursions. Recurring eradication campaigns for species such as Asian gypsy moths promise to become chronic, too, unless strict regulations are enforced on arrivals of timber, wood packaging, and vessels from moth-infested regions.[28]

When experts do determine, or admit, that an invader is too entrenched to eradicate, the question becomes what to do about it. This is not as straightforward as it sounds, and successful control programs today must be applied with social and political as well scientific understanding. Budgets for control of invasive species—with the exception of those targeting Medflies and a few other pests of agriculture—remain puny or nonexistent in most areas. Most methods available for controlling invaders were developed for use against the pests and weeds that plague agriculture and do not have a long history of use in natural ecosystems. Many management techniques, from poisoning animal invaders to spraying herbicides on weeds, are costly, chronic, and unpalatable to various segments of society and may themselves cause economic hardship or ecological harm. Since control methods are not always benign, the damage caused by the invader must be, or must threaten to become, greater than the disturbance, toxicity, or other untoward effects caused by the remedy.

A further complication is that the costs and consequences of control are sometimes borne by people who do not benefit directly. Uproars over the chopping of backyard fruit trees in Florida to halt the advance of citrus canker and the aerial spraying of pesticides over urban Los Angeles to prevent Medflies from threatening commercial fruit production provide striking examples. Any decision to chop, spray, release a biocontrol agent, or otherwise manage an invading plant or animal is a social and political decision based on how we view the effects of an invader. Scientists can help to predict or quantify those effects—water lost, native plants threatened, erosion intensified—but the decision to target a species for control should be made in consultation with all elements of society that have a stake in the outcome. Working for Water has been developing a process to ensure that all interested parties, from agriculturalists, foresters, and watershed managers to farmers and environmentalists, have a voice in the decision making, particularly when the intended targets are commercially valuable trees.

Finally, a proposed control effort should stand a good chance of success if undertaken. A classic cautionary tale is that of the all-out war against imported red fire ants in the southeastern United States, a $200 million fiasco that ecologist Edward O. Wilson has called "the Vietnam [War] of entomology." Despite two decades of heavy insecticide use, the fire ants expanded their territory severalfold, apparently aided by the poisoning of the native ant populations.[29]

Pesticides remain the most commonly used control measures against pests, weeds, and other invaders worldwide, despite growing concerns about their safety. Global pesticide sales, including herbicides and insecticides, have increased by about 250 percent since the 1980s. At the request of the member nations of the Convention on Biological Diversity, experts in the Global Invasive Species Programme (GISP) reviewed the available measures for controlling invaders and concluded that "solutions need to be based on combinations of methods, focusing on the minimum use of pesticides as far as possible." Because immediate action is often required to combat a pest incursion, "pesticide use has a role, but on a rational basis that maximizes impact while minimizing use." Physical control measures can also play an important role in integrated pest management (IPM). However, because biological approaches and land management practices have the potential to be environmentally benign, the GISP review concluded that these "should form the cornerstone of IPM programs in the longer term." Classical biocontrol—the introduction of exotic natural enemies to reduce the abundance of exotic invaders such as the wattles and willows in South Africa—requires much longer start-up times and greater start-up costs than physical and chemical control methods, but it is the only technique with the potential to be self-sustaining and thus cost-free in the long term.[30]

Enemies of Our Enemies

Biocontrol as a scientific discipline is little more than a century old, but its use has burgeoned recently as successes mount and global trade liberalization spawns new waves of invasions. The bulk of the research and introductions is taking place in the United States, Australia, New Zealand, Canada, and South Africa. These countries, along with international agencies such as CABI Bioscience, are also helping to develop and deploy biocontrol solutions in dozens of other countries, from Kenya to India to Papua New Guinea.

Half of the world's biodiversity is composed of plant-feeding insects and the insect predators and parasites that attack them, so it would seem at first glance that the pool of potential biocontrol agents available to target pests and weeds is almost unlimited.[31] However, modern codes of conduct formalized in the 1990s direct researchers to seek out single-minded natural enemies that attack only the intended target. That means scientists must travel to a pest's home range, whether that is South America for miconia, China for kudzu, or Russia for the wheat aphid, to search for highly specialized bugs, blight, or other attackers. When candidate agents are found, they must be collected, reared, and tested in a quarantine facility, often overseas, to see how effectively they attack the target species and also whether they are likely to stray and attack any nontarget crops or native species.

Wherever they are used, biocontrol agents are expected not to eliminate their targets but to suppress them and, it is hoped, eliminate their adverse effects. The strategy does not suit all invaders. Although every creature endures pests and pathogens, these enemies may not actually affect a species' numbers. Instead, its population may be kept in check in its original range by other factors, such as limited nutrients, food resources, or habitat. Practicalities also limit the application of biocontrol: each agent can take several years and millions of dollars to screen and develop. The attraction of biocontrol, as noted earlier, is that when it is successful, it can provide a permanent solution.

The first dramatic demonstration of that came in 1889, when an Australian ladybug called the vedalia beetle was released into the orange groves of southern California, which were being ravaged by insects known as cottony cushion scales. In barely a year, the ladybugs had reduced the scales to insignificant numbers. The same ladybug has since been set loose on cottony cushion scales in more than fifty other countries with the same result. An equally dramatic success story involving an invasive plant began in 1925, when *Cactoblastis,* a moth from Argentina, was set loose in eastern

Australia to help control prickly pear cactus across some 24 million hectares of land in Queensland and New South Wales. Within a decade, the cactus was rare.[32]

Since that time, some 5,000 introductions have been made of more than 1,000 species of biocontrol agents to combat insect pests, with about 10 to 15 percent of those releases considered successful.[33] On the invasive plant side, more than 1,100 releases have been made involving 365 species of insects and fungi targeted at 133 weed species. Perhaps 25 percent of those releases are reported to have contributed at some level to control of the weeds.[34] If those estimates of success sound fuzzy, it reflects the fact that monitoring of releases and standards for judging effectiveness and success historically have been anything but rigorous.

Buried in those numbers, however, are some tremendous triumphs that have occurred since the 1980s. Especially heartening successes have been achieved in the developing world, where countries have few resources to devote to chronic battles against invading weeds and pests, even when they threaten to devastate entire villages or create regional famine. Biocontrol programs targeting three invasive waterweeds from South America—water hyacinth, salvinia fern, and water lettuce—have been particular standouts. Take the case of salvinia fern. A weevil from Brazil has been able to clear choking infestations of this floating waterweed in a matter of months from lakes and rivers in Australia, Papua New Guinea, India, Sri Lanka, Malaysia, Kenya, South Africa, and Zambia.

In Papua New Guinea, salvinia fern literally threatened the future of village life along the vast Sepik River system, a largely roadless region in the northeastern part of the island, where 80,000 people rely on the river for food and travel. Salvinia fern had been introduced into Papua New Guinea in the 1970s, probably as an aquarium plant. Within a decade, it had clogged the Sepik with dense mats as much as a meter thick, causing people to abandon some villages as they became inaccessible by canoe. In the early 1980s, the Australian government and the United Nations Development Programme introduced the biocontrol weevil along the lower Sepik. As the waters there began to clear, villagers upstream, alerted by radio, traveled down by canoe to fetch bags of weevil-infested fern to take back and release in otherwise inaccessible reaches of the river. The weevil did its work rapidly, reopening the waters and literally restoring communities and livelihoods.[35]

An equally spectacular success, this one with biocontrol of an insect pest, came in Africa, where introduction of a wasp from South America headed off almost certain famine. The invader in this case, the cassava mealybug, had been accidentally introduced into West Africa in the early 1970s on illegally imported cassava planting stock. Cassava—also called manioc or tapioca—originated in South America and was brought to Africa some 500 years ago by the Spanish and Portuguese. It has since become a major staple for 200 million people across a swath of sub-Saharan Africa larger by half than the continental United States. Within less than two decades of the mealybug's arrival, the pest had spread through thirty African countries, destroying 80 percent or more of the cassava crop. In 1979, a team led by entomologist Hans R. Herren of the International Institute of Tropical Agriculture (IITA) in Benin began searching the American Tropics for an effective enemy of the mealybug. Luckily, their explorations and testing quickly turned up a wasp in Paraguay that attacks and kills the mealybug. The team bred huge numbers of the wasps in Africa and released them by airplane across the entire cassava-growing belt. Within less than a decade, the wasp had brought the invader under control and ended the famine threat. So immense were the human benefits that Herren was awarded the 1995 World Food Prize.[36]

Ecologist and biocontrol expert Jeff Waage of Imperial College at Wye believes international aid and development agencies ought to step in more often to fund the up-front costs of researching biocontrol solutions for these types of devastating and otherwise intractable weed and pest problems in the developing world. The affected countries cannot hope to pay the start-up costs of biocontrol, and their farmers often cannot afford the ongoing costs of spraying pesticides.

Researchers at IITA believe such biocontrol successes not only provide food security for people but also protect native biodiversity. When crop yields fall to weed or pest invasions, subsistence farmers in Africa often compensate by cultivating more land. Such clearing of natural habitat represents "the overwhelming threat to biodiversity in Africa," the researchers contend.[37]

Most past releases of biocontrol agents in developed and developing countries have been targeted at pests and weeds important to agriculture and forestry. Only recently has biocontrol been used specifically against invaders of conservation concern. One of the first efforts has targeted an alien scale insect from South America that appeared on the tiny South Atlantic island of St. Helena in the late 1980s. The scales, apparently carried in on imported plant material, quickly began to attack the few remnant patches of the island's unique tree daisies known as gumwoods. These gumwoods had survived centuries of land clearing and other human activities because they were isolated in steep terrain. But the invading scales ventured where humans had not. Within a few years, the scales had killed 10 percent of the remaining trees. The introduction of a specialized ladybug predator to eat the scales, however, quickly made the invader rare and has quite likely saved the gumwoods from extinction.[38]

The first weed of exclusively environmental concern targeted for biocontrol has been purple loosestrife, a serious wetland invader across the United States.[39] Other weeds in the crosshairs now include a notorious rogue's gallery of invaders of forests and wildlands as well as croplands. Among them are tamarisk, lantana, chromolaena, blackberry, parthenium, kahili ginger, banana poka vine, Brazilian pepper tree, fire tree, strawberry guava, kudzu, melaleuca, mesquite, miconia, mimosa, gorse, and giant reed.

Biocontrol targets of conservation concern also include rabbits and other pest animals that cause damage to native species as well as agriculture. Researchers are exploring a viral disease of snakes, for example, as a possible weapon against the brown tree snakes of Guam.[40] Other ideas are afloat for uses of biocontrol agents in the seas and coastal waters: parasitic barnacles to halt invading European green crabs along the West Coast of the United States,[41] a Caribbean sea slug to eat the caulerpa seaweed carpeting large expanses of the Mediterranean seabed, and introductions of Atlantic butterfish to gobble up predatory comb jellies in the Black Sea.

Rapid embrace of biocontrol as a conservation tool, however, has been hampered by an image problem. It stems, in part, from the ghosts of biocontrol past, those blunt-force releases of voracious predators with opportunistic diets such as mongooses, stoats, foxes, cane toads, and snails that still wreak havoc on native plants and animals as well as crops and livestock in many parts of the world. (One of these, the cane toad, introduced in the 1930s in a highly ill-advised and unsuccessful effort to control beetles in the sugarcane fields of Queensland, is now itself the target of a biocontrol program. Scientists had tried before but given up on finding a biological agent that would attack the toad without killing native species. Then, in 2001, when the toad hopped into Kakadu National Park, one of Australia's cultural and natural treasures, the government handed scientists at the Commonwealth Scientific and Industrial Research Organisation [CSIRO] money and told them to try again. Their strategy: find a toad gene that can be altered to prevent these heavyset amphibians from reaching maturity, and then load the genes into a virus that can infect toads. Stay tuned.[42])

The days of using vertebrates with broad appetites for biocontrol are not all behind us. The "supremely polyphagous," eat-practically-anything grass carp is still widely promoted and loosed into ponds for control of aquatic plants. Mosquito fish continue to be dumped into waterways worldwide for control of mosquitoes, despite the fact that the fish often inflict more damage on native fish than on mosquitoes. And in Mississippi in 1999, commercial catfish farmers convinced the state to allow them to use nonsterile black carp from China for the first time to control an outbreak of snail-borne fish parasites in their ponds. Fisheries managers and conservation officials in twenty-eight states throughout the Mississippi River basin protested, fearing the black carp would escape into the river and threaten its rich diversity of native mussels, clams, and snails—some of them already being pushed toward the brink by invading zebra mussels.[43]

Vertebrates are not the only biocontrol agents with an image problem, however. Since the 1990s, ecologists have increasingly taken classical biocontrol to task for the real or potential damage that invertebrate agents have inflicted on native species. The debate over collateral damage came to the fore in the United States in the mid-1990s, sparked by the behavior of a Eurasian flowerhead weevil that had been introduced into Canada and the United States three decades earlier to control weedy Eurasian musk thistle on rangelands. Svata Louda of the University of Nebraska, a researcher studying native thistles in the Midwest, noticed this biocontrol weevil feeding on her study plants and began to look further. She discovered that the weevil had begun to attack at least four native thistles, including some narrowly distributed endemics in parks and reserves, and could attack a threatened species called Pitcher's thistle.[44] Whether feeding by this exotic enemy will significantly depress populations of any native thistles remains to be seen.

More recently, biologists in Massachusetts have linked the steep decline of North America's largest native silk moth, the cecropia moth, to a European fly that was repeatedly released into northeastern forests from 1906 to 1986 to try to control invading gypsy moths. When these researchers set cecropia moth caterpillars out in the forests, eight in ten were killed by the flies, which lay their eggs on the caterpillars to provide food for their hatching maggots. The same fly is known to attack 180 other species of insects.[45] And this fly, it turns out, is only one among five dozen parasites and predators introduced in the failed effort to halt the gypsy moth.[46] What the others may be doing now, if they survived, is anyone's guess.

A number of agents released in other historic control efforts are being investigated for suspicious behavior, too. For instance, after the Russian wheat aphid—a sap-sucking insect that can stunt or kill plants and severely reduce grain yields—was discovered in Texas in 1986, the United States released a blitz of twenty-nine exotic insect predators and parasitoids to combat it. One of these, the seven-spot ladybug, is a suspect in the decline of native ladybugs across the Midwest.[47] In Hawaii, researchers reported in 2001 that three species of parasitic wasps released more than fifty years earlier for biocontrol of sugarcane pests have taken up residence in an isolated wilderness preserve on Kauai and become dominant in the invertebrate food web. Because no one documented the original community half a century earlier, it is impossible to tell what effects the newcomers have had.[48]

Despite a flurry of such reports, direct evidence of nontarget damage by biocontrol agents has been hard to come by, especially because there are not many people looking, certainly not the agencies that released the bugs in the first place. Past biocontrol practice has often involved little record keeping and no follow-up.

Today, candidate agents for weed biocontrol are extensively tested for their host specificity, the degree to which they attack only the target weed throughout their life stages. Safety testing has a long tradition in weed biocontrol because of the need to ensure that plant-feeding insects will not attack crops. Extending the testing to native plants has been a relatively straightforward prospect. Traditions are different, however, in insect biocontrol. Because economically valuable insects are few, practitioners traditionally considered little beyond threats to bees and silkworms, if that, when they chose agents to attack insect pests. That mind-set has been slow to change. Host testing remains much less common for biocontrol agents imported to combat insects.[49] The United States, for example, does not require such testing, a gap that reflects the scant protections or appreciation afforded to the country's smallest native fauna, for their roles either as natural enemies of economic pests or as vital players in natural food webs.[50]

As it happens, the case of the flowerhead weevil that is attacking native thistles in the United States is not an example of failed host testing. The weevil's potential affinity for native thistles was apparent when it was evaluated, but this risk did not arouse concern or follow-up from the agricultural bureaucracy that develops, regulates, and applies biocontrol solutions.[51] Values in a society may change over the decades, but creatures once released cannot be recalled. The same factors that can render biocontrol so effective—the ability of introduced agents to establish and persist in a new environment, that is, to invade—also create a risk that they will harm species and communities other than the target.

Now that biocontrol has become potentially valuable—and accountable—to a wider array of interests, including conservation organizations and development agencies, these sectors are raising "greater concerns about environmental safety and non-target effects than traditional stakeholders in the agricultural sector," Waage points out. Biocontrol has always met the safety expectations of agriculture—there are no records of insects or pathogens released after standard testing having become significant crop pests. Waage sees no reason why the technique cannot be made equally safe for the environment. The safety problems of biocontrol have been widely recognized by practitioners, and as a result, new international codes of conduct and best practices were drawn up in the 1990s to minimize the risks.[52]

In a room full of ecologists and biocontrol specialists, especially in the United States, you can still get a heated argument going about whether newer practices have truly reduced the risks to native species and ecosystems. In many ways, however, the debate is more political than scientific because the United States has been slow to reform its regulatory system to encourage wide consultation on release decisions or create a public review process that includes a range of stakeholders beyond agriculture.

Ecologist Don Strong of the University of California, Davis, and biocontrol specialist Robert Pemberton of the U.S. Agricultural Research Service, both critics of the nation's "archaic" regulatory system, commented: "Restraint is key to safe biological control. First must come judicious winnowing of potential targets. Not every alien species is a threat. Biological control is not the appropriate response to every pest, especially to native species perceived as pests. Second, not every available enemy promises relief. Importing multiple agents in a lottery search for one that might do the job increases the probability of non-target attacks upon the native biota. Restraint can come only from open discussion of risks versus benefits of biological control."[53]

Entomologist Peter McEvoy of Oregon State University also has criticized "profligate control programs" such as those that have released more than a dozen agents each for knapweed and leafy spurge in the western United States, with little apparent success. This "lottery model," in which more and more agents are released in hopes that a few will be effective, invites "revenge effects," he points out: "In our rush to solve local and acute pest problems we may be creating diffuse and chronic problems that are even harder to solve."[54]

Not only should fewer agents be released, he argues, but each should also be followed up for an extended period to determine its fate and its effects, both on and off target. Without knowledge of why releases succeed or fail, the selection of biocontrol agents will continue to be largely hit and miss. Ecologist Dan Simberloff and a number of colleagues advocate a perceptual shift to "guilty until proven innocent" for biocontrol agents, as well as for all other deliberate introductions of plants and animals.[55] Others believe, however, that such a shift—at least in the treatment of biocontrol agents—has already taken place outside the United States.

"I would argue that 'guilty until proven innocent' is exactly how screening is applied to most biocontrol agents right now," CSIR ecologist Brian van Wilgen told me. The trouble is, few countries apply the same standard to introductions of new plants and animals, some of which will inevitably invade and become targets for biocontrol. "Most governments are quite happy to introduce new plants without any screening whatsoever. So we keep the gates wide open to let in new problems, but shut them tightly for bringing in possible solutions. These double standards drive me crazy!"

At IITA in West Africa, some biocontrol researchers worry that the push for tougher safety standards in the developed world may work against biodiversity preservation in Africa. So little is known about bug and beetle biodiversity on that continent, for instance, that it would be an immense undertaking to host-test potential agents against native insects that might be at risk. Meanwhile, pest- and weed-plagued farmers push out into the bush, clearing more of the habitat that sustains these native species.[56] "With invasive species just as with many environmental issues, doing nothing is not neutral," Strong and Pemberton note. "Imported natural enemies are the last best hope to parry some of the most damaging exotic pests in natural areas as well as in agriculture."[57]

Focusing on a Goal, Not a Pest

The success of a biocontrol program, like any other effort to control a species invasion, must be evaluated not just in terms of "body counts" of pests killed but also in the context of how the invaded ecosystem as a whole is faring. Since the 1990s, an increasing number of ecologists and land management agencies have begun to criticize the approach of battling one invader at a time without giving enough thought to what is to be accomplished besides its defeat. They call instead for a more comprehensive or holistic approach to managing invaded ecosystems. Thus, neither biocontrol nor any other tactic should be considered in isolation.

"An expensive and successful program against one alien invasive weed may only result in another filling its place," Waage points out. "Or it may prove that a weed cannot be controlled without also controlling the alien vertebrates which are spreading it. Sound ecological research is needed to determine, for any conservation program, which alien species are genuine threats to biodiversity (many are not), how they interact, and how best they should be managed collectively."[58]

John Randall, invasive plant specialist for The Nature Conservancy (TNC), puts it this way: "Too often people get emotional about weeds, especially nasty, prickly ones, and killing them becomes the goal. I have to say, 'No, the goal is preserving native species.' If a weed doesn't interfere with this goal, even if it's nasty, it's not a priority."

In Australia, a biodiversity conservation law passed in 1999 takes a similar goal-oriented approach to the problem of feral animals. "In the past, the approach was to demonize the invasive species and measure success by the pile of bodies," Gerry Maynes, director of the invasive species section of Environment Australia, told me. "Now the measure of success is improving the problem. If you've spent twenty years controlling foxes and you don't see any improvement in threatened species, the funding agencies are going to get skeptical."

Waage believes this emerging approach of focusing on desired outcomes rather than individual pests can be usefully applied to agriculture, too. The tactic of reacting individually to each new pest arose from agriculture, as did most of today's control techniques. As an example of the new approach that is needed, Waage points to the pests and weeds that are being carried around the world in the grain trade. "It's clear that the global cereal agroecosystem is rapidly becoming more uniform in terms of pest species," he says. Western corn rootworms, golden apple snails, itchgrass, and rice water weevils are becoming cosmopolitan invaders. "We should be asking, what are the pests doing to the integrity of those systems? What do we want those agroecosystems to look like in the future? And from that perspective, how do we work control of invasive species into our management strategies rather than taking an ad hoc, fire-brigade approach to the appearance of each new pest?"

Whether the goal for a particular piece of land or aquatic system is preserving native plant and animal communities and ecological processes or ensuring sustained yields of wheat or supplies of water, resources seldom allow us to deal with all invasions with equal vigor. Taking a systems approach to a site thus means looking at trade-offs and setting priorities for action on the basis of what we hope to achieve.

Even after priorities are set, the order in which problems are tackled is critical to the outcome. Sometimes removal or control of a dominant invader is a necessary starting point before control of other pests and restoration of native species and habitats can begin. In the case of Elim, native plants were able to reclaim the land quickly once the pines and hakeas were removed. Fencing off sections of Hawaii Volcanoes National Park and removing pigs has proved to be a necessary first step before weed control and native plant restoration can even begin. In Yellowstone National Park, sustaining healthy populations of native cutthroat trout in Yellowstone Lake now depends on continuing efforts to reduce (by netting and fishing) populations of illegally planted lake trout that prey on young cutthroat.

In other systems, removing or knocking down populations of a long-established, dominant invader may cause hardship for some native species, at least in the short term. In the United States, deployment of a Chinese leaf beetle to attack invasive tamarisk trees was delayed by more than a year by the U.S. Fish and Wildlife Service while it made sure that other agencies were serious about restoring the native willows and cottonwoods the tamarisks had replaced. In the absence of those native stream bank trees, birds such as the endangered southwest willow flycatcher had taken to nesting in tamarisk. Conservation officials wanted assurances that the larger goal of restoring the native riparian community would not be forgotten after the beetles were released.

Another problem with battling the worst invaders one at a time is that the campaign may simply create an opening for another invader to fill.[59] Predicting how exotic sleepers or minor players in the habitat will respond to the removal of a dominant invader is no easy task, however. Control of water hyacinth has sometimes led to the spread of another lake- and river-choking weed called water lettuce—although countries can now release effective biocontrol agents against both to prevent this—and control of purple loosestrife in U.S. wetlands has sometimes been followed by invasion of reed canary grass.[60] On several islands in New Zealand, successful eradication of Norway rats or Pacific rats has been followed by eruptions of mice or exotic snails.[61]

On rugged Santa Cruz Island off the coast of California, TNC's attempts to restore the native plant communities in the lowlands have set off an unexpected cascade of woes. TNC removed cattle from former ranchlands in the island's central valley and, together with the National Park Service, eliminated thousands of feral sheep that had denuded vast tracts of the island. In response, exotic fennel exploded into a near monoculture within a few years. When crews burn, spray herbicide on, or dig up the fennel, noxious yellow starthistle and exotic grasses often move in. Treating the fennel with herbicide seems to improve the percentage of cover of native grasses and forbs, but Randall reports that even in these treated areas, the extent of ground reclaimed by natives has been "disappointingly low" so far. To further complicate restoration efforts, feral pigs—responding in part to the resurgence of greenery following elimination of the sheep—are now burgeoning across the island, helping to foster invading plants just as they do in Hawaii.[62]

Cases such as these need not cause us to throw up our hands in the face of invasions, but they certainly call for a commitment to what is known as adaptive management. In bare-bones form, that means designing control projects as experiments, monitoring the results of our actions, and rethinking our plans if things begin to go sour.

Sweeping changes in the way we manage land and water may be necessary for successful long-term control of invasions as well as restoration of healthy native plant and animal communities. That's because many invaders get their start by taking advantage of human disturbances such as suppression of wildfires, damming or channelizing of rivers, draining of wetlands, and heavy livestock grazing. Sometimes simply reversing the underlying disturbance (although economically and socially, that is seldom simple) by restoring more natural pulsed flows to a dammed river, for example, or by using prescribed burns can eliminate or diminish an invasion. In other cases, the system is not so easily tipped back to its former dynamics. Take cheatgrass in the American West, for instance. Although heavy grazing by cattle helped initiate the cheatgrass takeover, removal of cattle does not weaken that invader's hold on the land.[63]

When rangeland ecologist Roger Sheley showed me the spotted knapweed advancing through bunchgrass habitat in his research plots in Montana (see chapter 3), he described his search for the critical point at which a rancher or land manager could still intervene, knock down the weeds with carefully managed grazing or herbicide use, and let the native grassland reassert itself. There must be a point in any invasion when simply knocking back the invader will restore the native fynbos, as happened in Elim, instead of opening an opportunity for alien fennel, as on Santa Cruz Island—or, in the Montana case, trading alien spotted knapweed for alien leafy spurge or sulfur cinquefoil. Sheley believes that beyond some point, however, when there are too few of the desired species left in the system or too many changes to the land, replanting with a carefully chosen set of species must be part of any long-term solution.

With some entrenched invaders, the goal of managers may be to restrict their range or keep them from reaching sensitive areas rather than to reduce the size of current infestations. One exclusion effort, for example, is the U.S. Army Corps of Engineers' project to build an underwater electric "fence" to prevent invaders in the Great Lakes from transiting through a Chicago-area canal and into the Mississippi River drainage. The targets for exclusion include fishes such as the round goby and the ruffe, neither of which is being controlled in the lakes. (Unfortunately, gobies were detected downstream of the barrier site before the project was in place.) Containment often requires the active cooperation of the public to avert the spreading of an invader, either deliberately or inadvertently. To try to prevent the westward spread of the barnacle-sized zebra mussel, for instance, a number of agencies have launched the 100th Meridian Initiative (the meridian runs from Manitoba, Canada, south through the middle of the Dakotas and on to the border between the Texas panhandle and Oklahoma) to educate fishermen and boaters traveling west from the Great

Lakes about the need to clean their gear and trailered boats of any hitchhiking mussels.

An all-out battle to keep an invader in check on one piece of land is obviously doomed unless neighboring landowners—or even upstream or upslope land managers—cooperate to control sources of seeds or animals that can reinvade. Feral pig populations nurtured on behalf of hunters by state game agencies will always work against efforts to keep unfenced neighboring lands, such as those in Great Smoky Mountains National Park, pig-free.[64] Invasive species, like air or water pollutants, can easily spread among countries separated only by land borders or narrow waterways. Thus the *Cactoblastis* moth, released to clear prickly pear cactus on the Caribbean island of Nevis, made its way to Florida and is now expanding southwest across the United States, creating a potential hazard for dozens of endemic cactus species and for commercial growers in neighboring Mexico.

Finally, in some cases, the only practical response to an invasion may be to learn to live with it or to let it run its course, hoping that over time the native species and community will develop an accommodation with the newcomer that does not involve extinction or degradation of ecological processes or social and economic values. Managers can also attempt to mitigate the adverse effects of the invader on native species or human enterprises, whether control is attempted or abandoned. One example is the predator-proof nest boxes erected to save the chicks of the critically endangered Seychelles black parrot from being devoured by rats after rat control failed. Another is the attempt to fashion snake-proof barriers for utility poles to reduce the number of power outages caused by pole-climbing brown tree snakes on Guam.

It is easy to find people in every society who would opt for this passive response even when active control of an invader is feasible. The conservation movement in Western society has long had a philosophical preference for a hands-off approach that "lets nature take its course." It is grounded in the belief that humans create problems (true enough) and that our attempts to fix things only make them worse. If so, I believe that obliges us to work harder at improving our repair skills. Reliance on nature's healing powers is not likely to restore the communities and processes we value on the land. It is at best a convenient fiction that lets us off the hook for having to do unpalatable things: kill animals, spray herbicides, change our ways. Nature does not optimize; she makes do, and what she cobbles together in the face of the stresses we place on her will often disappoint and impoverish us. Invasive plant specialist John Randall points out that TNC, like many other stewards of conservation lands, believed in the 1950s that it could accomplish its mission by buying land, fencing it, and letting it be. Instead, the organization found over time that with a hands-off approach, "the values they acquired the land for would be lost to invaders." Now Randall and others at TNC consider control of invaders such as those that plague Santa Cruz Island an integral part of their stewardship. In wildlands the world over, we, not nature, will be responsible for the consequences of inaction.

As the pace of invasion increases, there will be few circumstances in which a single control or management technique will be adequate for restoring the economic or ecological conditions we desire in a landscape. There also will be no one-size-fits-all solution to managing invaded lands. But we have a solid and growing tool kit of remedies that, applied intelligently, can do much to preserve or restore the living communities and processes we value in our lands and waters. The goals we set for restoration and conservation, in a park, on an island, or across a region, reflect our values, and as such they are inherently political. Science can provide tools and insights, but only society can define the vision that will guide their application. Bioinvasions are a global problem, but we will need to develop local solutions that reflect local culture and values, just as South Africa has done so creatively through Working for Water.

ELEVEN Islands No Longer

"To read of the old whaling days in this archipelago is to see a vision of shouting men, wild in the freedom of a day ashore after long months of seafaring, striking one of these isolated Edens like a pestilence and rushing aboard ship again with dozens of creatures which, aside from the giant tortoises, served no purpose except the amusement of the moment and which in a few hours were tossed overside, limp bundles of feathers or sprawling scaly limbs."

—William Beebe, *Galápagos, World's End,* 1924

"We may infer from these facts what havoc the introduction of any new beast of prey must cause in a country, before the instincts of the aborigines become adapted to the stranger's craft or power."

—Charles Darwin, *Journal of Researches,* 1839

"We bring strangers together to make strange bedfellows, and we remake the beds they lie in, all at once."

—Jonathan Weiner, *The Beak of the Finch,* 1994

Not long after Charles Darwin visited the Galápagos Islands in 1835, Herman Melville stopped through on a whaling ship. He was unimpressed: "Take five-and-twenty heaps of cinders dumped here and there in an outside city lot; imagine some of them magnified into mountains, and the vacant lot the sea; and you will have a fit idea of the general aspect. . . . Little but reptile life is here found; tortoises, lizards, immense spiders, snakes, and that strangest anomaly of outlandish nature, the aguano [iguana]. No voice, no low, no howl is heard; the chief sound of life here is a hiss."[1]

If Melville could only sail into Academy Bay today and see the bustling town of Puerto Ayora on Santa Cruz Island in the Galápagos, he might be quite pleasantly surprised. The tin-roofed cafes and T-shirt shops of pastel stucco and cinder block run along a waterfront street now lined with flamboyant royal poinciana trees, coconut palms, bougainvilleas, and hibiscus. It is a postcard setting, a largely imported and landscaped cliché that says "tropical paradise" from Miami and Cancún to the beaches of Costa Rica and Hawaii. Of the more than 60,000 visitors who now sail into this bay each year, most on expensive cruises or private yachts, I imagine few are offended by, or even aware of, the town's makeover. Most probably never give it a thought. Certainly most of the 16,000-plus people who have taken up residence on the four inhabited islands in the Galápagos to profit from the booming tourist traffic enjoy the imported greenery, as well as newly paved roads and round-the-clock electricity.

But tourists seldom book cruises to the Galápagos to see palms and hibiscus, and certainly not to hear the low of cattle, the howl of dogs, or the voices of the cats, rats, goats, and pigs that on some islands threaten to overwhelm the legendary hiss. There are far more accessible and less expensive faux tropical paradises for a sun-sand-and-disco vacation. After tourists have spent a day or two here, visiting the Charles Darwin Research Station and the headquarters of Galápagos National Park and perhaps touring the relatively lush volcanic highlands of Santa Cruz Island, the cruise ships take them away to see the 97 percent of the Galápagos that is protected from human settlement and remains largely as Darwin beheld it: an equatorial anomaly, bathed by dueling ocean currents that nourish both penguins and flamingos, fur seals and sea turtles; an archipelago of surreal volcanic landscapes filled with giant tortoises, sunflowers and cacti that grow like trees, iguanas that graze on land and iguanas that graze at sea, and a world of other astonishing creatures molded by long isolation.

Clearly, that isolation is getting harder to maintain, although the Galápagos have held out longer than most of the world's islands. In the past 500 to 1,000 years, as we've seen, the unique living communities on most other oceanic islands have been devastated by the arrival of people and their plant and animal followers. Now Galápagos is facing a similar dilemma—how to preserve its still-intact biodiversity, the very drawing card that has recently rendered this harsh realm so emphatically habitable, in the face of a growing human population. In the three decades since organized tourism arrived in the Galápagos, population in the inhabited 3 percent of the islands has quadrupled. The lure of the Galápagos has been heightened not only by the tourism boom but also by chronic political and financial turmoil in mainland Ecuador, which annexed the islands in 1832.

Along with the growing traffic in people and goods have come hundreds of new plants and pest animals, both as hitchhikers and as deliberate imports. These new invaders have joined a cadre of animal intruders that have harassed the native life of these islands since the days when buccaneers and whalers careened their ships here and filled their bellies with the flesh of giant tortoises and land iguanas. The new invaders—not palms and hibiscus but guava and quinine trees, blackberry and lantana thickets, scale insects and fire ants—are neither as visibly threatening as the larger animals nor as attention-grabbing. They cannot compete for international headlines with oil spills and the bitter clashes over illegal fishing that remain the greatest threats to the rich marine life of the Galápagos. Yet the invading alien plants, pests, and pathogens now spilling out of villages and farms and overwhelming the protected parts of the inhabited islands constitute the greatest threat to the land-based life of this place and the human economy that has sprung up around it.

Darwin Station's director, Rob Bensted-Smith, who came to this Pacific outpost in 1996 after a decade of conservation work in East Africa, says it has taken three or four years "to get everybody focused on introduced species as the problem." Now, he says, "rather than talking about many different things—pollution and resource use and population growth—all as separate problems, they're now seen much more in terms of what they contribute to the introduced species problem. Because that's the one that could actually cause the extinction of the terrestrial biodiversity."

It can be daunting to recognize that most other island groups have failed the test that the Galápagos now face. From Hawaii to Mauritius, remnants of native plant and animal communities thrive only behind well-guarded fences in patches that must be periodically weeded and purged of intruders. (Indeed, many conservation areas on our continents, turned into veritable islands by the roads, plowed fields, cities, and suburbs that encircle them and serve as beachheads for invasive alien plants and animals, now require the same intensive management.)

"The data from across the globe indicate that when you put people on islands, you lose anywhere between 30 and 60 percent of the native biological diversity," says ecologist Howard Snell, program leader of vertebrate ecology and ecological monitoring at the station. "It's almost a law."

During the 1990s, scientists at the station, as well as administrators of Ecuador's Galápagos National Park Service, recognized that the islands were reaching a critical juncture in their history and that continuing to battle invaders with ad hoc and sporadic campaigns would not stop the slide to ecological degradation. They have been struggling ever since to establish the overall vision as well as to put in place the programs, safeguards, and funding the Galápagos will need in order to avoid the fate of other oceanic islands. Starting in 2000 with $4 million in pilot projects funded by Cable News Network founder Ted Turner's United Nations Foundation, the park and station launched an unprecedented effort to battle invasive species on all fronts at once. The program is now being bankrolled by $18 million from the United Nations– and World Bank–run Global Environment Facility (GEF), the funding mechanism for the Convention on Biological Diversity, and other sources. The projects include strategically designed control and eradication programs for the most damaging plant and animal invaders, quarantines on goods from the mainland to reduce future introductions, restrictions on immigration to the islands, public education, and active engagement of everyone from farmers to tour operators in preventing new introductions of unwanted species. In addition, tough enforcement of the designated marine reserve around the islands will be required to help end rampant illegal harvesting of sea urchins, sea cucumbers, and spiny lobsters by local fishermen as well as overexploitation of tuna and sharks by industrial fishing fleets.

It was the ambitious nature of this effort that drew me to the Galápagos, and I soon came to realize that the lessons learned here in pulling together all the pieces needed for an effective protection system may serve as a model for other regions.

One of the most troublesome challenges the Galápagos face in their effort to preserve their unique character and living communities is stemming rampant growth of the human population. Ecuador passed in 1998 and implemented in January 2000 a special law for the Galápagos with a provision that could reduce further in-migration, although it has seldom been invoked.

"Stopping human population growth in the Galápagos is not a very popular position, so there's an honest hope that if you could break the connection between human presence and invasive species, then you could have a sort of Utopia where the human population can coexist with biodiversity," Snell explained one evening as I joined him and his wife, Heidi Snell, at an open-air café in Puerto Ayora. It was late January, and the rain drumming on the tin roof testified to the start of the wet season. Even the arid scrub of the lowlands had taken on a greenness I had not expected as leaves burst from the gaunt, gray-barked Palo Santo trees.

"The idea is that if quarantine works perfectly, and we are able to devise eradication and control programs for species that are already here, we'll be the first archipelago on which people don't spell doom for native biodiversity," continued Snell, who is also a professor of biology at the University of New Mexico. He is not at all convinced the idea will work if the human population in the islands continues to mushroom, but it would be wrong to mistake his skepticism for resignation. Both of the Snells have devoted most of their professional lives to studying and protecting the life of these islands since first coming here as Peace Corps volunteers in 1977.

On that January evening, Heidi Snell, captain of the research sailing vessel *Prima,* had just returned from tiny South Plaza Island off the eastern coast of Santa Cruz Island, where the couple and their colleagues have been following the fate and breeding habits of land iguanas for more than twenty years. Young iguanas on South Plaza take fifteen or sixteen years to mature, so studying multiple generations is a lifetime enterprise. These meter-long jaundice-colored beasts that so repulsed Melville and fascinated Darwin can live for eighty years, at least when there are no dogs, pigs, or hungry humans around.

When the Snells first arrived here, South Plaza was one of the few islands where such a life span for land iguanas was still possible. Iguanas had already vanished from Santiago, an island about twenty-five kilometers northwest of Santa Cruz where Darwin was barely able to pitch a small tent because of the density of iguana burrows. It seemed quite likely by the 1970s that the rest of the land iguanas on the Galápagos would follow those on Santiago to oblivion. The iguanas' peril mirrored that of many other Galápagos natives and reflected the legacy of the earliest human influences on these islands.

Buccaneers and Castaways, Cattle and Rats

Dealing with the devastation wrought by invading mammals and direct human exploitation was the first task that faced conservationists on the Galápagos, and it is far from finished. It is a legacy not of recent human colonization but of centuries of hit-and-run human encounters. Those encounters began in 1535, when a Spanish bishop from Panama drifted off course on his way to Peru and ended up out here, almost 1,000 kilometers off the coast of Ecuador, badly in need of freshwater. "Everyone who ever came to the Galápagos arrived thirsty," a later explorer wrote, and the bishop and most others left as thirsty as they had arrived.[2] There are few permanent sources of water amid the nineteen named islands and forty or so islets of the Galápagos, which are spread over almost 8,000 square kilometers of ocean. Spanish soldiers drifting by in the mists called these islands Las Islas Encantadas, the Enchanted Isles, but the bishop's account of giant tortoises caused a mapmaker in 1574 to christen them the Galápagos, or Islands of the Tortoises.

For more than two centuries after they acquired a name on a map, the

Galápagos served as little more than a haven for British buccaneers who haunted these waters to prey on Spanish treasure ships and periodically sack port cities such as Guayaquil. Among them for a time was Alexander Selkirk, the real-life Robinson Crusoe, after his rescue from Juan Fernández Island, more than 4,000 kilometers to the south. On the beaches of Isabela, Santiago, and a few other islands, the buccaneers careened their wooden ships for repair and feasted on the behemoth tortoises they dubbed "Galápagos mutton." Undoubtedly, black rats jumped ship and colonized a number of islands during these stopovers. The only native land mammals the black rats would have encountered were seven species of mostly vegetarian rice rats (five now vanished) that had managed to raft here from the mainland on drifting mats of vegetation in eons past.

By the opening of the nineteenth century, demand for whale oil to lubricate the machines of the new industrial age brought another fleet to these islands, whose waters at that time served as a feeding ground for sperm whales. British and American whaling ships, including one that carried Herman Melville to the Galápagos, enjoyed a short-lived heyday harpooning whales and filling their holds with casks of oil. They also filled their holds with tortoises, as many as 500 at a time. The beasts served as an ideal fresh meat supply because they could survive for months marooned on their backs without food or water, making no complaint humans could hear. The whalers were soon joined in Galápagos plunder by fur seal hunters. As a result, sperm whales were eliminated from these waters, fur seals were all but driven extinct, some 100,000 to 200,000 tortoises were killed, and races of tortoises on three islands went extinct. In turn, more rats went ashore, along with dogs, cats, goats, pigs, donkeys, horses, and cattle.

It was amid this swashbuckling scene that Darwin arrived, and the four islands he visited during his monthlong stay (San Cristóbal, Floreana, Isabela, and Santiago Islands) were far from pristine by then. In the highlands of Floreana, the southernmost island, Darwin encountered a short-lived penal colony where 200 or 300 convicts raised sugar, sweet potatoes, plantains, and cattle. Darwin watched some of the finches that he would later make famous in *On the Origin of Species* raid seeds from the convicts' fields. He wrote in *The Voyage of the Beagle* that the convicts lived like Robinson Crusoe, hunting wild pigs and goats and eating the unresisting tortoises. Cats and rats, dogs and donkeys also roamed about. The largest of the famous finches Darwin collected on Floreana died out soon afterward, apparently with the help of feral livestock that knocked over and ate the tree cactus on which that finch species depended.

In the century after Darwin came a dozen more scientific expeditions and more short-lived colonies and enterprises on the islands, from harvesting salt, guano, and dyer's moss to mining sulfur from volcanic craters. When an expedition from the California Academy of Sciences arrived on southern Isabela in 1905, leader Rollo Beck reported finding thousands of feral cattle on the volcanic slopes and packs of ferocious wild dogs that ate young tortoises as fast as they hatched. Men from the mainland had set up camp, too, killing hundreds of adult tortoises for their oil and slaughtering feral cattle for their hides. Beck believed that there was little more hope for the survival of the Galápagos' tortoises than for America's bison, which by the end of the nineteenth century had been reduced to near extinction by hide hunters. (His grim prediction did not keep Beck, like many other scientists before and after him, from collecting and dissecting dozens more of the beleaguered tortoises, some more than a meter and a half in length.)[3]

A key event that helped to romanticize the islands and bring them to world attention—and eventual protection—was publication of a book called *Galapagos, World's End,* biologist and explorer William Beebe's bestselling 1924 account of his brief expedition here. Soon after that, a steady stream of luxury yachts began to call in the islands, and a handful of European eccentrics and castaways set up court on Floreana, titillating the international press with petty intrigues, deaths, and disappearances that are still recounted at cocktail hour on the tourist boats. For the first time, settlers moved to Santa Cruz to farm the highlands, where the rains of the wet season and the garua mists of the dry season make farming possible. Soon citrus, coffee, blackberries, guava, red quinine trees, elephant grass, and other species planted by settlers began to invade, adding to the threat posed by feral animals.

In 1935, the centennial of Darwin's visit, Ecuador designated some of the islands sanctuaries, and scientists began to talk of the need for an international effort to preserve the integrity of the unique animal and plant communities of the Galápagos. The coming of World War II delayed their efforts, but talks resumed in the 1950s. It was the centennial of *On the Origin of Species* in 1959 that finally sparked the beginning of significant protection of the islands. That year, Ecuador declared all Galápagos lands not already settled a national park, and the Charles Darwin Foundation was formally established as an independent nongovernmental organization under the auspices of the government of Ecuador, the United Nations Educational, Scientific and Cultural Organization (UNESCO), and the World Conservation Union (IUCN) to found a research station in the islands. There were fewer than 1,500 people on the Galápagos then, scattered among four inhabited islands—Santa Cruz, Floreana, San Cristóbal, and the southern end of Isabela. Paradoxically, it was the drive to protect the Galápagos that helped open the way for the current tourism and population boom that has accelerated the threat of invasions.

I mentioned four occupied islands in the Galápagos, but actually there is a fifth, Baltra Island, smaller than the rest and not part of the park. Baltra lies north across a narrow channel from Santa Cruz and serves as the major airport in the islands. Its airstrip, first developed by the U.S. military during World War II to guard the air approach to the Panama Canal, is now occupied by the Ecuadoran armed forces. Herman Melville's choleric opinion of the Galápagos probably would not have improved had he been able to trade his berth on the whaling ship for a flight to Baltra, a flat expanse of lava desert blotched here and there with remnants of concrete foundations and pilings from its war service. Passenger jets from Quito and Guayaquil land on Baltra (or at a newer, smaller airstrip on San Cristóbal) daily, bringing waves of tourists to meet cruise boats anchored nearby. Other passengers en route to Puerto Ayora once had to travel another four hours by small boat around to Academy Bay, on the southern side of Santa Cruz. Now a short bus and ferry ride takes them across the channel, and buses waiting on the Santa Cruz side make the final forty-kilometer journey across the top of the island to the southern coast. The land route, opened in the early 1970s with the advent of organized tourism, cuts the journey to an hour and a half and gives riders who know what to look for a preview of some of the challenges the Galápagos face. The original cinder road was paved in the mid-1990s, and finches, yellow warblers, and other wildlife, too naïve to be wary of humans or other predators, are easy targets for vehicles that often hurtle at high speed across the island. But the most serious threats posed by the road are far less visible.

"Arguably, one of the worst things anybody ever did in Galápagos was to build that road across the top of Santa Cruz instead of around the edge," Bensted-Smith told me. The coast route would have traversed only the arid zone, the least hospitable environment in the island for arriving plants and animals to colonize. As it is, the road cuts up from the arid scrub and through a wooded transition zone, the humid Scalesia (giant sunflower tree) forest, and the miconia shrub zone to the fern and sedge zone that on Santa Cruz tops out at approximately 850 meters above the sea.

"Whatever insects and seeds you happen to be carrying on your bus or the load of rubbish you're taking to the tip, you're dispersing through all the possible habitats to make absolutely sure it can find a place to survive," Bensted-Smith explained with more than a hint of sarcasm.

In Puerto Ayora itself, the same tourism boom that got the road paved has sent rebar and cinder block sprouting skyward as new homes, shops, and tourism companies proliferate. Residents who enjoyed a slower pace of life were not pleased when electrical service was extended around the clock a few years back, making it possible, among other things, for discos to keep the village rocking into the wee hours. Signs of the universal youth culture are popping up everywhere. In a town with only sporadic sidewalks and cinder-block streets, for example, I spotted two ambitious youngsters with skateboards under their arms.

The other three islands peopled by civilians are experiencing many of the same development pressures, creating scenes you will seldom see featured in television documentaries or coffee-table books on the Galápagos. Isabela now has an airstrip for light planes at Puerto Villamil, and some would like to see it expanded to host direct passenger flights from the mainland. There is talk in Puerto Ayora, too, of developing an airstrip nearby to avoid the ninety-minute drive to Baltra. A more immediate concern for conservation is the pressure to expand the road network on the inhabited islands so that small-scale local tour operators who do not own boats can take tourists to see land iguana and blue-footed booby colonies, sea lion and sea turtle beaches, tortoise watering holes, and volcanic features.

In the late 1960s, the development of an organized tourism industry was considered vital to bringing world attention and funding to the new conservation effort in the Galápagos. Scientists worried even then, however, that the influx of tourists would spread new pests and weeds and exacerbate the historic human-caused destruction in the uninhabited parts of the archipelago. So far, that has not happened. Tour boats take the place of onshore hotels, restaurants, and scuba diving centers here, and visitors go ashore on uninhabited islands only at specified landing beaches and travel on well-marked footpaths shepherded by licensed naturalist guides. Compared with most of the world's islands, the Galápagos have managed to keep the direct consequences of tourism to a minimum. But Galápagos residents are feeling shortchanged.

"Galápagos tourism started as local fishermen taking people out in boats, but now most boat owners are rich people on the continent," Snell told me one morning in his office at the station, a compound of white stucco and lava rock buildings sited in the coastal scrub vegetation east of Puerto Ayora. "Local people receive very little from tourism or from preserving the biodiversity here, so the idea of building more roads is to give them a piece of the pie," he explained. Fair enough, everyone agrees. "But the saving grace of Galápagos tourism has been keeping people on boats and on a few foot trails. More roads mean more pathways for invasion and more opportunities for poaching and other illegal activities."

The special law for the Galápagos implemented in 2000 requires that invasive species concerns be incorporated into policies such as road and airport plans. But the station, the park, and local governments will have to tread new ground to implement that, finding practical ways to incorporate knowledge about invasion pathways and vectors into every aspect of development planning. If there's a head-on collision between conservation and development, Bensted-Smith notes, the "law says take a precautionary approach."

The complex side effects of tourism, especially the population growth it has generated, are clearly some of the most pressing issues facing the Galápagos. But the park and research station must also continue to battle direct threats inherited from the past.

Eliminating the Howl, Reviving the Hiss

One of the most urgent tasks of the station, which opened in 1961, and the Galápagos National Park Service, established in 1968, was to halt the human slaughter of giant tortoises, the most recognizable international symbol of the Galápagos. Tortoise poaching on southern Isabela, for example, has continued ever since the establishment of the park, and in 1994 it took a sharp upturn with the killing of seventy-three tortoises in just an eight-month period.[4] Besides the direct human threat to adult tortoises, invasive feral animals were preventing the survival of eggs and young on many islands when the park and station began operation. Pigs, black rats, dogs, and cats devoured eggs and hatchlings, goats competed for vegetation, and cattle and donkeys trampled nests dug in the soft sand. On islands such as Española, the few remaining adults were so scattered that scientists feared males and females would not find each other to breed.

Starting in 1965, eggs were collected and adult tortoises from Española, Pinzón, Santiago, Santa Cruz, and San Cristóbal Islands were taken to breeding pens at Darwin Station or, beginning in 1994, at Puerto Villamil on Isabela, where their offspring could be raised for three to four years until they reached rat- and pig-proof size. By 2000, some 2,500 young tortoises had been repatriated throughout the islands.[5]

The land iguanas, as noted earlier, were also in need of direct intervention thanks to invading animals and past human hunting. Just before the Snells arrived in 1977, a grisly series of attacks by feral dog packs had exterminated nearly 500 iguanas on the western coast of Santa Cruz and devastated another large iguana colony on Isabela. Station and park service personnel managed to rescue fewer than 100 survivors from these two sites, and the Snells were handed the task of managing a captive breeding program for these iguanas. Few of the nitty-gritty details of reptile husbandry were known then, such as how to get males to breed with females instead of killing them, and how long and at what temperature to incubate iguana eggs. It was studies of the free-living animals on South Plaza Island that helped provide the answers. Later, iguanas from other colonies were brought to rearing pens at the station, too. As the wild dogs have been eliminated from Santa Cruz and Isabela, and goats, pigs, cats, and rats have come under control on other islands since the 1990s, hundreds of young iguanas have been returned to their ancestral islands.

From the beginning, park and station personnel had made uneven efforts, as money allowed, to eliminate animal invaders from various islands. By 1990, for example, intensive hunting by park service teams had eliminated goats from six small islands—Española, Santa Fe, Rabida, Marchena, South Plaza, and Pinta. On a few small islets, black rats were eradicated by poisoning. Many other feral animal problems, however, have remained intractable, especially on the larger islands. Sporadic efforts to eliminate pigs from Santa Cruz, for example, have failed. Now, armed with the GEF funding and the potential for sustained campaigns against the most destructive feral animals, station personnel are attempting to introduce greater scientific rigor into eradication and control efforts.

"We want to be less macho and more oriented to the ecological goal of sustaining native populations," says Howard Snell, who oversees the station's vertebrate animal control projects. That means evaluating whether large-scale eradication projects are actually doable, whether the park and station can commit the resources necessary to see them through, and whether knocking invader populations back periodically could achieve the same benefits for native creatures as a costlier eradication effort. "Still, there are some animals so obviously destructive and difficult to control that we are going for eradication where possible," Snell adds.

One such effort now under way is unprecedented in geographic scope: eliminating goats from the northern end of Isabela. The uninhabited northern arm of the island is isolated from the populated southern part by a tortuous expanse of lava known as the Perry Isthmus, twelve kilometers long and twelve across from coast to coast. Feral goats have long been abundant south of the isthmus, but in 1979 they were spotted in the north, on the slopes of Alcedo Volcano, for the first time. The goat population around Alcedo has since exploded to 100,000 animals, and goats have also turned up on two volcanoes farther north.[6] The consequences for biodiversity could be severe. Isabela accounts for more than half the landmass in the Galápagos and has more endemic species of plants and animals than any other island. Trees high on the volcanic slopes comb moisture from the garua mists in the dry season, creating drip-water pools where half of all surviving tortoises in the Galápagos hang out to drink and stay cool. When goats strip the trees bare, the pools dry up and plant life dies.

The station and park service recognized the threat and launched the Isabela Project in 1995 to remove the goats and restore the island to as close to a pristine condition as possible. Isabela, however, is thirty-five times the size of the largest island from which goats have been eliminated in the past, and the volcanic terrain creates a brutal working environment. Until recently, crews have fought a holding action, concentrating on hunting to control goats in critical tortoise habitat and fencing or wrapping protective chicken wire around drip trees, tree ferns, and other vulnerable plant species.[7] In early 2002, however, with GEF funding, a veritable army of as many as 100 riflemen were scheduled to begin a systematic attack on the goats, employing helicopters, global positioning devices, trained tracking dogs, and radio-collared "Judas" goats whose social instincts will help them—and the hunters—locate any surviving goats.

"It's something that has never been done at this scale," comments Marc Patry, who with Felipe Cruz directs the campaign. "If we succeed, we'll end up with a proven methodology that can be useful everywhere." Conservation efforts in the Galápagos enjoy strong international interest and support, Patry pointed out, and the islands have the advantage of being somewhat isolated from animal rights activists who would object to killing goats or pigs, even to preserve unique plant or animal species. "If you can't turn the tables on introduced species in a situation like [that in the] Galápagos, where can you do it?" he asked.

Besides these grand-scale efforts, Snell's team is planning a series of campaigns against smaller but no less problematic animals. One goal:

develop methods to eliminate feral cats from Baltra, a move that would give repatriated land iguanas a better shot at survival and perhaps provide a strategy for controlling feral cats on larger islands such as Isabela. Another goal, recommended since the 1980s, is to get rid of the city pigeons that have become established in the four populated areas and have strong potential to spread parasites and diseases to native birds.[8]

Most of the invaders that have arrived since the park was established are plants and insects, but a few new vertebrates have slipped into the islands, too. Aggressive Norway or brown rats invaded in the early 1980s and are now spreading across Santa Cruz and San Cristóbal.[9] Two species of frogs—creatures intolerant of salt water and unable to drift to the islands naturally aboard flotsam—arrived on cargo and succeeded in establishing populations in Puerto Ayora and Puerto Villamil during the rainy El Niño season of 1997–1998, becoming the first amphibians in the Galápagos. Also during that El Niño, big black birds known as smooth-billed anis, introduced into Santa Cruz by a farmer, were able to get a toehold on Genovesa and Fernandina Islands, the only two major islands in the archipelago that had until then remained free of alien vertebrates.[10]

Scientists realize that for many widespread and long-established invaders, eradication efforts would be costly and futile. Instead, park personnel spend a significant amount of their time mitigating the adverse effects of these invaders on vulnerable native species, especially the myriad creatures that nest on the ground. Now, Snell's team wants to test whether these control efforts provide the most cost-effective way of improving the lot of native species. A particularly vulnerable species, for example, is the endangered Hawaiian or dark-rumped petrel. In Hawaii, rats, cats, pigs, and mongooses have drastically reduced its numbers by ambushing both adults and hatchlings in their nesting burrows on the volcanic slopes. (The parents, mated for life, must leave their single chick alone and vulnerable for several days at a time while they fish at sea.) This seabird's only other nesting grounds are in the Galápagos, and the same cast of characters—fortunately absent the mongoose—takes a toll on its populations here as well. For almost two decades, park service crews have been poisoning rats around the petrel colonies on Floreana, and more recently in the highlands of Santa Cruz, to knock down predator numbers during the nesting season.

Breeding success has improved markedly since the 1960s, when the petrel looked bound for extinction. But there has been little effort to examine the methods that lead to success. For instance, what proportion of the rat population must you kill to get an acceptable survival of petrel chicks? At what point do you gain little in petrel breeding success by eliminating more rats? What is the most cost-effective way to control rats to the level needed? The same questions can be asked of efforts to protect tortoise and sea turtle eggs on the nesting beaches from rats and pigs. In coming years, control efforts will be designed and monitored to answer just such questions.

"When we control rats around petrel or sea turtle nesting areas, we'll evaluate success not by the number of pigs or rats killed but by increases in the number of young petrels," Snell says. "Even when we achieve positive results, we want to know whether they are cost-effective. Will some other measure produce as many young petrels or tortoises at less cost?"

I heard a similar message from Bensted-Smith: "We need to know, are we going to spend $100,000 a year wasting our time? Because undoubtedly you could do that with a poorly researched campaign, spend a lot of money and have no noticeable long-term effect. As a research station, we have to have a better capability to provide the park with good advice. And when we undertake control or eradication campaigns, we need to do it in a way that we learn from it. If we get rid of cats, do the rats suddenly increase? Is there a better survival of juvenile land iguanas? Would we be able to achieve the same thing on a bigger island?"

Cost is critical because despite the new influx of money, the number of battles that need fighting continues to grow, and the new battles are more complex than ever. It is no longer just feral animals that plague nesting petrels, for instance, and eliminating rats alone will not ensure their survival. Galápagos National Park Service technical officer Carlos Carvajal told me that even though his crew must poison rats around the nesting grounds on the summit of Floreana five times each year, they also spend at least one-fourth of their time each month pulling up, chopping, or carefully daubing herbicides on invasive plants such as the lantana thickets and dense mats of an ornamental known as mother-of-thousands that threaten to overgrow the area and cause petrels to abandon their nests. The same goes for clearing the quinine trees that have invaded the petrel nesting area on Media Luna, a half-moon-shaped volcanic crater in the highlands of Santa Cruz. In most cases, green invaders such as quinine, lantana, and others have proven more intractable than the four-legged ones, but scientists here hope the new GEF-funded projects will give them their first morale-boosting successes.

The Growing Green Threat

Besides the two dozen exotic animals that have invaded the Galápagos, there are at least 300 introduced insects and other invertebrates, an uncounted array of pathogens, and more than 600 plant species. Although the arrival of new animals has been severely curtailed in recent years, new plants, insects, and pathogens have continued to turn up on these islands with disheartening regularity.[11]

The pace of new plant introductions, for instance, has paralleled the rise in the human population. In 1971, there were 77 exotic plants in the Galápagos. By 1999, there were 471 in the station database, and by 2001, the number of alien plants had overtaken the 560 native plants.[12] The nature of plant introductions in the Galápagos fits the pattern I reported earlier for many other parts of the world: a good three-fourths were brought here deliberately. Among the worst invaders, such as blackberry, quinine, lantana, guava, Cuban cedar, citrus, elephant grass, and passion vine, 84 percent were imported deliberately. About half of the new arrivals have escaped cultivation, and some 40 species, including those just listed, have already become aggressive invaders. Another 150 or so of those naturalized plants are suspected to be sleepers or incipient weeds—still uncommon but capable of running amok. Undoubtedly there are many more that will behave badly at some point in the future. Most newly arrived plants take hold in the high humid zones on the inhabited islands, where the conditions are more amenable and where the farms and settlements that supply nearly all the new plants are located. Anywhere from 60 to 95 percent of the natural habitat in these zones (the highest percentage being on San

Cristóbal) has already been lost to agriculture, which makes the concentration of new invaders doubly troubling.[13]

Land managers in the Galápagos, as in most other parts of the world, have tended to regard invasive plant problems as less urgent and serious than the very visible destruction caused by animals. Fewer resources have thus been devoted to the development of control techniques that are appropriate for severe weeds in sensitive island ecosystems, says Alan Tye, head of botany at Darwin Station. Also lacking has been a system for prioritizing weed problems and channeling the limited available funds to the worst problems. Yet Tye believes that over time, "plants are the more insidious, and almost certainly the greater threat, certainly in the Galápagos."[14]

"There have been control trials for invasive plants since the 1980s, but at that time there was little recognition of the scale of the problem and no real effort at coordinated control," Tye says. Like Bensted-Smith, he came to the Galápagos from conservation work in East Africa. "For about ten years, the park tried to keep Media Luna free of quinine. Then those efforts were abandoned as people were pulled off to shoot goats. And control trials were poorly set up and monitored, so it's hard to tell what methods really work." The current park director, however, recognizes the seriousness of plant invasions, Tye says, and "this seriousness is working its way down."

Indeed, the director of Galápagos National Park, Eliecer Cruz, recently created a separate department to handle invasive plants, putting that effort on the same level as feral animal control. "We need a very strong response to control introduced plants, but it's going to be a very difficult problem," he explains—more difficult because "it's not very exciting" and it has been hard until recently to interest donors in paying for it.

Plant control can be even less exciting than usual when the best approach is not quick and dramatic like chopping down trees, Tye says. Often, chopping or pulling creates such a disturbance that new invaders move in quickly. In many cases, a "hack and squirt" approach—a few strategic hacks and twists with a machete and a squirt of herbicide in the cut—works better in restoring the native community because it kills trees such as guava or cedar slowly, giving the native vegetation time to fill in the gaps.

Documenting what works and finding the most cost-effective ways to achieve control are a key part of the GEF-funded effort. But Tye believes that invasive plant management in the Galápagos also needs a botanical equivalent of the ambitious Isabela goat eradication project. "We don't have anything like that strategy or goal for plants or insects," he says. The Galápagos need a morale-boosting success in plant eradication to show people the effort can be worthwhile, he believes. And he has a couple of targets in mind: quinine and a selection of 30 of the 150 suspected sleepers—species known to be serious invaders in other parts of the world but still present as only small infestations in the Galápagos.[15]

Quinine is an invader that both farmers and conservationists despise. "Quinine is present on only one island—Santa Cruz; it is very conspicuous, and no one wants it," Tye says. Further, it is threatening to transform several of the most imperiled native vegetation zones in the Santa Cruz highlands, including the petrel nesting areas. Tye hopes to make quinine a poster child, the first established invasive plant to be targeted for eradication in the Galápagos. Station and park teams are examining the feasibility of eradication by testing the effectiveness of environmentally safe and minimally toxic herbicides in clearing quinine from the petrel nesting areas.

At the other end of the invasion scale, Tye is hoping over the next six years to eradicate thirty sleeper weeds from the archipelago before they awaken and invade on a large scale. If that project succeeds, he says, "it might be possible to report, for the first time since 1535, a decrease in the number of introduced plants in the islands."

Equally novel challenges are facing personnel who must deal with newly arriving insects such as wasps, fire ants, scale insects, and biting black flies that threaten native biodiversity and human health and livelihoods in the islands.

Calling In the Ladybugs

Pulgón is a name Galápagos Islanders affix to all little insects, but when they say it these days they are usually referring to the little white globs, ranging in size from a peppercorn to a pine nut, that I saw as soon as I arrived at my hotel in Puerto Ayora. The white mangroves along the hotel entrance and the walkway below them were coated with sooty black mold, and ranks of puffy white pulgóns were lined up along the branches and under the leaves. I soon learned that these pulgóns are cottony cushion scales, the same sap-sucking Australian exports that were the target of the world's first wildly successful biocontrol effort, in 1889 (described in chapter 10). This scale is known to attack more than 200 plant species and has invaded more than eighty countries. The Galápagos are one of its most recent conquests, and the scale's arrival has prompted the first introduction of an exotic biocontrol agent into these islands.

Of the nearly 300 exotic insects identified in the Galápagos, six are considered "highly aggressive and a threat to the flora and fauna" of the islands. This pulgón is one.[16] First seen in 1982, the scale apparently had arrived on a shipment of imported acacia trees. In 1996 it exploded across the archipelago, probably lofted between islands on warm El Niño air currents. It now infests thirteen islands and attacks at least eighty plant species.[17] Scale insects seldom attract the kind of attention we give to wasps, biting flies, and fire ants, much less marauding goats or rats, but the pulgón quickly became an in-your-face invader by creating unsightly messes in the mangroves and other trees about town. Scale insects suck vital nutrients from trees and exude a honeydew that attracts ants (including fire ants), which tend and protect the scales. The honeydew also nourishes sooty molds that coat leaves and reduce photosynthesis, further weakening the trees.

Biocontrol expert Charlotte Causton arrived in 1997 to head the station's quarantine efforts and invertebrate program, and the pulgón became her concern. As we saw in chapter 10, an Australian ladybug known as the vedalia beetle has been released in more than fifty countries over the past century to combat this scale. But many people, including Howard Snell, were concerned that the cure might be worse than the problem. Sure, the vedalia beetle had been widely used, mostly for protecting citrus and other crops, but virtually no one had ever bothered to learn what else besides cottony cushion scales the vedalia beetle might attack. Decision makers in the Galápagos needed to know how much of a threat the pulgón posed to native plants and whether bringing in the vedalia beetle to stop that threat would, in turn, pose a problem for native insects. So, for the first time in its career, the vedalia beetle was put through the kind of character and behavioral screening ecologists hope to see for all biocontrol agents in the future.

Collecting native scale insects and shipping them to overseas test facilities to confront the vedalia beetle was not feasible, Causton realized. Just finding them was difficult, and when researchers went looking they found new scale species that had never been named. The alternative, bringing the vedalia beetle to the islands for testing, was not a simple task either. It required building a quarantine facility that meets international standards. The result is a little tin-roofed stucco building that sits outside Causton's office window at the station.

"It's pretty basic, but it does the job," she says. "A room within a room, with a refrigerated cool-room door between and a black passageway with an insect-killing light." The vedalia beetle arrived there in March 1999 to meet the island scales, sixteen species from nine families.

Meanwhile, Tom Poulsom, who had worked on biocontrol of prickly pear cactus in South Africa's Kruger National Park, was testing the effect of the pulgón on native plants. By the time he walked me out to some empty rearing pens for land iguanas to show me rows of scale-infested shrubs growing in plastic pots, the count of Galápagos plants that could be debilitated or killed by the pulgón was nearing sixty, including some endangered species. A quick look at the uprooted white mangrove seedlings from a just-completed trial was telling. The ones grown from seed to seven months of age without scales were visibly more vigorous in every way, from their lush tangles of roots to their taller height, than the scale-infested seedlings. Not far away, at the top of an escarpment, Poulsom pointed out native paloverdes and thorn trees marked with white plastic tape. The pulgón had found these trees on its own, attacking at densities as great as 400 scale insects per meter and a half of branch.

By mid-2001, the researchers had determined that although adult vedalia beetles might nibble on native scales occasionally, this biocontrol agent would not survive without the pulgón to attack. And the pulgón itself had been shown to pose a clear threat, not only to many native plants but also to the specialized moths and other invertebrates that depend on them. By the end of the year, the park had approved the release of the vedalia beetle starting in 2002. If the vedalia beetle does its work well, this will be the first, some hope, of a number of biocontrol efforts in the Galápagos.

Control and eradication programs, whether they use herbicides, beetles, or guns, cannot hope to halt ecological degradation in the Galápagos unless something is done to prevent the arrival of new invaders. "I think the first goal is to stop things from getting worse, and then start reversing the process," Bensted-Smith says. "Because right now, it is getting worse." The invaders already on the islands are spreading, and more are arriving.

Keeping things from getting worse will require action on a number of fronts, including quarantine and inspection of goods headed for the Galápagos, a monitoring and rapid response system for new arrivals, environmental education, and perhaps some indirect tactics such as increasing local farm production to reduce the need for importation of potentially pest-ridden produce from the mainland.[18]

In the early 1990s, Ecuador began setting up a legal structure—the System of Inspection and Quarantine of the Galápagos Islands (SICGAL)—to do what few countries have done. That is to set up an internal quarantine structure to protect the islands from pests, pathogens, and weeds carried over from mainland Ecuador. It is the sort of protection from the mainland that many in Hawaii have sought unsuccessfully for years from the U.S. quarantine bureaucracy.

There have been many frustrating fits and starts in trying to get the program up and running, from getting inspectors trained and certified, procedures and policies defined, equipment purchased, and facilities built to setting user fees and penalties. With the support of international donor agencies, a pilot program was launched in May 1999 that spread six inspectors throughout the Galápagos to meet arriving airplanes and cargo boats. It was a token force at best, and they concentrated on searching hand luggage and commercial cargo. (Park inspectors check outgoing cargo to intercept illegally harvested goods such as shark fins and sea cucumbers.)

Still, by the time I visited eight months later, these cursory inspections had intercepted thirty-three species of exotic moths, flies, ants, beetles, and weevils as well as ninety plant and animal products, including a fighting cock, a live white rabbit, and a rooted clump of sedge carried by a Swiss tourist. One especially worrisome insect found on two cargo boats was a geometrid moth that is already defoliating mangroves along the coast of mainland Ecuador.[19] The concern about rabbits escaping on islands will most likely be obvious to anyone who has read this far in the book. The cock raised the specter of another outbreak of a highly contagious avian plague called Marek's disease that hit the islands in 1995 and 1996, threatening not only poultry but also the Galápagos finches. Marek's is one of at least eight diseases of birds detected in the Galápagos so far.[20] (The fighting cock was indeed infected, but with another extremely virulent and deadly bird virus, the agent that causes Newcastle disease.)

This pattern of interceptions in just eight months of token searches provides a sobering indication of what else must have been arriving in all the unmonitored luggage and cargo. By late 2001, SICGAL had thirty-six inspectors and was covering all ports of exit and entry.

Importing new plants and animals into the islands has been prohibited for years, but the law has been neither routinely enforced nor obeyed. As a result, the Galápagos also experienced an isolated outbreak of foot-and-mouth disease and the arrival of canine parvovirus in dogs during the 1990s.[21] In early 2001, canine distemper broke out, threatening not only domestic dogs but also native sea lions.[22] What's more, farmers have felt free to carry in whatever plants or seeds looked useful. In 1996, a botanist from the station, out spraying blackberries in the agricultural zone, got curious about a test plot of a new cattle fodder a farmer had planted. The farmer showed him a two-and-a-half-kilogram bag of kudzu seeds he had bought on the mainland.[23] Luckily, the farmer agreed to allow the plants to be destroyed, and after four years of treatment and monitoring, kudzu was considered eradicated—unless, of course, another farmer still has some, undetected, elsewhere in the Galápagos.[24]

In early 2000, the first detailed list of permitted, prohibited, and restricted products was completed in a collaborative effort by the station, park, key stakeholders, and government groups. I saw the new lists publicized in everything from posters and television spots to interactive games for children. The prohibited goods range from any live animals and genetically modified organisms to fresh milk and cheese, cherries, apricots, fresh flowers, and sugarcane. The restricted items include kiwi fruit, lemons, and mangos, which must be cleared for fruit flies, and bananas, which might carry black Sigatoka and other diseases. During the year 2000, SICGAL inspectors intercepted more than 800 items, most of them prohibited.

It was hardly surprising to learn that the highest number of pest interceptions in the Galápagos have been made on imported fruits and vegetables. The growing urban population and growing wealth in the islands have created a demand for fresh produce that local farmers are not set up to supply. Whether local production can or should be increased is much debated. "The problem with farming here is that it's not most people's primary income," Tye says. Almost anyone can make more money in fishing or tourism. There are no farmworkers to hire, returns are low, little water is available at the surface, landowners do not own rights to subsurface water, and weed invasions increase the labor and expense of farming. The percentage of agricultural land overrun by invasive plants ranges from 12 percent on Santa Cruz to 22 percent on Isabela.[25] Yet farmers have little interest in working hard at weed control. As one resident of Puerto Ayora told me, "Owning land is attractive, but farming isn't." Most farmland sits unused.

On a visit to a 500-hectare farm in the highlands, for instance, I walked through a beautiful vegetable garden full of giant cabbages, watermelons, cantaloupes, broccoli, tomatoes, and corn, all watered by one of the few permanent springs on the island. Most of the seeds and a great deal of technical help had been supplied by the station's agricultural outreach program. Yet the vegetable plot covered less than a hectare while 90 percent of the farm sat unused, rank with ungrazed elephant grass and beset by invading blackberry thickets.

"It is totally unclear what should happen in the agricultural zone, but critically important," Bensted-Smith says. Since there is so little tradition of farming here, the current outreach project "serves to get farmers organized and bring them into the issue. It has a value beyond any specific effects on production right now. It will help feed into the overall strategy for the agricultural zone, whatever that becomes."

The goals for the quarantine system are clearer. More inspectors have already been certified and stationed in the Galápagos. Some inspection of Galápagos-bound passengers and goods is taking place at departure points in Quito and Guayaquil, and more is in the works. Galápagos authorities are pressing for construction of a special dock and treatment facility in Guayaquil to handle produce and other goods bound for the islands. For all the fits and starts, quarantine in the Galápagos is more advanced than in many far wealthier parts of the world. And invasions and biodiversity loss are much further behind. These facts, and the recent infusion of capital and expertise into the battle against invaders, have created a spirit of optimism here.

"The park is very pleased with the results we've had," Park Director Cruz comments, "and we're not going to let things like [the biodiversity loss] in Hawaii happen here. We are optimistic for the future."

The Galápagos Islands face extreme challenges in their struggle to avoid the fate of Hawaii and other oceanic islands, where onslaughts of invasive alien species have devastated much of the unique natural heritage. Yet leaders in this archipelago are facing the threat not with sporadic and reactive efforts but with a strategic vision, seeking both to eliminate or blunt the adverse effects of current invaders and to prevent the arrival of new ones. Whether the task is chopping lantana bushes, hunting goats, checking insect traps, or inspecting baggage, the goal is to ensure that the clamor of new arrivals never drowns out the legendary hiss. If the Galápagos succeed in remaining a fabled land of anomaly and isolation, the strategy pioneered here may truly guide other regions as they struggle to protect their human economies and native biodiversity from the threat of invasive species.

TWELVE

Can We Preserve Integrity of Place?

"Unless one merely thinks man was intended to be an all-conquering and sterilizing power in the world, there must be some general basis for understanding what it is best to do. This means looking for some wise principle of co-existence between man and nature, even if it has to be a modified kind of man and a modified kind of nature."

—Charles Elton, *The Ecology of Invasions by Animals and Plants,* 1958

"For one species to mourn the death of another is a new thing under the sun. . . . To love what was is a new thing under the sun."

—Aldo Leopold, *Sketches Here and There,* 1949

On a small island east of Auckland one gray fall day, Sarah Lowe and I peered under a scrubby thicket, hoping to catch sight of a kokako, New Zealand's largest surviving native songbird. The kokako belongs to an ancient family of wattlebirds that exist only in New Zealand and survive only tenuously there. I had heard the haunting, organ-like call of the kokako once, but not from a living bird. The kokako's song reverberates through the sound track of director Jane Campion's 1994 Oscar-winning movie *The Piano,* which portrays British colonists carving out a settlement in New Zealand's primeval forests in the 1850s. In that era, male and female kokako regularly greeted the dawn with resounding and complex duets. These days, seeing or hearing a kokako in the remnants of those forests is rare. The legendary dawn chorus of native birdsong is largely gone from the country's major landmasses, the North and South Islands, having fallen victim not only to massive forest clearing but also to possums, rats, stoats, ferrets, and dozens of other alien predators and competitors loosed by human colonists. On that fall day, however, Lowe—a member of the Invasive Species Specialist Group of the World Conservation Union (IUCN)—had taken time out to accompany me on a forty-minute ferry ride from Auckland Harbor to an island called Tiritiri Matangi to show me a place where kokako and other beleaguered native birds are making a comeback.

Tiritiri is a 220-hectare island of rolling terrain edged by coastal cliffs. Its native coastal forests were long ago cleared for farming and sheep grazing. Remnants of the forests remained in the gullies, however, along with small populations of bellbirds and other native birds that were growing increasingly rare on the North Island. In 1984, an active restoration effort began on Tiritiri using seedlings grown from these forest remnants and from nearby mainland vegetation. Over the next decade, thousands of volunteers planted more than a quarter of a million seedlings, from native flax to cabbage trees and tree ferns. But restoring habitat was not sufficient. Lowe remembers spending a night on the island in 1992 and finding "the ground just heaving" with the stirring of Pacific rats, a species that arrived with the first Maori 800 years ago. A single island-wide drop of poison pellets in 1993 eliminated the rats. Since then, surviving bird populations have rebounded and a number of endangered birds, including the kokako, have been successfully reintroduced on Tiritiri.[1]

As we walked the trails of the island, I learned that bird-watching here often means looking down instead of up, thanks to the same aversion to flying that has rendered New Zealand's birds literal sitting ducks for introduced predators. Red-crowned parakeets, or kakariki, hopped ahead of us on the trail under a dense canopy of tree ferns and an immense, gnarled New Zealand Christmas tree, a scarlet-flowering relative of Hawaiian ʻohiʻa trees. From the bush came an overwrought commotion of bellbirds and equally noisy eruptions of whiteheads. Above us darted a large tui with a single downy white feather dangling from its dark throat. The special pride of Tiritiri, beloved of volunteers and schoolchildren, are the takahe, rare and endangered flightless swamp hens released to the safety of this island. The turkey-sized, red-faced takahe, cloaked in brilliant blue and green feathers, strutted before us on a lawn beneath the island's historic lighthouse. Along a coastal trail, we saw volunteer-built rock and cement nesting cairns recently vacated by blue penguins. All around us, the shrubs were alive with tiny stitchbirds and robins, fantails and saddlebacks—the latter the only other surviving member of New Zealand's wattlebird family. Then there were the kokako, only eight on the island at the time, the first having been released in 1997.

Kokako of both sexes are gray-blue, their eyes strikingly masked in black and their throats hung with bright blue wattles (a South Island variety, which is feared to be extinct, had orange wattles). They are poor fliers and prefer to walk and hop about the treetops on their powerful legs. That day, however, we heard the sound of a kokako not from the canopy but from beneath a thicket. Lowe heard it first, not a melodious burst of song but a kittenish mewing. Dropping to our hands and knees, we saw a single kokako hopping through the low branches. It was afternoon by then, not the most auspicious time for a kokako to sing. Yet it was thrilling to listen to this bird's busy thrumming and mewing and to realize that in this place it has potential mates to sing with and no alien predators to attack a female or her chicks on the nest. What's more, sanctuaries such as Tiritiri are already supplying birds for reintroduction on other islands and may one day provide birds to mainland areas where ongoing control programs are keeping predator populations low.

What New Zealanders have chosen to re-create on Tiritiri is not an authentic prehuman tableau but a lush, protected setting that supports key elements of the lost native community and allows restoration of processes and possibilities that earlier had been stifled. No attempt has been made to remove all alien species, only the harmful invaders. Indeed, the kokako we saw was making itself at home under a thicket of Australian acacias, coincidentally known as wattles. The mewing and singing of the kokako mingles here with the voices of European skylarks and song thrushes, and the takahe feed amid visiting starlings that fly over from the mainland. You can either lament this human-driven commingling or take pleasure in how much of the unique natural heritage of New Zealand survives, according to your sensibilities and values. Lowe, like many others, is unabashedly thrilled at what has been achieved on Tiritiri. It is, I believe, what Charles Elton envisioned as a "modified kind of nature"—different, undoubtedly diminished in the eyes of many, yet no less a thing of value or source of optimism than the primeval communities of the 1850s.

Unlike their forebears, however, today's New Zealanders recognize that the survival of what remains can never be taken for granted. Along the black cobble beaches of Tiritiri, No Dogs signs warn visiting boaters not to allow their pets onto the island. And as we walked out onto the pier to meet the return ferry, we passed one of numerous yellow rattraps that stand guard against accidental reinvasion around the island. The signs and traps are reminders that only constant vigilance keeps the reconstituted dawn chorus in voice here.

Tiritiri is one of dozens of predator-free sanctuaries that New Zealanders have created amid the more than 700 islands that dot their coastline, and the first to be opened to the perils that accompany unrestricted public access. These island sanctuaries represent more than national treasures. They also symbolize a conservation reality in much of the world, on continents as well as remote islands: unique and valued native plant and animal communities, increasingly beset by habitat loss, surrounded by a sea of human development, and beleaguered from within by invading alien predators and competitors, will persist only in places where they are actively managed and protected. From Mick Clout's patch of primal forest outside Auckland, with which I began this book, to Tiritiri, Hawaii, the Galápagos Islands, and parks and reserves from Kruger National Park in South Africa to Yellowstone National Park in the continental United States, management increasingly involves eliminating or controlling damaging invaders and maintaining constant vigilance against reinfestation. The same is true of working landscapes and waterways everywhere, from African cassava fields to the North American Great Lakes. Hard-won victories on an island or in a continental park, a river, or a cornfield cannot be sustained unless they are accompanied by precautionary actions at local, national, and international levels to prevent the arrival or detect and quickly blunt the adverse effects of potential new invaders that now ride so freely on the conveyor belt of global trade.

Invasions are an international problem, driven by the global movement of people, goods, vessels, and creatures. The nature and severity of the consequences for biodiversity, human health, and economic life, however, fall unevenly across countries and regions. As a consequence, some aspects of the problem call for international cooperation and collaboration while others require solutions tailored to specific local values and needs. Awareness of the accelerating threat of bioinvasions is growing among the world's governments and peoples, but action at all levels lags far behind what is required. To stem the tide of invasions, vigilance and action must take a number of different forms, from global agreements to our personal decisions about everything from pets and garden plants to out-of-season produce.

Acting Globally

Almost every country on the earth, through participation in the 1992 Convention on Biological Diversity (Biodiversity Treaty) or through its own laws protecting native plants and animals, has declared a commitment to safeguard biodiversity and ensure responsible stewardship and sustainable use of its local share of the earth's living riches. Yet until the mid-1990s, only a few countries were keenly aware that invasive species were a major threat to those commitments. Even fewer were taking concerted action. The same was true of major conservation organizations. Since then, participants in the Global Invasive Species Programme (GISP)—scientists, economists, lawyers, natural resource managers, and policy makers—as well as others have been spreading the message that the goals of the Biodiversity Treaty cannot be achieved unless the damaging effects of invasive plants and animals already present are minimized and further invasion threats are thwarted. Through workshops, reports, technical documents, and consultation, these efforts have succeeded in raising the visibility of the issue and pushing it to the top of the action agenda of the Biodiversity Treaty and a growing number of other international treaty organizations.

The activities of GISP and others have also heightened awareness among national governments and international agencies.

Well-funded efforts at the international level are vital because of the fundamental link between species introductions and the growth of global trade. It is critical, for example, to ensure that the cost of establishing and operating needed quarantine and treatment facilities and surveillance and rapid response programs be borne by those who import, export, and transport the world's growing volume of goods and commodities.

In September 2000, after three years of international collaboration, GISP participants met in Cape Town, South Africa, to draw up a global strategy on invasive alien species, identifying priorities for national and international action that offer the greatest practical payoff in halting the escalating problem of invasions.[2] One key priority is to develop economic policies and tools to reduce the risk of harmful introductions, such as insurance and liability requirements for importers and users of potential invasive species, or economic incentives for sectors such as shipping and tourism. Since most invasions are spin-offs of economic activity, market prices must begin to reflect the costs such activities impose on society in the form of damaging invasions.

Some of the other identified priorities include the following:

- Promoting international cooperation and resolving gaps and conflicts among the mandates of various international treaties and conventions, from the Biodiversity Treaty to the World Trade Organization (WTO).
- Building the capacity and will of nations to deal with the problem through border control and quarantine efforts, rapid response programs, and other management initiatives.
- Raising public awareness of and support for policies and activities that reduce the threat of invasions, and engaging key segments of the public in implementing solutions.
- Building research capacity so that nations can make more accurate projections of which species have the potential to invade, which ecosystems are most vulnerable, how best to exclude potential invaders from trade,

how to eradicate and control invasive species, and how to restore sustainable ecosystems.

- Applying risk analysis to the prevention, eradication, and control of invaders as well as to activities such as land-use planning and development. Risk analysis should shift the burden of proof to individuals who propose to move potentially invasive species within or between countries.
- Promoting the sharing of information about invasive species and their management through further development of a global information and early warning system.
- Integrating the issue of invasions into consideration of other human-driven global changes such as climate warming, alterations in the atmosphere, and land-use changes that are expected to exacerbate the adverse effects of invasive species.

Many of these priorities are reflected in a set of fifteen broad guiding principles recommended to the Conference of Parties to the Convention on Biological Diversity in March 2001 by its technical advisory group.[3] One of the most important roles for these nonbinding guiding principles will be to help governments incorporate invasive species concerns into the national biodiversity strategies and action plans that each treaty nation has committed to develop.

The draft guiding principles lead off with a call for nations to take a precautionary approach in efforts to prevent unintentional introductions, in decisions about purposeful imports of new plants and animals, and in eradication or control measures for established invaders. They also recommend that prevention, including border controls and quarantine measures, be given priority because prevention is usually "far more cost effective and environmentally desirable" than actions taken after an invasion.[4] Wherever possible, however, rather than just intercepting species as they arrive, governments should assess and take steps to minimize the risks of unintentional introductions that occur through key commercial pathways such as import and export, shipping, ground and air transport, construction projects, landscaping, tourism, fisheries, agriculture, horticulture, and the pet trade. The guidelines also consider it vital for nations to develop a capacity for early detection and rapid response, since the ability to react quickly and with adequate resources to eradicate a potential invader can save enormously on both environmental damage and long-term control costs.

For purposeful imports of new species, the draft guidelines recommend, "No first-time intentional introduction of an alien species should take place without authorization from a competent authority unless it is known that an alien species poses no threat to biological diversity."[5] Risk analyses and environmental impact assessments are recommended before decisions to authorize an introduction are made.

Such comprehensive prescreening has been attempted so far only by New Zealand and Australia. It remains to be seen whether other nations will follow suit. Most other countries still deal with invasive alien species in piecemeal fashion, if at all, with the most effort directed at excluding or controlling known pests of agriculture such as fruit flies. Only a handful of governments, such as those of the United States,[6] South Africa, and Ecuador's remote Galápagos Islands, have begun to develop comprehensive, well-coordinated policies to protect their biodiversity and economies from the threat of invasive alien species.

As I write, the United States' National Research Council has, ironically, just recommended that work begin soon on designing, building, and testing a quarantine facility of unprecedented complexity—in this case, to receive a cargo of rocks and soil samples from Mars that are expected to arrive aboard an unmanned spacecraft around 2014.[7] "Although the probability is extremely low that these samples will contain hazardous organisms, prudence dictates that all material must be rigorously quarantined at first," the council's press release stated. The goal is to protect both the samples and the earth's environment from cross-contamination. It will be interesting to see whether, by 2014, the United States and its trading partners will be using equal prudence in dealing with the much higher probability that hazardous life-forms are arriving from around planet Earth daily by ship, plane, and vehicle.

Bioinvasions are what social scientists call a weakest-link problem. That means the country with the least capacity or motivation to address the problem determines the threat level faced by nearby countries and even distant trading partners. Water hyacinth will float downriver from the weakest-link country to others, fruit flies will disperse across its borders, pigs across the landscape, nutrias through the wetlands, weed seeds on the wind. What's more, untold numbers of stowaway and hitchhiker species will ride on the vehicles, vessels, crates, and bulk goods the weakest-link country exports around the globe.

The draft guiding principles urge Biodiversity Treaty nations to recognize and "take appropriate individual and cooperative actions to minimize" the risks they pose to other states as sources of potential invaders.[8] Few governments as yet have this capacity, however, and few of the ones that do have enacted stringent procedures to minimize export of potentially invasive species. Nations will not be able to depend on their trading partners to help keep trade pathways clean unless those partners develop the capacity to control invaders that reach their shores, to identify their own native species that could cause harm if exported, and to manage the condition of goods and vessels that leave their territory. At the receiving end, even the most sophisticated border protection systems cannot keep out newly arriving weeds, pests, and pathogens if goods, crates, packing materials, vehicles, and vessels continue to arrive heavily infested. Solving the problem will require increased funding for developing countries, including many of the small island states that face the greatest threats from invasive species. Developed countries also will need to share information, research, and technology with their trading partners to help build capacity that will ultimately protect the environment, human health, and economic well-being along all points on the trade routes.

Biodiversity Treaty nations have shown little interest to this point in drafting a new binding protocol on invasive alien species, one that would parallel the Cartagena Protocol on Biosafety, which treaty members drafted in 2000 to set standards for international trade in genetically modified organisms. That leaves the issue of invasion threats governed by a patchwork of international accords and agreements that have significant gaps and inconsistencies. For example, some agreements are binding on signatory nations and others are not; most are limited in coverage (say, to plant pests or species imported for aquaculture); and most establish their own bureaucracies and procedures and have little tradition of consultation or cooperation with others.[9]

Heightened concern about damaging economic and environmental effects of invasive species, however, has recently led to attempts to close some of the identified gaps in international standards, policies, and procedures. The International Maritime Organization (IMO), for instance, adopted voluntary guidelines in 1997 to minimize the risk of introducing pathogens and other harmful organisms in ships' ballast water. Now the IMO is being urged by some of its member nations to develop a legally binding standard for ballast water management. Meanwhile, the IMO has launched a capacity-building program called GloBallast to help developing countries protect themselves from ballast-borne invasions.

Work on filling gaps has also begun among parties to the International Plant Protection Convention (IPPC), a multilateral treaty that sets standards for regulating the movement in international trade of pests, pathogens, weeds, and animals that could harm plants, either wild or cultivated. The IPPC in 2000 began developing new standards for incorporating environmental effects of invasive species into pest risk assessments. However, the IPPC's traditional species-specific, commodity-specific, and plant-centered approach to risk assessment still falls short of the mark set by GISP, the Biodiversity Treaty, and others for a broad precautionary approach to analyzing risks posed by all types of invaders that might be moving in a trade pathway.

An incident that occurred in Los Angeles in the summer of 2001 illustrates the inadequacies of a narrowly focused approach to risk assessment. When agricultural inspectors at the Port of Los Angeles opened a shipment of a trendy plant from South China known as lucky bamboo, mosquitoes flew out. The U.S. Centers for Disease Control identified them as Asian tiger mosquitoes, those efficient human disease carriers that first entered the southeastern United States in shipments of used tires from Asia in the mid-1980s and have now spread throughout the eastern half of the country. But Asian tiger mosquitoes had not been seen on the West Coast before. The trigger for this incursion, it turns out, was a change in the way lucky bamboo was packed and shipped—a risk not usually subject to scrutiny when import requests are assessed. Lucky bamboo—not really a bamboo but a species of dracaena—was first imported from China in small quantities packed dry, as airfreight. But the popularity of the plant in home decorating has soared along with interest in the Chinese art of feng shui, which involves the auspicious placement of objects. Lucky bamboo is usually sold bound into artful bundles of cut stalks that sprout leaves from the sides when kept in a vase of pebbles and water. To keep up with the burgeoning demand, dealers had begun importing six-meter cargo containers loaded with plants. To keep the plants alive during the two-week sea journey from China, they were packed in shallow trays of water. The result was seagoing mosquito incubators.[10] The mosquitoes hatching in these freight containers provide a dramatic illustration of why we must consider trade as a package.

Elsewhere on the international front, invasive species issues have come to prominence under the 1971 Ramsar Convention on Wetlands, and a working group on invasive species has been formed under the Convention on International Trade in Endangered Species of Wild Fauna and Flora (CITES). From the United Nations Framework Convention on Climate Change to aid and development agencies such as the United Nations Development Programme and the World Bank, organizations are being urged to take into account the risk of alien invasions when considering everything from adaptation to climate change to famine relief or development projects.

Paradoxically, while the physical and economic connectedness of the world is accelerating invasions, our electronic connectedness is enhancing our ability to share information and solutions. Every day on an Internet list server known as Aliens-L set up by IUCN's Invasive Species Specialist Group, the message traffic reflects what is happening on the front lines of the battle against invasions: an official in Portugal needs information about a Peruvian frog and some tropical butterflies that someone wants to import; a British diplomatic mission in Swaziland wants to know how best to use its limited funds to deal with parthenium, chromolaena, and lantana bushes overrunning a nature reserve; a New Zealander alerts colleagues to the discovery of a nest of alien red fire ants in Auckland; scientists in Palau need to know the weed potential of a legume being touted for revegetating road cuts; someone reports that an on-line auction site is refusing to police sales of noxious weeds such as purple loosestrife.

GISP, with new and broadened participation, is continuing the development of global information-sharing and early warning systems for invasive species. A great deal of information already exists about the invasive tendencies of various plants and animals, the rate at which they colonize and spread, the damage they cause, and methods for controlling or eradicating them. But this information is scattered about in scientific journals and specialized computer databases or, too often, sitting hidden in agency file cabinets. GISP is also working to advance other priorities that can minimize invasion threats: furthering the effort to fill gaps and enhance coordination between international treaties and organizations; supporting research and furthering the development and dissemination of tools, techniques, and best practices in the prevention and management of invasions; and facilitating regional cooperation on invasive species issues. Since early 2001, for example, GISP has coordinated a series of regional workshops, funded by the U.S. Department of State and various host country governments, to raise awareness of the problem of invasive species and the need for regional solutions in Europe, Asia, Africa, and tropical America.

Changing Business as Usual

Another critical GISP priority is encouraging and assisting in the development of voluntary codes of conduct that can reduce the likelihood of invasions through specific trade pathways or industries. For ships, airlines, trucking companies, freight importers, crating and container devanning operations, and many other trade and tourism sectors, unwanted species introductions are an inevitable side effect of business as usual. As public and government concern about invasion risk grows, many sectors should recognize the benefit of voluntarily cleaning up their standards and procedures before governments step in to impose regulations and penalties.

The IMO, for example, developed its voluntary guidelines on ballast water management for the maritime shipping industry in the wake of several highly publicized and costly ballast introductions, such as the zebra mussel invasion into the Great Lakes.

Too many invasive species have also been introduced, deliberately or accidentally, during well-meaning but shortsighted development projects and famine relief efforts. Whether agencies are giving away grain and other foodstuffs directly, attempting to create self-sufficiency by supplying crop plants and livestock, or financing construction of roads or water projects, they must begin to consider the danger that if they ignore the threat of damaging plant and animal invasions, they may be creating more problems over the long term than they solve. The introduction of the larger grain borer into Africa in aid grain, mentioned in chapter 3, is a sobering example.

Many industries and economic sectors in various countries still rely heavily on intentional importation of novel species from distant lands and waters: examples are forestry and agricultural research, fish farming, game ranching, fish and game agencies, zoological and botanical collections, horticulture, and aquarium and pet suppliers. Some imported species are intended for planting or for release into fields, forests, and waterways, whereas others inevitably escape or are turned out by their owners. In most countries, import and release decisions continue to be left to the sectors that benefit. Without regulations or economic incentives to do otherwise, these sectors may give little weight to the risks they impose on the public or on publicly owned resources. Even vital institutions such as agricultural research centers, which shelter the world's plant diversity in seed banks and develop new crop cultivars, have been lax in assessing the weediness potential of species they plant in field trials or release for commercialization.

Horticultural introductions, as we've seen, have supplied the majority of plant invaders in many parts of the world. New Zealand and Australia, besides clamping down on new plant imports, have also begun efforts to restrict cultivation and sales of invasive plants already widely sold in nurseries and garden centers. Unfortunately, most other countries not only place few restrictions on new imports but also allow unrestricted sales of plant material already within their borders, even species that have proven themselves invasive in forests, waterways, or wildlands. While governments ponder the need for new regulations, changes in industry attitudes and practices could go a long way toward reducing the problem of plant invasions.

To help remedy the threat from horticulture, GISP has already begun to organize workshops and provide technical support to leaders in the botanical world—including scientists from universities, botanical gardens, and conservation groups as well as representatives of the nursery and seed trades—who have been working since the 1990s to change traditional thinking and standards. Because botanical gardens often serve as mentors or pacesetters for the horticultural world, ethical standards developed there can help to influence the conduct of retail nurseries, landscape architects, and gardeners. In 1999, at a meeting of the American Association of Botanical Gardens and Arboreta held in British Columbia, ecologist Peter White issued a proposed code of ethics called the Chapel Hill Challenge, which is based on policies he instituted as director of the North Carolina Botanical Garden:

Do no harm to plant diversity and natural areas.
Perform risk assessment for introductions.
Remove invasives from plant collections.
Control invasives in natural areas.
Develop noninvasive and native alternative plant material.
Do not distribute plants and seeds that will be invasive elsewhere.
Educate the public.
Become partners with conservation organizations.[11]

Notice that the challenge does not mandate a natives-only policy, although White's own garden does emphasize North Carolina native plants. White is firm that "being anti-invasive is not being anti-exotic" because most non-native plants pose little or no risk to native species or natural areas. Critics who ignore that point have lumped anti-invasive species initiatives together with the back-to-natives gardening movement and targeted both with lurid charges of xenophobia or "ethnic cleansing."[12]

(Although virtually all anti-invasion initiatives target only the small minority of alien species that are invasive, passionate rhetoric combined with sloppy use of terms can confuse the point and provide grist for critics.) Pulling all invasive plants from the seed racks and seedling flats for sale at garden centers would still leave myriad well-behaved exotics available to gardeners everywhere. Further, by encouraging gardeners to root out and replace invasive plants, savvy nurserymen could actually increase sales of both natives and well-behaved exotics.[13]

Although the Chapel Hill Challenge does not mandate natives only, there is much to be said for promoting the cultivation of native plants, not only in local or regional botanical gardens but also on roadsides, in parks, and in public and private gardens. Every meadow plowed, wetland filled, or forest bulldozed for roads, homes, or shopping malls leaves fewer refuges for native plants and animals. It would be a shame to evict them from our gardens and public spaces, too, even for lovely and well-mannered exotics. As we saw earlier, there is also the possibility that some seemingly noninvasive exotics are sleepers that will one day awaken to spread explosively. This is not an argument against the use of all exotics but a further example of the need for vigilance in the form of surveillance and monitoring programs.

The code proposed by White challenges botanical gardens and others to help develop native (as well as noninvasive exotic) alternatives to the use of invasive plants. Plant hunters and horticulturalists have traditionally expended more effort in seeking out and developing novel exotics than in exploring the capabilities of native plants. Agricultural and forestry scientists, for their part, have concentrated their efforts on a small number of crops and timber trees that are now grown the world over, to the neglect of traditional food plants and native forest trees.[14] The Biodiversity Treaty mandates both conservation and enhanced sustainable use of native species, and that in itself should serve as a challenge to nations to spend more effort investing in the horticultural, landscape, and even forestry and cropping potential of native trees, shrubs, and flowers. Already in the western United States, the demand for native seeds and plants for restoration efforts as well as general landscaping has made once-unvalued native plants into cash crops.

Choices, Values, and Payoffs

Changing business as usual will prove a slow job in many arenas. But importers and exporters, conservation workers, land and water managers, ranchers, farmers, public health officials, and many others on the front lines will not be able to halt the rising tide of damage caused by invaders without such an effort. Well-funded initiatives at the global level are vital. So are new initiatives at national government and industry levels. There is also a vital role for individuals to play, however, if we are ever to get control of the invasive alien species problem. Engaging individuals throughout the world in meaningful solutions is perhaps the toughest hurdle we face.

Each new chronic environmental ill that comes to light sparks the message that each of us must change the way we think, act, and live—in this case, how we travel, what we buy, what we eat, the pets and garden plants we keep. The need to change our ways is never a popular message. In this case, it is made more difficult by the fact that the problem remains invisible to so many of us. Those of us who live in cities seldom see the ravages caused by the nematode worms that arrive on the crates that deliver our cell phones and crystal, the fruit flies that hatch from the out-of-season plums and melons we buy, or the smothering vines that spring up in the woods from seeds that blow from our gardens. Those who work the land or depend for their livelihoods directly on natural resources and processes are well aware of the threats posed by new weeds, pests, and plant and animal diseases. Even they, however, usually do not see the larger picture that links the depredations of misplaced knapweed, beetles, and pines with that of misplaced zebra mussels, West Nile virus, Formosan termites, possums, and pigs. We are all in this together. The same global movement of people and goods brings invaders to all countries, to the detriment of all ecosystems and every economic sector.

In many cases, people will need help and guidance to become direct participants in the solution. Even when we want to plant gardens or see that our local parks are landscaped with natives or well-behaved exotics, we will need reliable information about suitable alternatives. Fortunately, such help is increasingly available. In almost every state in the United States, for example, Exotic Plant Pest Councils can provide consumers with lists of locally invasive plants to exclude. In Australia, the Nursery and Garden Industry Australia has teamed up with government agencies to choose fifty-two "weeds of national significance" to withdraw from sale and discourage gardeners from cultivating. In areas where such help is not yet available, consumers or garden clubs could usefully urge government agencies, ecologists, botanists, or even nurserymen to seek out and provide alternatives.

Perhaps the threat of damaging invasions, along with the other stresses that plague our wild places, will also serve as a challenge to more of us to take up the mantle of amateur naturalists and cultivate the pride and enjoyment of place too often sacrificed in a mobile, global society. For example, New Zealanders, while not at all renouncing sheep and English-style cottage gardens, have increasingly come to see themselves as kiwis and Pacific Islanders rather than British expatriates, I'm told. The result has been a growing pride and interest in the survival of native forest birds like the kokako.

Educational and media campaigns by conservation groups, government, and industry can engage individuals in other efforts to prevent invasions. Many people remain unaware of the potentially harmful consequences of such thoughtless or misguided actions as "freeing" unwanted pets—whether dogs, cats, ferrets, or tropical fish—into wildlands and waterways or flaunting airport quarantines. Fishermen and hunters need to understand the consequences for native wildlife of demanding that exotic fish and game be stocked in streams and natural areas. Educated consumers would be more likely to support quarantines and other efforts to sanitize the onrush of free trade, even if it raises the price of imported goods. Greater understanding also might lead more of us to consider reducing the frenzy of consumption that puts so many goods and fossil fuel–driven vehicles in motion around the earth. "Buying local" may seem a quaint notion in this global age, but where feasible it would help reduce many environmental stresses, including the problem of bioinvasions, which is so fundamentally tied to trade.

Despite the international scope of the invasive species problem, most direct action against invaders is taken at the local level, and often it requires the support of citizens and the community. In such cases, land managers and local constituents must agree on the end point to be achieved and have a clear vision of what they value and expect from the land and its living communities. Both managers and citizens must be willing to listen, learn, and help find locally acceptable solutions, even if those involve unpleasant activities such as spraying chemicals, releasing natural enemies, or removing harmful animals. Both action and inaction in the face of invasions have consequences. We cannot preserve the earth's biological wealth if we continue to mix species mindlessly. Tiritiri could not long sustain both rats and kokako.

If we are willing to act, as individuals and as institutions, the payoff can be a reduction in the pace of troublesome new arrivals and in the damaging effects of current invaders. If instead we remain relatively lax or outright sloppy in our approach to both deliberate and incidental movements of species, the cost to our health, our homes, and our industries and enterprises will escalate. So will the toll on native plant and animal communities and the ecological processes so vital to the fertility, productivity, and health of our lands and waters. The human species, ingenious enough to invent airplanes and sentient enough to value fresh flowers jetted daily between hemispheres, surely has the capacity to learn to use the world's biological wealth without diminishing it.

Exotic species remain vital to every modern culture, and their roles may become more important in the future. Outspoken ecologist James Brown believes scientists should approach exotics with balance and realism because "exotic species will sometimes be among the few organisms capable of inhabiting the drastically disturbed landscapes that are increasingly covering the earth's surface."[15] Certainly there are a growing number of sites degraded by human activities, from dammed rivers and polluted coastal waters to heavily logged forests and eroded croplands, where few natives may be able or willing to set up shop. Human-driven changes in the atmosphere and global warming will also make some habitats unsuitable for natives. All these stresses also strain the capacity of our lands and waters to produce food and fiber and deliver the natural services that our economies and growing populations require. Ultimately, it is a sad reflection on our environmental stewardship that we must consider importing creatures and greenery—or genetically engineering pollution- or salt-tolerant plants and other modified species—simply because we have killed off the natives or rendered their habitats inhospitable.

It is also unfortunately true that societies have been too quick to give up on the vigor and staying power of native species. In Hawaii, for example, the periodic natural diebacks of ʻohiʻa trees were interpreted in the early twentieth century as a sign of their "inability to survive in the modern world," and many alien trees were subsequently introduced to replace them.[16] Ecologist Edward O. Wilson can imagine a day when we might have the wisdom and the need to create "synthetic faunas and floras, assemblages of species carefully selected from different parts of the world and introduced into impoverished habitats." Yet he cautions: "Ecology is still too primitive a science to predict the outcome of the synthesis of predesigned biotas. No responsible person will risk dumping destroyers into the midst of already diminished communities. Nor should we delude ourselves into thinking that synthetic biotas increase global diversity."[17]

A few ecologists and evolutionary biologists find comfort in the exceedingly long view, pointing out that the number of species has not dropped in most invaded habitats and that, in eons hence, the newcomers will evolve into a diverse array of new species.[18] It is quite possible that if New Zealand loses the kokako and the takahe to invading rats, the rats and starlings will keep the local species count as high as ever. Globally, however, Wilson is correct. The world's tally of creatures will drop if kokako slip away, and the presence of rats and starlings in New Zealand adds nothing to global biodiversity because those species exist in many places. This homogenization of the life of the earth leaves us poorer, whatever the head count. Besides, biodiversity is not just a numbers game. Our key concern is for relationships among species and the roles plant and animal species play in sustaining the structure and functioning of ecosystems. It would be hard to claim on purely philosophical grounds that a kokako is intrinsically better, more natural, or more valuable in New Zealand than a starling or a rat. Yet we humans have chosen for sound reasons to place a higher value on kokako than on rats and starlings in New Zealand, a choice reflected in our laws, customs, and agreements among nations. Theoretical detachment and a long-term view are even harder to maintain when the focus is not kokako but weeds, pests, predators, and diseases that destroy human lives and livelihoods.

As for life's ability to generate new diversity over the ages, it is true that species will continue to evolve and new communities will come together from whatever plants and creatures survive on this increasingly human-dominated planet. But that renewal will take place on a timescale irrelevant to human hopes and history. We and our descendants in the foreseeable future will have to leave it to some post–*Homo sapiens* Darwin stopping through on an intergalactic *Beagle* to discover what sundry new forms and associations the starlings of New Zealand, South Africa, or North America have adopted. In the span of time meaningful to us as a species, starlings will be starlings wherever we have moved them, and kokako, once lost, will not be replaced.

We will always have to walk a tightrope between the benefits that exotics provide and the dangers posed by the few invaders among them. We have learned hard lessons since yaks were first delivered to the Jardin des Plantes in Paris, but we have not yet put that knowledge to effective use. We cannot undo most of the mistakes of the past, but we also cannot afford to keep repeating them. It is time to replace our ad hoc approaches with coherent policies that recognize both the fact of our scrambled biota and the perils of continuing to allow the haphazard movement of species. We should aim not just to preserve or restore what we value but also to ensure that we do not foreclose the options of future generations. It would be sad to imagine our descendants making do amid only the plants and animals that seemed pleasing or useful to us at the opening of the third millennium—or the weeds, wanderers, and vermin too wily for us to eliminate.

APPENDIX A 100 of the World's Worst Invasive Alien Species

The "100 of the World's Worst Invasive Alien Species" list presented here is designed to enhance awareness of the invasive alien species issue. Species were selected for the list according to two criteria: their serious adverse effects on biodiversity and/or human activities and their illustration of important issues surrounding biological invasion. To ensure the inclusion of a wide variety of examples, only one species from each genus was selected. There are many other invasive alien species in addition to those in this list of examples. *Absence from the list does not imply that a species poses a lesser threat.*

Aquatic Plants

caulerpa seaweed	(*Caulerpa taxifolia*)
common cordgrass	(*Spartina anglica*)
wakame seaweed	(*Undaria pinnatifida*)
water hyacinth	(*Eichhornia crassipes*)

Land Plants

African tulip tree	(*Spathodea campanulata*)
black wattle	(*Acacia mearnsii*)
Brazilian pepper tree	(*Schinus terebinthifolius*)
chromolaena (Siam weed, triffid weed)	(*Chromolaena odorata*)

Land Plants *(cont.)*

cluster pine	(*Pinus pinaster*)
Cogon grass	(*Imperata cylindrica*)
fire tree	(*Myrica faya*)
giant reed	(*Arundo donax*)
gorse	(*Ulex europaeus*)
hiptage	(*Hiptage benghalensis*)
Japanese knotweed	(*Polygonum cuspidatum*)
kahili ginger	(*Hedychium gardnerianum*)
Koster's curse	(*Clidemia hirta*)
kudzu	(*Pueraria lobata*)
lantana	(*Lantana camara*)
leafy spurge	(*Euphorbia esula*)
leucaena	(*Leucaena leucocephala*)
melaleuca	(*Melaleuca quinquenervia*)
mesquite	(*Prosopis glandulosa*)
miconia	(*Miconia calvescens*)
mile-a-minute weed	(*Mikania micrantha*)
mimosa (giant sensitive plant)	(*Mimosa pigra*)
prickly pear cactus	(*Opuntia stricta*)
privet	(*Ligustrum robustum*)
pumpwood	(*Cecropia peltata*)
purple loosestrife	(*Lythrum salicaria*)
quinine	(*Cinchona pubescens*)
shoebutton ardisia	(*Ardisia elliptica*)
strawberry guava	(*Psidium cattleianum*)
tamarisk (saltcedar, Athel pine)	(*Tamarix ramosissima*)
wedelia (Singapore daisy)	(*Wedelia trilobata*)
yellow Himalayan raspberry	(*Rubus ellipticus*)

Amphibians

bullfrog	(*Rana catesbeiana*)
cane toad	(*Bufo marinus*)
Caribbean tree frog	(*Eleutherodactylus coqui*)

Fishes

brown trout	(*Salmo trutta*)
common carp	(*Cyprinus carpio*)
large-mouth bass	(*Micropterus salmoides*)
mosquito fish	(*Gambusia affinis*)
Mozambique tilapia	(*Oreochromis mossambicus*)
Nile perch	(*Lates niloticus*)
rainbow trout	(*Oncorhynchus mykiss*)
walking catfish	(*Clarias batrachus*)

Birds

Indian myna	(*Acridotheres tristis*)
red-whiskered bulbul	(*Pycnonotus cafer*)
starling	(*Sturnus vulgaris*)

Reptiles

brown tree snake	(*Boiga irregularis*)
red-eared slider turtle	(*Trachemys scripta*)

Mammals

black or ship rat	(*Rattus rattus*)
brushtail possum	(*Trichosurus vulpecula*)
cat	(*Felis catus*)
crab-eating macaque monkey	(*Macaca fascicularis*)
goat	(*Capra hircus*)
gray squirrel	(*Sciurus carolinensis*)
mouse	(*Mus musculus*)
nutria (coypu)	(*Myocastor coypus*)
pig	(*Sus scrofa*)
rabbit, European	(*Oryctolagus cuniculus*)
red deer	(*Cervus elaphus*)
fox	(*Vulpes vulpes*)
small Indian mongoose	(*Herpestes auropunctatus*)
stoat	(*Mustela erminea*)

Aquatic Invertebrates

Chinese mitten crab	(*Eriocheir sinensis*)
comb jelly	(*Mnemiopsis leidyi*)
green crab	(*Carcinus maenas*)
marine clam	(*Potamocorbula amurensis*)
Mediterranean mussel	(*Mytilus galloprovincialis*)
northern Pacific seastar	(*Asterias amurensis*)
spiny water flea	(*Cercopagis pengoi*)
zebra mussel	(*Dreissena polymorpha*)

Land Invertebrates

Argentine ant	(*Linepithema humile*)
Asian long-horned beetle	(*Anoplophora glabripennis*)
Asian tiger mosquito	(*Aedes albopictus*)
big-headed ant	(*Pheidole megacephala*)
common wasp	(*Vespula vulgaris*)
crazy ant	(*Anoplolepis gracilipes*)
cypress aphid	(*Cinara cupressi*)
flatworm	(*Platydemus manokwari*)
Formosan subterranean termite	(*Coptotermes formosanus shiraki*)
giant African snail	(*Achatina fulica*)
golden apple snail	(*Pomacea canaliculata*)
gypsy moth (Asian and European)	(*Lymantria dispar*)
khapra beetle	(*Trogoderma granarium*)
little fire ant	(*Wasmannia auropunctata*)
malaria mosquito	(*Anopheles quadrimaculatus*)
red imported (tropical) fire ant	(*Solenopsis invicta*)
rosy wolf snail	(*Euglandina rosea*)
sweet potato whitefly	(*Bemisia tabaci*)

Disease Agents

avian malaria	(*Plasmodium relictum*)
banana bunchy top	(banana bunchy top virus)
chestnut blight	(*Cryphonectria parasitica*)
crayfish plague	(*Aphanomyces astaci*)

Dutch elm disease	(*Ophiostoma ulmi*)
frog chytrid fungus	(*Batrachochytrium dendrobatidis*)
phytophthora root rot	(*Phytophthora cinnamomi*)
rinderpest	(paramyxovirus)

The World Conservation Union's (IUCN's) Invasive Species Specialist Group (ISSG) is developing a Global Invasive Species Database as part of the Global Invasive Species Programme (GISP), coordinated by the Scientific Committee on Problems of the Environment (SCOPE). IUCN, CAB International, and the United Nations Environment Programme (UNEP) are partners in GISP. The "100 of the World's Worst Invasive Alien Species" list and database constitute an integrated subset of the Global Invasive Species Database. Development of the list and database was made possible by the generous contribution of the Fondation d'Entreprise TOTAL.

APPENDIX B Index to Scientific Names of Cited Invasive Species

Aquatic Plants

Brazilian elodea	(*Elodea densa*)
caulerpa seaweed	(*Caulerpa taxifolia*)
cordgrass (spartina grass)	(*Spartina* spp.)
Eurasian watermilfoil	(*Myriophyllum spicatum*)
giant reed	(*Arundo donax*)
hydrilla	(*Hydrilla verticillata*)
reed canary grass	(*Phalaris arundinacea*)
salvinia fern (giant salvinia, kariba weed)	(*Salvinia molesta*)
water hyacinth	(*Eichhornia crassipes*)
water lettuce	(*Pistia stratiotes*)

Land Plants

Aleppo pine	(*Pinus halepensis*)
asparagus fern	(*Asparagus scandens*)
banana poka vine	(*Passiflora mollissima*)
Bathurst burr	(*Xanthium spinosum*)
beardgrass	(*Andropogon glomeratus*)
bitou bush (boneseed)	(*Chrysanthemoides monilifera*)
blackberry	(*Rubus* spp.)
black wattle	(*Acacia mearnsii*)
blue thunbergia	(*Thunbergia grandiflora*)

boxthorn	(*Lycium ferocissimum*)
Brazilian pepper tree	(*Schinus terebinthifolius*)
bridal creeper	(*Asparagus asparagoides*)
broom	(*Cytisus* spp.)
broomsedge	(*Andropogon virginicus*)
buckthorn	(*Rhamnus cathartica*)
buffelgrass	(*Cenchrus ciliaris*)
bugweed	(*Solanum mauritianum*)
Cape ivy	(*Delairea odorata*)
capeweed	(*Arctotheca calendula*)
cardoon (wild artichoke)	(*Cynara cardunculus*)
casuarina (Australian pine, ironwood, she-oak)	(*Casuarina equisetifolia*)
catchweed bedstraw	(*Galium aparine*)
cheatgrass	(*Bromus tectorum*)
Chinese tallow tree	(*Triadica sebifera*)
chromolaena (Siam weed, triffid weed)	(*Chromolaena odorata*)
clidemia (Koster's curse)	(*Clidemia hirta*)
climbing glory lily	(*Gloriosa superba*)
cluster pine	(*Pinus pinaster*)
cocklebur	(*Xanthium strumarium*)
Cogon grass (alang-alang)	(*Imperata cylindrica*)
cotoneaster	(*Cotoneaster* spp.)
crabgrass	(*Digitaria* spp.)
Cuban cedar	(*Cedrela odorata*)
cut-leaved teasel	(*Dipsacus laciniatus*)
dalmation toadflax	(*Linaria genistifolia*)
dandelion	(*Taraxacum officinale*)
diffuse knapweed	(*Centaurea diffusa*)
dodder	(*Cuscuta* spp.)
drumstick tree (moringa)	(*Moringa oleifera*)
dyer's woad	(*Isatis tinctoria*)
elephant grass	(*Pennisetum purpureum*)
English ivy	(*Hedera helix*)

Land Plants *(cont.)*

eucalyptus	(*Eucalyptus* spp.)
European stone pine	(*Pinus pinea*)
fennel	(*Foeniculum vulgare*)
filaree	(*Erodium cicutarium*)
fire tree	(*Myrica faya*)
fireweed	(*Senecio madagascariensis*)
fountain grass	(*Pennisetum setaceum*)
garlic mustard	(*Alliaria petiolata*)
giant hogweed	(*Heracleum mantegazzianum*)
giant thistle	(*Silybum marianum*)
gorse	(*Ulex europaeus*)
groundsel	(*Senecio vulgaris*)
guava, common	(*Psidium guajava*)
guava, strawberry	(*Psidium cattleianum*)
hakea shrub	(*Hakea* spp.)
Indian laurel fig	(*Ficus microcarpa*)
itchgrass	(*Rottboellia cochinchinensis*)
ivy gourd	(*Coccinia grandis*)
jacaranda	(*Jacaranda mimosifolia*)
Japanese barberry	(*Berberis thunbergii*)
Japanese honeysuckle	(*Lonicera japonica*)
Jerusalem thorn tree	(*Parkinsonia aculeata*)
Johnson grass	(*Sorghum halepense*)
jointed goatgrass	(*Aegilops cylindrica*)
kahili ginger	(*Hedychium gardnerianum*)
Kikuyu grass	(*Pennisetum clandestinum*)
knotweed	(*Polygonum* spp.)
kochia	(*Kochia scoparia*)
kudzu	(*Pueraria lobata*)
lantana	(*Lantana camara*)
laurel tree	(*Cordia alliodora*)
leafy spurge	(*Euphorbia esula*)
leucaena	(*Leucaena leucocephala*)

lodgepole pine	(*Pinus contorta*)
Lolium ryegrass	(*Lolium rigidum*)
longleafed wattle	(*Acacia longifolia*)
medusahead	(*Taeniatherum caput-medusae*)
melaleuca	(*Melaleuca quinquenervia*)
mesquite	(*Prosopis* spp.)
Mexican poppies	(*Argemone mexicana*)
Mexican weeping pine	(*Pinus patula*)
miconia	(*Miconia calvescens*)
mile-a-minute weed	(*Mikania micrantha*)
mimosa (giant sensitive plant)	(*Mimosa pigra*)
molasses grass	(*Melinis minutiflora*)
Monterey pine	(*Pinus radiata*)
morning glory	(*Ipomoea indica*)
mother-of-thousands	(*Kalanchoe pinnata*)
mullein	(*Verbascum thapsus*)
multiflora rose	(*Rosa multiflora*)
musk thistle	(*Carduus nutans*)
nutsedge	(*Cyperus* spp.)
oleander	(*Nerium oleander*)
oriental bittersweet	(*Celastrus orbiculatus*)
pampas grass	(*Cortaderia* spp.)
parthenium (wild carrot)	(*Parthenium hysterophorus*)
passion fruit vine	(*Passiflora ligularis*)
passion vine	(*Passiflora edulis*)
periwinkle	(*Vinca minor*)
pigweed	(*Amaranthus retroflexus* and spp.)
pond apple tree	(*Annona glabra*)
Port Jackson willow	(*Acacia saligna*)
potatovine	(*Anredera cordifolia*)
prickly pear cactus	(*Opuntia* spp.)
privet	(*Ligustrum* spp.)
purple loosestrife	(*Lythrum salicaria*)
quackgrass	(*Elytrigia repens*)

Land Plants *(cont.)*

quinine	(*Cinchona pubescens*)
ragwort (tansy ragwort)	(*Senecio jacobaea*)
raspberry (blackberry)	(*Rubus* spp.)
red brome grass	(*Bromus rubens*)
rhododendron	(*Rhododendron ponticum*)
rubbervine	(*Cryptostegia grandiflora*)
rush skeletonweed	(*Chondrilla juncea*)
Russian olive tree	(*Elaeagnus angustifolia*)
Russian thistle	(*Salsola kali* or *S. iberica*)
St. John's wort (Klamath weed)	(*Hypericum perforatum*)
spotted knapweed	(*Centaurea maculosa*)
sulfur cinquefoil	(*Potentilla recta*)
tamarisk (saltcedar, Athel pine)	(*Tamarix* spp.)
tree of heaven	(*Ailanthus altissima*)
tropical soda apple	(*Solanum viarum*)
wandering jew	(*Tradescantia albiflora*)
wattles	(*Acacia* spp.)
wild oat	(*Avena fatua*)
willows	(*Salix* spp.)
witchweed	(*Striga asiatica*)
yellow starthistle	(*Centaurea solstitialis*)

Amphibians

bullfrog	(*Rana catesbeiana*)
cane toad	(*Bufo marinus*)

Fishes

alewife	(*Alosa pseudoharengus*)
Asian swamp eel	(*Monopterus albus*)
Atlantic salmon	(*Salmo salar*)
black carp	(*Mylopharyngodon piceus*)
brook trout	(*Salvelinus fontinalis*)
brown trout	(*Salmo trutta*)

common carp (*Cyprinus carpio*)
grass carp (*Ctenopharyngodon idella*)
lake trout (*Salvelinus namaycush*)
large-mouth bass (*Micropterus salmoides*)
mosquito fish (*Gambusia affinis*)
Nile perch (*Lates niloticus*)
Nile tilapia (*Oreochromis niloticus*)
pike (*Esox lucius*)
rainbow trout (*Oncorhynchus mykiss*)
round goby (*Neogobius melanostomus*)
ruffe (*Gymnocephalus cernuus*)
sea lamprey (*Petromyzon marinus*)

Birds

house finch (*Carpodacus mexicanus*)
house sparrow (English sparrow) (*Passer domesticus*)
Indian myna (*Acridotheres tristis*)
Japanese white-eye (*Zosterops japonica*)
mallard duck (*Anas platyrynchos*)
monk parakeet (*Myiopsitta monachus*)
pigeon (rock dove) (*Columba livia*)
red-whiskered bulbul (*Pycnonotus cafer*)
ruddy duck (*Oxyura jamaicensis*)
smooth-billed anis (*Crotophaga ani*)
starling (*Sturnus vulgaris*)

Reptiles

brown tree snake (*Boiga irregularis*)
red-eared slider turtle (*Trachemys scripta*)

Mammals

beaver (*Castor canadensis*)
brown hare (European hare) (*Lepus europaeus*)
brushtail possum (*Trichosurus vulpecula*)

Mammals *(cont.)*

camel (dromedary)	(*Camelus dromedarius*)
cat	(*Felis catus*)
cattle	(*Bos taurus*)
crab-eating macaque monkey	(*Macaca fascicularis*)
dog	(*Canis familiaris*)
donkey	(*Equus asinus*)
European rabbit	(*Oryctolagus cuniculus*)
ferret	(*Mustela furo*)
fox	(*Vulpes vulpes*)
goat	(*Capra hircus*)
gray squirrel	(*Sciurus carolinensis*)
Himalayan tahr	(*Hemitragus jemlahicus*)
horse	(*Equus caballus*)
Indian mongoose	(*Herpestes auropunctatus*)
mouse	(*Mus musculus*)
muskrat	(*Ondatra zibethicus*)
North American elk (wapiti)	(*Cervus elaphus*)
nutria (coypu)	(*Myocastor coypus*)
pig	(*Sus scrofa*)
rat, black or ship	(*Rattus rattus*)
rat, Norway or brown	(*Rattus norvegicus*)
rat, Pacific	(*Rattus exulans*)
red deer	(*Cervus elaphus*)
sheep	(*Ovis aries*)
sika deer	(*Cervus nippon*)
stoat	(*Mustela erminea*)
water buffalo	(*Bubalus bubalis*)
weasel	(*Mustela nivalis*)

Aquatic Invertebrates

Asian clam	(*Corbicula fluminea*)
black-striped mussel	(*Mytilopsis sallei*)
Chinese mitten crab	(*Eriocheir sinensis*)
comb jelly	(*Mnemiopsis leidyi*)

green crab	(*Carcinus maenas*)
gribble	(*Limnoria* spp.)
Japanese crab	(*Hemigrapsus sanguineus*)
northern Pacific seastar	(*Asterias amurensis*)
rusty crayfish	(*Orconectes rusticus*)
shipworm	(*Teredo navalis*)
zebra mussel	(*Dreissena polymorpha*)

Land Invertebrates

Africanized honeybee	(*Apis meillifera scutellata*)
Argentine ant	(*Linepithema humile*)
Asian honeybee	(*Apis cerana*)
Asian long-horned beetle	(*Anoplophora glabripennis*)
Asian tiger mosquito	(*Aedes albopictus*)
avocado thrips	(*Scirtothrips perseae*)
balsam wooly adelgid	(*Adelges piceae*)
banana skipper moth	(*Erionota thrax*)
bee tracheal mite	(*Acarapis woodi*)
big-headed ant	(*Pheidole megacephala*)
biting black fly	(*Simulium bipunctatum*)
Bostrychid beetle	(*Bostrychid* spp.)
cassava mealybug	(*Phenacoccus manihoti*)
common wasp	(*Vespula vulgaris*)
corn rootworm, western	(*Diabrotica virgifera*)
cotton boll weevil	(*Anthonomus grandis*)
cottony cushion scale	(*Icerya purchasii*)
crazy ant	(*Anoplolepis gracilipes*)
fire ant, little	(*Wasmannia auropunctata*)
fire ant, red imported (tropical)	(*Solenopsis invicta*)
Formosan termite	(*Coptotermes formosanus*)
geometrid moth	(*Thyrinteina arnobia*)
German wasp	(*Vespula germanica*)
giant African snail	(*Achatina fulica*)
glassy-winged sharpshooter	(*Homalodisca coagulata*)
golden apple snail	(*Pomacea canaliculata*)

Land Invertebrates *(cont.)*

gypsy moth (Asian and European)	(*Lymantria dispar*)
hibiscus mealybug	(*Maconellicoccus hirsutus*)
khapra beetle	(*Trogoderma granarium*)
larger grain borer	(*Prostephanus truncatus*)
malaria mosquito	(*Anopheles quadrimaculatus*)
mango leafhopper	(*Idioscopus nitidulus*)
Mediterranean fruit fly (Medfly)	(*Ceratitis capitata*)
nun moth	(*Lymantria monacha*)
painted apple moth	(*Teia anartoides*)
papaya fruit fly	(*Bactrocera papayae*)
phylloxera aphid	(*Daktulosphaira vitifoliae*)
pine wood nematodes	(*Bursaphelenchus* spp.)
red-banded mango caterpillar	(*Nooda albizonalis*)
rice water weevil	(*Lissorhoptrus oryzophilus*)
rosy wolf snail	(*Euglandina rosea*)
Russian wheat aphid	(*Diuraphis noxia*)
screwworm fly	(*Cochliomyia hominivorax*)
silver-leaf whitefly	(*Bemisia argentifolii*)
southern house mosquito	(*Culex quinquefasciatus*)
spiraling whitefly	(*Aleurodicus dispersus*)
spruce bark beetle	(*Ips typographus*)
sugarcane stem borer	(*Sesamia grisescens*)
two-spotted leafhopper	(*Sophonia rufofascia*)
Varroa mite (bee mite)	(*Varroa jacobsoni*)
white-spotted tussock moth	(*Orgyia thyellina*)
yellow fever mosquito	(*Aedes aegypti*)
yellow jacket wasp	(*Vespula squamosa*)

Disease Agents

acquired immune deficiency syndrome (AIDS)	(human immunodeficiency virus)
annosus root disease	(*Heterobasidion annosum*)
avian malaria	(*Plasmodium relictum*)

banana bunchy top	(banana bunchy top virus)
black Sigatoka of banana	(*Mycosphaerella fijiensis*)
blood disease of banana	(*Pseudomonas celebensis*)
bluetongue	(orbivirus)
brucellosis	(*Brucella abortus*)
bubonic plague	(*Yersinia pestis*)
canine distemper	(paramyxovirus)
canine parvovirus	(parvovirus)
chestnut blight	(*Cryphonectria parasitica*)
cholera	(*Vibrio cholerae*)
chytrid fungus	(*Batrachochytrium dendrobatidis*)
citrus canker	(*Xanthomonas campestris*)
Cyclospora intestinal parasite	(*Cyclospora cayetanensis*)
dengue fever	(flavivirus)
dogwood anthracnose	(*Discula destructiva*)
Dutch elm disease	(*Ophiostoma ulmi*)
foot-and-mouth disease	(aphthovirus)
geminiviruses	(begomoviruses)
grapevine leaf rust	(*Phakopsora euvitis*)
infectious salmon anemia	(orthomyxovirus)
influenza	(influenza virus)
Japanese encephalitis	(flavivirus)
Karnal bunt fungus	(*Tilletia indica*)
larch canker	(*Lachnellula* spp.)
malaria	(*Plasmodia* spp.)
Marek's disease	(chicken herpes virus)
measles	(paramyxovirus)
Newcastle disease	(paramyxovirus)
Nipah encephalitis	(paramyxovirus)
Panama disease of banana	(*Fusarium oxysporum* f. sp. *cubense*)
papaya ringspot	(apotyvirus)
phytophthora root rot	(*Phytophthora cinnamomi*)
pine pitch canker	(*Fusarium subglutanins*)

Disease Agents *(cont.)*

potato blight	(*Phytophthora infestans*)
red tide organisms	(various dinoflagellates)
Rift Valley fever	(phlebovirus)
rinderpest	(paramyxovirus)
Ross River virus (epidemic polyarthritis)	(togavirus)
St. Louis encephalitis	(flavivirus)
shigellosis	(*Shigella* spp.)
smallpox	(variola virus)
sudden oak death	(*Phytophthora* unnamed sp.)
sweet potato little leaf disease	(mycoplasma-like organism)
swine fever	(pestivirus)
syphilis	(*Treponema pallidum*)
tomato bushy stunt virus	(tombusvirus)
West Nile encephalitis	(flavivirus)
whirling disease	(*Myxobolus cerebralis*)
white pine blister rust	(*Cronartium ribicola*)
yellow fever	(flavivirus)

NOTES

Chapter 2. Reuniting Pangaea

1. Michael A. Osborne, *Nature, the Exotic, and the Science of French Colonialism* (Bloomington: Indiana University Press, 1994), 11.
2. Ibid.
3. Richard N. Mack, "Motivations and Consequences of the Human Dispersal of Plants," in *The Great Reshuffling: Human Dimensions of Invasive Alien Species,* ed. Jeffrey A. McNeely (Gland, Switzerland, and Cambridge, England: IUCN, 2001).
4. Alfred W. Crosby, *Ecological Imperialism: The Biological Expansion of Europe, 900–1900* (Cambridge, England: Cambridge University Press, 1986).
5. Tyler Whittle, *The Plant Hunters* (New York: PAJ Publications, 1970).
6. David J. Mabberley, "Where Are the Wild Things?" in *Paradisus: Hawaiian Plant Watercolors by Geraldine King Tam* (Honolulu: Honolulu Academy of Arts, 1999).
7. Sidney W. Mintz, "Pleasure, Profit, and Satiation," in *Seeds of Change: A Quincentennial Commemoration,* ed. Herman J. Viola and Carolyn Margolis (Washington, D.C.: Smithsonian Institution Press, 1991), 112–129.
8. Katharina Dehnen-Schmutz, "Medieval Castles as Centers of Spread of Non-native Plant Species," in *Plant Invasions: Ecological Mechanisms and Human Responses,* ed. U. Starfinger et al. (Leiden, Netherlands: Backhuys Publishers, 1998), 307–312.
9. Christopher Lever, *The Naturalized Animals of the British Isles* (London: Hutchinson and Company, 1977).
10. James T. Carlton, "Marine Bioinvasions: The Alteration of Marine Ecosystems by Nonindigenous Species," *Oceanography* 9, no. 1 (1996): 36–43.
11. William H. McNeill, *Plagues and Peoples* (Garden City, N.Y.: Anchor Books, 1976).

12. Crosby, *Ecological Imperialism.*
13. Mintz, "Pleasure, Profit, and Satiation."
14. P. S. Martin and R. G. Klein, *Quaternary Extinctions* (Tucson: University of Arizona Press, 1984).
15. Richard N. Mack et al., "Biotic Invasions: Causes, Epidemiology, Global Consequences, and Control," *Ecological Applications* 10, no. 3 (2000): 689–710.
16. Raymond Sokolov, *Why We Eat What We Eat: How the Encounter between the New World and the Old Changed the Way Everyone on the Planet Eats* (New York: Summit Books, 1991).
17. Francis L. Black, "Why Did They Die?" *Science* 258 (1992): 1739–1740; Alfred W. Crosby, "Metamorphosis of the Americas," in Viola and Margolis, *Seeds of Change,* 70–89.
18. Ralph T. Bryan, "Alien Species and Emerging Infectious Diseases: Past Lessons and Future Implications," in *Norway/UN Conference on Alien Species,* ed. Odd Terje Sandlund, Peter Johan Schei, and Aslaug Viken (Trondheim, Norway: Directorate for Nature Management and Norwegian Institute for Nature Research, 1996), 74–80.
19. Daniel Defoe, *The Life and Strange Surprising Adventures of Robinson Crusoe of York, Mariner,* (1719; reprint, New York: Heritage Press, 1930), 103.
20. Richard N. Mack, "The Motivation for Importing Potentially Invasive Plant Species: A Primal Urge?" in *People and Rangelands, Building the Future: Proceedings of the Sixth International Rangeland Congress,* ed. David Eldridge and David Freudenberger (Townsville, Queensland, Australia, 1999), 557–562.
21. Tim Low, *Feral Future: The Untold Story of Australia's Exotic Invaders* (Victoria, Australia: Viking, 1999).
22. Brian J. Huntley, "South Africa's Experience regarding Alien Species: Impacts and Controls," in Sandlund, Schei, and Viken, *Norway/UN Conference on Alien Species,* 182–188; quote, 182.
23. Richard N. Mack, "Temperate Grasslands Vulnerable to Plant Invasions: Characteristics and Consequences," in *Biological Invasions: A Global Perspective,* ed. J. A. Drake et al., SCOPE 37 (Chichester, West Sussex, England: John Wiley and Sons, 1989), 155–179; quotes, 160–161.
24. Christopher Lever, *They Dined on Eland: The Story of the Acclimatisation Societies* (London: Quiller Press, 1992), 3.
25. Osborne, *Nature, the Exotic, and the Science of French Colonialism.*
26. Lever, *They Dined on Eland,* 26–27.
27. R. H. Groves, "Plant Invasions of Australia: An Overview," in *Ecology of Biological Invasions: An Australian Perspective,* ed. R. H. Groves and J. J. Burdon (Canberra: Australian Academy of Science, 1986), 137–148.
28. Whittle, *Plant Hunters;* Toby Musgrave, Chris Gardner, and Will Musgrave, *The*

Plant Hunters (London: Ward Lock, 1998); Vernon H. Heywood, "Patterns, Extents, and Modes of Invasions by Terrestrial Plants," in Drake et al., *Biological Invasions,* 31–55.

29. Lever, *They Dined on Eland,* 97.
30. Ibid.
31. Crosby, *Ecological Imperialism.*
32. Lever, *They Dined on Eland,* 77–78.
33. Ibid.
34. Jared Diamond, "Blitzkrieg against the Moas," *Science* 287 (2000): 2170–2171; R. N. Holdaway and C. Jacomb, "Rapid Extinction of the Moas (Aves: Dinornithiformes): Model, Test, and Implications," *Science* 287 (2000): 2250–2254; Michael N. Clout, "Biological Conservation and Invasive Species: The New Zealand Experience," in Sandlund, Schei, and Viken, *Norway/UN Conference on Alien Species,* 161–166.
35. Clout, "Biological Conservation and Invasive Species."
36. Lever, *They Dined on Eland.*
37. George Laycock, *The Alien Animals: The Story of Imported Wildlife* (Garden City, N.Y.: Natural History Press, 1966).
38. Huntley, "South Africa's Experience."
39. Lloyd L. Loope, "Hawaii and the Pacific Islands," in *Status and Trends of the Nation's Biological Resources,* ed. M. J. Mac et al. (Reston, Va.: U.S. Department of the Interior, U.S. Geological Survey, 1998), 747–774.
40. Richard N. Mack, "The Commercial Seed Trade: An Early Disperser of Weeds in the United States," *Economic Botany* 45, no. 2 (1991): 257–273; Richard N. Mack, "Catalog of Woes: Some of Our Most Troublesome Weeds Were Dispersed through the Mail," *Natural History* (March 1990): 45–52.
41. Musgrave, Gardner, and Musgrave, *Plant Hunters.*
42. Mack, "Commercial Seed Trade."
43. Mack, "Motivation for Importing," 559.
44. W. Mark Lonsdale, "Inviting Trouble: Introduced Pasture Species in Northern Australia," *Australian Journal of Ecology* 19 (1994): 345–354.
45. Cal Kaya, "Arctic Grayling in Yellowstone: Status, Management, and Recent Restoration Efforts," *Yellowstone Science* 8, no. 3 (2000): 12–17.
46. Leo G. Nico and Pam L. Fuller, "Spatial and Temporal Patterns of Nonindigenous Fish Introductions in the United States," *Fisheries* 24, no. 1 (1999): 16–27; quote, 19.
47. Heywood, "Patterns, Extents, and Modes of Invasions," 42.
48. Ibid.
49. David Pimentel et al., "Environmental and Economic Costs of Nonindigenous Species in the United States," *BioScience* 50, no. 1 (2000): 53–65.

50. David Pimentel et al., "Economic and Environmental Threats of Alien Plant, Animal, and Microbe Invasions," *Agriculture, Ecosystems, and Environment* 84 (2001): 1–20.
51. Randy G. Westbrooks, *Invasive Plants: Changing the Landscape of America: Fact Book* (Washington, D.C.: Federal Interagency Committee for the Management of Noxious and Exotic Weeds, 1998).
52. Clout, "Biological Conservation and Invasive Species."
53. Wendy Strahm, "Invasive Species in Mauritius: Examining the Past and Charting the Future," in Sandlund, Schei, and Viken, *Norway/UN Conference on Alien Species,* 167–181.
54. Richard N. Mack, "Alien Plant Invasion into the Intermountain West: A Case History," in *Ecology of Biological Invasions of North America and Hawaii,* ed. Harold A. Mooney and James A. Drake (New York: Springer-Verlag, 1986), 191–213.
55. Mack, "Temperate Grasslands Vulnerable to Plant Invasions."

Chapter 3. Wheat and Trout, Weeds and Pestilence

1. Richard N. Mack, "Alien Plant Invasion into the Intermountain West: A Case History," in *Ecology of Biological Invasions of North America and Hawaii,* ed. Harold A. Mooney and James A. Drake (New York: Springer-Verlag, 1986), 191–213; Jeffrey C. Mosley, Stephen C. Bunting, and Mark E. Manoukian, "Cheatgrass," in *Biology and Management of Noxious Rangeland Weeds,* ed. Roger L. Sheley and Janet K. Petroff (Corvallis: Oregon State University Press, 1999), 175–188.
2. Mack, "Alien Plant Invasion."
3. Randy G. Westbrooks, *Invasive Plants: Changing the Landscape of America: Fact Book* (Washington, D.C.: Federal Interagency Committee for the Management of Noxious and Exotic Weeds, 1998).
4. Roger L. Sheley, James S. Jacobs, and Michael L. Carpinelli, "Spotted Knapweed," in Sheley and Petroff, *Biology and Management of Noxious Rangeland Weeds,* 350–361.
5. Frank Forcella and Stephen J. Harvey, "Relative Abundance in an Alien Weed Flora," *Oecologia* 59 (1983): 292–295.
6. R. F. Bucher, *Potential Spread and Cost of Spotted Knapweed on Range Uses* (Bozeman: Montana State University Extension Service, 1984).
7. David Pimentel, "Pest Management in Agriculture," in *Techniques for Reducing Pesticide Use: Economic and Environmental Benefits,* ed. David Pimentel (Chichester, West Sussex, England: John Wiley and Sons, 1997), 1–11; David Pimentel et al., "Economic and Environmental Benefits of Biodiversity," *BioScience* 47, no. 11 (1997): 747–757.
8. H. A. Mooney and J. A. Drake, "The Ecology of Biological Invasions," *Environment* 29, no. 5 (1987): 12.

9. Agriculture and Resource Management Council of Australia and New Zealand, Australian and New Zealand Environment and Conservation Council, and Forestry Ministers, *The National Weeds Strategy: A Strategic Approach to Weed Problems of National Significance* (Canberra: Commonwealth of Australia, 1999).
10. Pimentel, "Pest Management in Agriculture."
11. David Pimentel et al., "Environmental and Economic Costs of Nonindigenous Species in the United States," *BioScience* 50, no. 1 (2000): 53–65.
12. David Pimentel et al., "Environmental and Economic Costs of Pesticide Use," *BioScience* 42, no. 10 (1992): 750–760.
13. U.S. Department of Agriculture, Animal and Plant Health Inspection Service (USDA APHIS), *APHIS FY2001 Reference Book: Agricultural Quarantine Inspection* (Washington, D.C.: USDA APHIS, 2001), on-line at http://www.aphis.usda.gov/.
14. Kevin Hackwell and Geoff Bertram, *Pests and Weeds: The Cost of Restoring an Indigenous Dawn Chorus* (Wellington: New Zealand Conservation Authority, 1999), 45.
15. David Pimentel et al., "Economic and Environmental Threats of Alien Plant, Animal, and Microbe Invasions," *Agriculture, Ecosystems, and Environment* 84 (2001): 1–20.
16. L. G. Holm et al., *The World's Worst Weeds* (Honolulu: University Press of Hawaii, 1977).
17. T. P. Tomich et al., "*Imperata* Economics and Policy," *Agroforestry Systems* 36 (1997): 233–261.
18. Vandana Shiva, "Species Invasions and the Displacement of Biological and Cultural Diversity," in *Norway/UN Conference on Alien Species*, ed. Odd Terje Sandlund, Peter Johan Schei, and Aslaug Viken (Trondheim, Norway: Directorate for Nature Management and Norwegian Institute for Nature Research, 1996), 47–52.
19. Vernon H. Heywood, "Patterns, Extents, and Modes of Invasions by Terrestrial Plants," in *Biological Invasions: A Global Perspective*, ed. J. A. Drake et al., SCOPE 37 (Chichester, West Sussex, England: John Wiley and Sons, 1989), 31–55.
20. Rudiger Wittenberg and Matthew J. W. Cock, eds., *Invasive Alien Species: A Toolkit of Best Prevention and Management Practices* (Wallingford, Oxon, England: CAB International, 2001).
21. Shiva, "Species Invasions and the Displacement of Biological and Cultural Diversity"; O. P. Sharma, H. P. S. Makkar, and R. K. Dawra, "A Review of the Noxious Plant *Lantana camara*," *Toxicon* 26 (1988): 975–987.
22. C. Ronald Carroll, "The Interface between Natural Areas and Agroecosystems," in *Agroecology*, ed. C. Ronald Carroll, John H. Vandermeer, and Peter Rosset (New York: McGraw-Hill, 1990), 365–383.

23. Adama Daou, "Grappling with Water Weeds in Mali," *World Conservation* (IUCN journal) 28, no. 4, and 29, no. 1 (April 1997–January 1998): 29–30; Brian Ligomeka, "World's Most Abundant Lake Threatened by Weeds," Environment News Service, 8 November 2000.
24. Victor Kasulo, "The Impact of Invasive Species in African Lakes," in *The Economics of Biological Invasions,* ed. Charles Perrings, Mark Williamson, and Silvana Dalmazzone (Cheltenham, England: Edward Elgar Publishing, 2000).
25. U.S. Congress, Office of Technology Assessment, *Harmful Non-indigenous Species in the United States,* OTA-F-565 (Washington, D.C.: U.S. Government Printing Office, 1993).
26. Erika Zavaleta, "Valuing Ecosystem Services Lost to *Tamarix* Invasion in the United States," in *Invasive Species in a Changing World,* ed. Harold A. Mooney and Richard J. Hobbs (Washington, D.C.: Island Press, 2000), 261–300.
27. M. Lynne Corn et al., *Harmful Non-native Species: Issues for Congress* (Washington, D.C.: Congressional Research Service, 1999), on-line at http://cnie.org/NLE/CRSreports/Biodiversity/biodv-26.cfm.
28. Jean-Yves Meyer, "French Polynesia: A Natural Paradise . . . for Invasive Species," *Aliens* (IUCN newsletter) 6 (1997): 5–6.
29. Jean-Yves Meyer and Jacques Florence, "Tahiti's Native Flora Endangered by the Invasion of *Miconia calvescens* DC. (Melastomataceae)," *Journal of Biogeography* 23 (1996): 775–781; Jean-Yves Meyer, "Polynesia's Green Cancer," *World Conservation* (IUCN journal) 28, no. 4, and 29, no. 1 (April 1997–January 1998): 7.
30. John J. Ewel et al., "Deliberate Introductions of Species: Research Needs," *BioScience* 49, no. 8 (1999): 619–630.
31. Carroll, "Interface between Natural Areas and Agroecosystems."
32. Alan Holt, "An Alliance of Biodiversity, Agriculture, Health, and Business Interests for Improved Alien Species Management in Hawaii," in Sandlund, Schei, and Viken, *Norway/UN Conference on Alien Species,* 155–160.
33. Frank G. Zalom and Joseph G. Morse, "Expanded Efforts Needed to Limit Exotic Pests," *California Agriculture* (March–April 1999): 2.
34. Kathryn Brown, "Florida Fights to Stop Citrus Canker," *Science* 292 (2001): 2275–2276.
35. Environment News Service, "Invasive Leafhopper Creates Agricultural Emergency in California," 26 June 2000; Tara Weingarten, "Attack of the Wine Bugs," *Newsweek,* 2 July 2001, 52.
36. Robin Meadows, "Medfly: Going but Not Gone," *California Agriculture* (March–April 1999): 6.
37. Holt, "Alliance of Biodiversity, Agriculture, Health, and Business Interests."
38. Jane E. Polston and Pamela K. Anderson, "The Emergence of Whitefly-Transmitted Geminiviruses in Tomato in the Western Hemisphere," *Plant Disease*

81, no. 12 (1997): 1358–1367; Anne Simon Moffat, "Geminiviruses Emerge as Serious Crop Threat," *Science* 286 (1999): 1835.

39. William E. Fry and Stephen B. Goodwin, "Resurgence of the Irish Potato Famine Fungus," *BioScience* 47, no. 6 (1997): 363–371.
40. Clive M. Brasier, "Rapid Evolution of Introduced Plant Pathogens via Interspecific Hybridization," *BioScience* 51, no. 2 (2001): 123–133.
41. Charles S. Elton, *The Ecology of Invasions by Animals and Plants* (1958; reprint, Chicago: University of Chicago Press, 2000).
42. Otis C. Maloy, "White Pine Blister Rust Control in North America: A Case History," *Annual Review of Phytopathology* 35 (1997): 87–109.
43. G. S. Gilbert and S. P. Hubbell, "Plant Diseases and the Conservation of Tropical Forests," *BioScience* 46 (1996): 98–106.
44. F. D. Podger and M. J. Brown, "Vegetation Damage Caused by *Phytophthora cinnamomi* on Disturbed Sites in Temperate Rainforest in Western Tasmania," *Australian Journal of Botany* 37 (1989): 443–480.
45. Random Samples, "Culprit Named in 'Sudden Oak Death,'" *Science* 289 (2000): 859.
46. Reuters, "Asian Beetle Quarantine Expanded in Illinois," 29 May 1999.
47. Piran C. L. White and Geraldine Newton-Cross, "An Introduced Disease in an Invasive Host: The Ecology and Economics of Rabbit Calcivirus Disease (RCD) in Rabbits in Australia," in Perrings, Williamson, and Dalmazzone, *Economics of Biological Invasions.*
48. Rosamond Naylor, "Invasions in Agriculture: Assessing the Cost of the Golden Apple Snail in Asia," *Ambio* 25, no. 7 (1996): 443–448; Robert H. Cowie, "Apple Snails as Agricultural Pests: Their Biology, Impacts, and Management," in *Molluscs as Crop Pests,* ed. G. M. Barker (Wallingford, Oxon, England: CAB International, 2002).
49. Asna B. Othman et al., "Increasing Rice Production in Malaysia: Department of Agriculture's Approach" (paper presented at seminar, "Sustainable Rice Production towards the Next Millennium," Langkawi, Kedah, Malaysia, 7–9 November 1998).
50. Cowie, "Apple Snails as Agricultural Pests."
51. Les Kaufman, "Catastrophic Change in Species-Rich Freshwater Ecosystems," *BioScience* 42, no. 11 (1992): 846–858.
52. Richard Ogutu-Ohwayo, "Nile Perch in Lake Victoria: Effects on Fish Species Diversity, Ecosystem Functions, and Fisheries," in Sandlund, Schei, and Viken, *Norway/UN Conference on Alien Species,* 93–98.
53. Yvonne Baskin, "Africa's Troubled Waters," *BioScience* 42, no. 7 (1992): 476–481.
54. Pimentel et al., "Economic and Environmental Benefits of Biodiversity."
55. Ed Stoddard, "Foot-and-Mouth Threatens South African Wildlife," Reuters, 14 November 2000.

56. Martin Enserink, "Barricading U.S. Borders against a Devastating Disease," *Science* 291 (2001): 2298–2300.
57. Erwin Northoff, "Animal Diseases Spreading Warns UN," *AQIS Bulletin* (Australian Quarantine and Inspection Service) 13, no. 8 (October 2000), on-line at http://www.affa.gov.au/.
58. David Sharp, "Salmon Destroyed to Stop Virus," Associated Press, 7 September 2001.
59. Rosamond L. Naylor et al., "Effect of Aquaculture on World Fish Supplies," *Nature* 405, no. 29 (June 2000): 1017–1024.
60. Corn et al., *Harmful Non-native Species;* Peter B. Moyle, "Effects of Invading Species on Freshwater and Estaurine Ecosystems," in Sandlund, Schei, and Viken, *Norway/UN Conference on Alien Species,* 86–92.
61. Stephen L. Buchmann and Gary Paul Nabhan, *The Forgotten Pollinators* (Washington, D.C.: Island Press, Shearwater Books, 1996).
62. Martin Enserink, "Biological Invaders Sweep In," *Science* 285 (1999): 1834–1836.
63. New Zealand Ministry of Agriculture and Forestry, *Varroa Management Programme: Questions and Answers,* on-line at http://www.maf.govt.nz/biosecurity.
64. Robert H. Boyle, "Flying Fever," *Audubon* (July–August 2000): 63–68.
65. Durland Fish, "Testimony by Durland Fish, Ph.D.," in *Senate Field Hearing on West Nile Virus* (Fairfield University, Fairfield, Conn., 14 December 1999), on-line at http://www.senate.gov/~epw/fis_1214.htm.
66. Martin Enserink, "New York's Lethal Virus Came from Middle East, DNA Suggests," *Science* 286 (1999): 1450–1451.
67. R. S. Lanciotti et al., "Origin of the West Nile Virus Responsible for an Outbreak of Encephalitis in the Northeastern United States," *Science* 286 (1999): 2333–2337.
68. Environment News Service, "West Nile Virus Spreads West of Mississippi River," 10 October 2001.
69. Sue Binder et al., "Emerging Infectious Diseases: Public Health Issues of the Twenty-first Century," *Science* 284 (1999): 1311–1313; Laurie Garrett, *The Coming Plague: Newly Emerging Diseases in a World Out of Balance* (New York: Farrar, Straus and Giroux, 1994).
70. Ralph T. Bryan, "Alien Species and Emerging Infectious Diseases: Past Lessons and Future Implications," in Sandlund, Schei, and Viken, *Norway/UN Conference on Alien Species,* 74–80.
71. Martin Enserink, "Malaysian Researchers Trace Nipah Virus Outbreak to Bats," *Science* 289 (2000): 518–519; See Yee Ai, "Profile of a Virus," *Star Online: Star Publications (Malaysia) Bhd.,* 10 April 2000.
72. Enserink, "Malaysian Researchers Trace Nipah Virus Outbreak."

73. Peter Daszak, Andrew A. Cunningham, and Alex D. Hyatt, "Emerging Infectious Diseases of Wildlife: Threats to Biodiversity and Human Health," *Science* 287 (2000): 443–449.
74. John S. Mackenzie, "Emerging Viral Diseases: An Australian Perspective," *Emerging Infectious Diseases* 5, no. 1 (January–March 1999), on-line at http://www.cdc.gov/ncidod/EID/vol5no1/mackenzie.htm; E. Wright, K. Dunn, and A. Brown, *Report to the Minister for Agriculture, Fisheries, and Forestry on a Review of the Northern Australia Quarantine Strategy* (Canberra, Australia: Quarantine and Exports Advisory Council, 1998).
75. Institute of Medicine, Board on International Health, *America's Vital Interest in Global Health* (Washington, D.C.: National Academy Press, 1997).
76. Eystein Skjerve and Yngvild Wasteson, "Ecological Consequences of Spreading of Pathogens and Genes through an Increasing Trade in Foods," in Sandlund, Schei, and Viken, *Norway/UN Conference on Alien Species*, 141–147.
77. Gregory M. Ruiz et al., "Global Spread of Microorganisms by Ships," *Nature* 408 (2000): 49–50.
78. R. V. Tauxe, E. D. Mintz, and R. E. Quick, "Epidemic Cholera in the New World: Translating Epidemiology into New Prevention Strategies," *Emerging Infectious Diseases* 1 (1995): 141–146.
79. Eduardo H. Rapoport, "Tropical versus Temperate Weeds: A Glance into the Present and Future," in *Ecology of Biological Invasion in the Tropics*, ed. P. S. Ramakrishnan (New Delhi: International Science Publications, 1991), 11–51.

Chapter 4. Elbowing Out the Natives

1. Lloyd L. Loope, "Hawaii and the Pacific Islands," in *Status and Trends of the Nation's Biological Resources*, ed. M. J. Mac et al. (Reston, Va.: U.S. Department of the Interior, U.S. Geological Survey, 1998), 747–774.
2. Alan Holt, "An Alliance of Biodiversity, Agriculture, Health, and Business Interests for Improved Alien Species Management in Hawaii," in *Norway/UN Conference on Alien Species*, ed. Odd Terje Sandlund, Peter Johan Schei, and Aslaug Viken (Trondheim, Norway: Directorate for Nature Management and Norwegian Institute for Nature Research, 1996), 155–160.
3. Lloyd L. Loope, Ole Hamann, and Charles P. Stone, "Comparative Conservation Biology of Oceanic Archipelagoes," *BioScience* 38, no. 4 (1988): 272–282.
4. United Nations Environment Programme, *Global Biodiversity Assessment* (Cambridge, England: Cambridge University Press, 1995).
5. P. R. Ehrlich and E. O. Wilson, "Biodiversity Studies: Science and Policy," *Science* 253 (1991): 758–762.
6. Peter M. Vitousek et al., "Human Domination of Earth's Ecosystems," *Science* 277 (1997): 494–499.

7. Edward O. Wilson, foreword to *Strangers in Paradise: Impact and Management of Nonindigenous Species in Florida,* ed. Daniel Simberloff, Don C. Schmitz, and Tom C. Brown (Washington, D.C.: Island Press, 1997), ix–x.
8. World Conservation Union (IUCN), *1996 IUCN Red List of Threatened Animals* (Gland, Switzerland, and Cambridge, England: IUCN, 1996).
9. United Nations Environment Programme, *Global Biodiversity Assessment.*
10. World Conservation Union (IUCN), *1997 IUCN Red List of Threatened Plants* (Gland, Switzerland, and Cambridge, England: IUCN, 1997).
11. David S. Wilcove et al., "Quantifying Threats to Imperiled Species in the United States," *BioScience* 48, no. 8 (1998): 607–615; quote, 614.
12. Yvonne Baskin, "Curbing Undesirable Invaders," *BioScience* 46, no. 10 (1996): 732–736; quote, 734.
13. James H. Brown, *Macroecology* (Chicago: University of Chicago Press, 1995), 218–219.
14. IUCN, *1997 IUCN Red List.*
15. Paul L. Angermeier, "Does Biodiversity Include Artificial Diversity?" *Conservation Biology* 8, no. 2 (1994): 600–602; National Research Council, *Ecological Indicators for the Nation* (Washington, D.C.: National Academy Press, 1999).
16. Jeffrey A. McNeely, "The Great Reshuffling: How Alien Species Help Feed the Global Economy," in Sandlund, Schei, and Viken, *Norway/UN Conference on Alien Species,* 53–59.
17. Richard N. Mack et al., "Biotic Invasions: Causes, Epidemiology, Global Consequences, and Control," *Ecological Applications* 10, no. 3 (2000): 689–710.
18. Mark Williamson, *Biological Invasions* (London: Chapman and Hall, 1996), 115.
19. Don C. Schmitz et al., "The Ecological Impact of Nonindigenous Plants," in Simberloff, Schmitz, and Brown, *Strangers in Paradise,* 39–61.
20. I. Kowarik, "Time Lags in Biological Invasions with Regard to the Success and Failure of Alien Species," in *Plant Invasions: General Aspects and Special Problems,* ed. P. Pysek et al. (Amsterdam: SPB Academic Publishing, 1995), 15–38.
21. Jeff Crooks and Michael E. Soulé, "Lag Times in Population Explosions of Invasive Species: Causes and Implications," in Sandlund, Schei, and Viken, *Norway/UN Conference on Alien Species,* 39–46.
22. Williamson, *Biological Invasions,* 33.
23. U.S. Congress, Office of Technology Assessment, *Harmful Non-indigenous Species in the United States,* OTA-F-565 (Washington, D.C.: U.S. Government Printing Office, 1993).
24. Marcel Rejmánek et al., "Ecology of Invasive Plants: State of the Art," in *Invasive Alien Species: Searching for Solutions,* ed. H. A. Mooney et al. (Washington, D.C.: Island Press, 2002).
25. Daniel S. Simberloff, "Biological Invasions: What Are They Doing to Us, and What Can We Do about Them?" (paper presented at conference, "Exotic

Organisms in Greater Yellowstone: Native Biodiversity under Siege," Yellowstone National Park, Wyo., October 1999).

26. Peter M. Vitousek et al., "Biological Invasion by *Myrica faya* Alters Ecosystem Development in Hawaii," *Science* 238 (1987): 802–804; Peter M. Vitousek and Lawrence R. Walker, "Biological Invasion by *Myrica faya* in Hawaii: Plant Demography, Nitrogen Fixation, Ecosystem Effects," *Ecological Monographs* 59, no. 3 (1989): 247–265.
27. Jean-Yves Meyer, "French Polynesia: A Natural Paradise . . . for Invasive Species," *Aliens* (IUCN newsletter) 6 (1997): 5–6.
28. Jean-Yves Meyer, "Polynesia's Green Cancer," *World Conservation* (IUCN journal) 28, no. 4, and 29, no. 1 (April 1997–January 1998): 7.
29. Wendy Strahm, "Invasive Species in Mauritius: Examining the Past and Charting the Future," in Sandlund, Schei, and Viken, *Norway/UN Conference on Alien Species,* 167–181.
30. Ian A. W. Macdonald et al., "Wildlife Conservation and the Invasion of Nature Reserves by Introduced Species: A Global Perspective," in *Biological Invasions: A Global Perspective,* ed. J. A. Drake et al., SCOPE 37 (Chichester, West Sussex, England: John Wiley and Sons, 1989), 215–255.
31. Schmitz et al., "Ecological Impact of Nonindigenous Plants"; Stella E. Humphries, "Invasive Plants in Australia," *Aliens* (IUCN newsletter) 1 (1995): 13–14.
32. David M. Richardson and Steven I. Higgins, "Pines as Invaders in the Southern Hemisphere," in *Ecology and Biogeography of Pinus,* ed. David M. Richardson (Cambridge, England: Cambridge University Press, 1998), 450–473.
33. Erika Zavaleta, "Valuing Ecosystem Services Lost to *Tamarix* Invasion in the United States," in *Invasive Species in a Changing World,* ed. Harold A. Mooney and Richard J. Hobbs (Washington, D.C.: Island Press, 2000), 261–300.
34. Surayya Khatoon, "Pakistan's Alien Forests," *World Conservation* (IUCN journal) 28, no. 4, and 29, no. 1 (April 1997–January 1998): 15.
35. Todd Esque and Cecil Schwalbe, "Non-native Grasses and Fires Create Double Jeopardy," *People, Land, and Water* (U.S. Department of the Interior) (July–August 2000): 26.
36. Flint Hughes, Peter M. Vitousek, and Timothy Tunison, "Alien Grass Invasion and Fire in the Seasonal Submontane Zone of Hawaii," *Ecology* 72, no. 2 (1991): 743–746.
37. Humphries, "Invasive Plants in Australia."
38. Ibid.; Agriculture and Resource Management Council of Australia and New Zealand, Australian and New Zealand Environment and Conservation Council, and Forestry Ministers, *The National Weeds Strategy: A Strategic Approach to Weed Problems of National Significance* (Canberra: Commonwealth of Australia, 1999).

39. Anonymous, "Kudzu on the Loose in Oregon," *Noxious Times* (California Department of Food and Agriculture) (fall 2000): 6.
40. Llewellyn C. Foxcroft, "A Case Study of Human Dimensions in Invasion and Control of Alien Plants in the Personnel Villages of Kruger National Park," in *The Great Reshuffling: Human Dimensions of Invasive Alien Species,* ed. Jeffrey A. McNeely (Gland, Switzerland, and Cambridge, England: IUCN, 2001), 127–134.
41. Alison J. Leslie and James R. Spotila, "Alien Plant Species Threatens Nile Crocodile (*Crocodylus niloticus*) Breeding in Lake St. Lucia, South Africa," *Biological Conservation* 98 (2001): 347–355.
42. Alexandre Meinesz, *Killer Algae: The True Tale of a Biological Invasion,* trans. Daniel Simberloff (Chicago: University of Chicago Press, 1999).
43. O. Jousson et al., "Invasive Alga Reaches California," *Nature* 408 (2000): 157–158; Lars Anderson, "Caulerpa: the Newest Aquatic Threat," *Noxious Times* (California Department of Food and Agriculture) (fall 2000): 6.
44. M. Lynne Corn et al., *Harmful Non-native Species: Issues for Congress* (Washington, D.C.: Congressional Research Service, 1999), on-line at http://cnie.org/NLE/CRSreports/Biodiversity/biodv-26.cfm.
45. Daniel S. Simberloff and Donald R. Strong, "Exotic Species Seriously Threaten Our Environment," *Chronicle of Higher Education* 47, no. 2 (8 September 2000): B20.
46. Williamson, *Biological Invasions;* Chris Bright, *Life Out of Bounds,* Worldwatch Environmental Alert Series (New York: W. W. Norton and Company, 1998).
47. J. H. Lawton and C. G. Jones, "Linking Species and Ecosystems: Organisms as Ecosystem Engineers," in *Linking Species and Ecosystems,* ed. C. G. Jones and J. H. Lawton (New York: Chapman and Hall, 1995), 141–150.
48. Williamson, *Biological Invasions,* 115.
49. I. M. Parker et al., "Impact: Toward a Framework for Understanding the Ecological Effects of Invaders," *Biological Invasions* 1 (1999): 3–19.
50. Mary E. Power et al., "Challenges in the Quest for Keystones," *BioScience* 46, no. 8 (1996): 609–620.
51. Robert T. Paine, "A Note on Trophic Complexity and Community Stability," *American Naturalist* 103 (1969): 91–93; Robert T. Paine, "Food Web Complexity and Species Diversity," *American Naturalist* 100 (1966): 65–75.
52. Macdonald et al., "Wildlife Conservation and the Invasion of Nature Reserves."
53. Adele Conover, "Foreign Worm Alert," *Smithsonian* (August 2000): 29–30.
54. Buddhi Marambe et al., "Human Dimensions of Invasive Alien Species in Sri Lanka," in McNeely, *The Great Reshuffling,* 135–142.
55. Lloyd Loope, "Aliens to Watch: Garden Slugs, *Milax gagates,*" *Aliens* (IUCN newsletter) 1 (1995): 19.
56. Williamson, *Biological Invasions.*

57. William Bond and P. Slingsby, "Collapse of an Ant–Plant Mutualism: The Argentine Ant (*Iridomyrmex humilis*) and Myrmecochorous Proteaceae," *Ecology* 65, no. 4 (1984): 1031–1037; Caroline E. Christian, "Consequences of a Biological Invasion Reveal the Importance of Mutualism for Plant Communities," *Nature* 413 (2001): 635–639.
58. William F. Laurance, "The Top End of Down Under," *Wildlife Conservation* (July–August 1990): 26–45.
59. David Pimentel et al., "Environmental and Economic Costs of Nonindigenous Species in the United States," *BioScience* 50, no. 1 (2000): 53–65.
60. David R. Klein et al., "Alaska," in Mac et al., *Status and Trends of the Nation's Biological Resources,* 707–745; John Hughes, "Aleutian Canada Geese Make Recovery," Associated Press, 29 July 1999.
61. Dick Veitch, "The Invader at Our Side," *World Conservation* (IUCN journal) 28, no. 4, and 29, no. 1 (April 1997–January 1998): 19.
62. Heritage and Aboriginal Affairs South Australia Department for Environment, *Flinders Range Bounceback: Progress Report, Stage 1* (Adelaide: Heritage and Aboriginal Affairs South Australia Department for Environment, 1999).
63. Western Australia Department of Conservation and Land Management, *Western Shield: Bringing Wildlife Back from the Brink of Extinction,* on-line at http://www.calm.wa.gov.au/projects/west_shield_article.html.
64. Mack et al., "Biotic Invasions."
65. Peter Daszak, Andrew A. Cunningham, and Alex D. Hyatt, "Emerging Infectious Diseases of Wildlife: Threats to Biodiversity and Human Health," *Science* 287 (2000): 443–449.
66. Peter Daszak et al., "Emerging Infectious Diseases and Amphibian Population Declines," *Emerging Infectious Diseases* 5, no. 6 (November–December 1999), on-line at http://www.cdc.gov/ncidod/EID/vol5no6/daszak.htm; Rick Speare and Lee Berger, *Global Distribution of Chytridiomycosis in Amphibians,* on-line at http://www.jcu.edu.au/school/phtm/PHTM/frogs/chyglob.htm.
67. Yvonne Baskin, "A Sickening Situation: Emerging Pathogens Pose a Threat to Wildlife," *Natural History* (April 2000): 24–27.
68. Robert H. Boyle, "Flying Fever," *Audubon* (July–August 2000): 63–68; Martin Enserink, "The Enigma of West Nile," *Science* 290 (2000): 1482–1484.
69. Michael N. Clout, "Biological Conservation and Invasive Species: The New Zealand Experience," in Sandlund, Schei, and Viken, *Norway/UN Conference on Alien Species,* 161–166; Jacqueline Beggs, "Exotic Wasps in New Zealand," *Aliens* (IUCN newsletter) 1 (1995): 3–4.
70. Joseph Kiesecker and Andrew Blaustein, "Effects of Introduced Bullfrogs and Smallmouth Bass on Microhabitat Use, Growth, and Survival of Native Red-Legged Frogs," *Conservation Biology* 12, no. 4 (1998): 776–787.
71. Wilson, foreword to Simberloff, Schmitz, and Brown, *Strangers in Paradise,* x.

72. Ettore Randi et al., "Mitochondrial DNA Variability in Italian and East European Wolves: Detecting the Consequences of Small Population Size and Hybridization," *Conservation Biology* 14, no. 2 (2000): 464–473.
73. Mack et al., "Biotic Invasions."
74. Baz Hughes and Gwyn Williams, "What Future for the White-Headed Duck?" *World Conservation* (IUCN journal) 28, no. 4, and 29, no. 1 (April 1997–January 1998): 27–28.
75. Kenneth Schmidt and Christopher Whelan, "Effects of Exotic *Lonicera* and *Rhamnus* on Songbird Nest Predation," *Conservation Biology* 13, no. 6 (1999): 1502–1506.
76. Bruce E. Coblentz, "Strangers in Paradise: Invasive Mammals on Islands," *World Conservation* (IUCN journal) 28, no. 4, and 29, no. 1 (April 1997–January 1998): 7–8

Chapter 5. The Good, the Bad, the Fuzzy

1. Gordon H. Rodda, Thomas H. Fritts, and David Chiszar, "The Disappearance of Guam's Wildlife," *BioScience* 47, no. 9 (1997): 565–574; quote, 570.
2. Craig Hoover, *The U.S. Role in the International Live Reptile Trade: Amazon Tree Boas to Zululand Dwarf Chameleons* TRAFFIC North America report (World Wildlife Fund and World Conservation Union [IUCN]) (September 1998), on-line at http://www.traffic.org/reptiles.
3. Brian P. Butterfield, Walter E. Meshaka Jr., and Craig Guyer, "Nonindigenous Amphibians and Reptiles," in *Strangers in Paradise: Impact and Management of Nonindigenous Species in Florida,* ed. Daniel Simberloff, Don C. Schmitz, and Tom C. Brown (Washington, D.C.: Island Press, 1997), 123–138.
4. Hoover, *U.S. Role in the International Live Reptile Trade.*
5. Richard Hartley, "Pet Trade Blues," *International Wildlife* (March–April 2000), on-line at http://www.nwf.org/internationalwildlife/2000/lears.html.
6. World Trade Organization, *International Trade Statistics, 2001* (Geneva, Switzerland: World Trade Organization, 2001), on-line at http://www.wto.org.
7. Michael E. Soulé, "Conservation: Tactics for a Constant Crisis," *Science* 253 (1991): 744–750.
8. World Trade Organization, *International Trade Statistics, 2001.*
9. United Nations Conference on Trade and Development (UNCTAD), *Review of Maritime Transport, 2001* (Geneva, Switzerland: UNCTAD, 2001), on-line at http://www.unctad.org.
10. International Air Transport Association (IATA), *World Air Transport Statistics, 2001* (London: IATA, 2001), on-line at http://www.iata.org.
11. World Tourism Organization, *Tourism Highlights, 2000* (Madrid: World Tourism Organization, 2000), 26, on-line at http://www.world-tourism.org/.

12. World Tourism Organization, "Tourism Industry Takes Action to End Crisis," news release (12 November 2001), on-line at http://www.world-tourism.org/newsroom/Releases/more_releases/R011112.html; International Air Transport Association, "IATA Traffic Confirms Effect of September 11," news release (30 October 2001), on-line at http://www.iata.org/pr/.
13. James T. Carlton, "The Scale and Ecological Consequences of Biological Invasions in the World's Oceans," in *Invasive Species and Biodiversity Management,* ed. Odd Terje Sandlund, Peter Johan Schei, and Auslaug Viken (Dordrecht, Netherlands: Kluwer Academic Publishers, 1999), 195–212.
14. Greg Ruiz et al., "Invasion Pathways of Marine Invertebrates in North America: Spatial and Temporal Patterns" (paper presented at conference, "GISP Invasion Pathways: Analysis of Invasion Patterns and Pathway Management," Smithsonian Environmental Research Center, Edgewater, Md., November 1999).
15. Charles F. Boudouresque, "The Red Sea–Mediterranean Link: Unwanted Effects of Canals," in *Norway/UN Conference on Alien Species,* ed. Odd Terje Sandlund, Peter Johan Schei, and Aslaug Viken (Trondheim, Norway: Directorate for Nature Management and Norwegian Institute for Nature Research, 1996), 107–115.
16. Cynthia S. Kolar and David M. Lodge, "Freshwater Nonindigenous Species: Interactions with Other Global Changes," in *Invasive Species in a Changing World,* ed. Harold A. Mooney and Richard J. Hobbs (Washington, D.C.: Island Press, 2000), 3–30.
17. Murray Gregory, "Pelagic Plastics and Marine Invaders," *Aliens* (IUCN newsletter) 7 (1998): 6–7.
18. Keizi Kiritani and Kohji Yamamura, "Exotic Insects and Their Pathways for Invasion" (paper presented at conference, "GISP Invasion Pathways: Analysis of Invasion Patterns and Pathway Management," Smithsonian Environmental Research Center, Edgewater, Md., November 1999).
19. Mary Palm, "Invasion Pathways of Terrestrial Plant-Inhabiting Fungi" (paper presented at conference, "GISP Invasion Pathways: Analysis of Invasion Patterns and Pathway Management," Smithsonian Environmental Research Center, Edgewater, Md., November 1999).
20. Soetikno S. Sastroutomo, "The Roles of Plant Quarantine in Preventive Control of Invasive Species" (paper presented at conference, "Global Invasive Species Programme: Workshop on Management and Early Warning Systems," Kuala Lumpur, Malaysia, March 1999).
21. N. Wace, "Assessment of Dispersal of Plant Species: The Car Borne Flora in Canberra," *Proceedings of the Ecological Society of Australia* 10 (1977): 166–186.
22. Francois Moutou, "Cautionary Tales: Bats Journey to England; Raccoon's Sojourn in France," *Aliens* (IUCN newsletter) 7 (1998): 14.

23. M. Lynne Corn et al., *Harmful Non-native Species: Issues for Congress* (Washington, D.C.: Congressional Research Service, 1999), on-line at http://cnie.org/NLE/CRSreports/Biodiversity/biodv-26.cfm.
24. Agriculture and Resource Management Council of Australia and New Zealand, Australian and New Zealand Environment and Conservation Council, and Forestry Ministers, *The National Weeds Strategy: A Strategic Approach to Weed Problems of National Significance* (Canberra: Commonwealth of Australia, 1999); James P. H. Paterson, "The Role of Warfare in Promoting the Introduction and Invasion of Alien Species," unpublished manuscript (2000), on-line at http://members.tripod.co.uk/WoodyPlantEcology/docs/war-ww2.rtf; James T. Carlton, "Pattern, Process, and Prediction in Marine Invasion Ecology," *Biological Conservation* 78 (1996): 97–106.
25. Buddhi Marambe et al., "Human Dimensions of Invasive Alien Species in Sri Lanka," in *The Great Reshuffling: Human Dimensions of Invasive Alien Species*, ed. Jeffrey A. McNeely (Gland, Switzerland, and Cambridge, England: IUCN, 2001), 135–142.
26. Martin Enserink, "Biological Invaders Sweep In," *Science* 285 (1999): 1834–1836.
27. Paterson, "Role of Warfare."
28. Simon V. Fowler, "Alien Weeds and Their Specific Problems regarding Management" (paper presented at conference, "Global Invasive Species Programme: Workshop on Management and Early Warning Systems," Kuala Lumpur, Malaysia, March 1999).
29. Anonymous, "NAQS Supports East Timor Response," *NAQS News* (Northern Australia Quarantine Strategy) (December 1999): 1, on-line at http://www.affa.gov.au/.
30. Christine Kopral and Katherine Marshall, *The Wildlife Industry: Trends and Diseases Issues* (Fort Collins, Colo.: U.S. Department of Agriculture, Animal and Plant Health Inspection Service, Centers for Epidemiology and Animal Health, Center for Emerging Issues, 2000), on-line at http://www.aphis.usda.gov/vs/ceah/cei/wildlife_industry.htm.
31. Ronald M. Jurek, *Domestic Ferret Issues in California*, on-line at http://www.dfg.ca.gov/.
32. Pet Industry Joint Advisory Council (PIJAC), *U.S. Ornamental Aquarium Industry* (Washington, D.C.: PIJAC Pet Information Bureau, 2000), on-line at http://petsforum.com/.
33. Crispian Balmer, "France Faces Up to Threat of Invasion of Apes," Reuters, 20 December 2000.
34. Agustin Iriarte Walton, "Chilean Program on Control of Invasive Wildlife Species" (paper presented at conference, "GISP Invasion Pathways: Analysis of

Invasion Patterns and Pathway Management," Smithsonian Environmental Research Center, Edgewater, Md., November 1999).
35. Marambe et al., "Human Dimensions of Invasive Alien Species."
36. Fred Kraus, "Pathways of Global Terrestrial Vertebrate Introductions, with an Emphasis on Reptiles and Amphibians" (paper presented at conference, "GISP Invasion Pathways: Analysis of Invasion Patterns and Pathway Management," Smithsonian Environmental Research Center, Edgewater, Md., November 1999).
37. Kopral and Marshall, *Wildlife Industry.*
38. R. K. Brooke, P. H. Lloyd, and A. L. de Villiers, "Alien and Translocated Terrestrial Vertebrates in South Africa," in *The Ecology and Management of Biological Invasions in Southern Africa,* ed. I. A. W. Macdonald, F. J. Kruger, and A. A. Ferrar (Cape Town, South Africa: Oxford University Press, 1986), 63–74.
39. Rosamond L. Naylor et al., "Effect of Aquaculture on World Fish Supplies," *Nature* 405, no. 29 (June 2000): 1017–1024.
40. John Baker, "Gourmet Invader," *Aliens* (IUCN newsletter) 1 (1995): 6.
41. U.S. Geological Survey, Biological Resources Division, "USGS Scientists Find New Population of Asian Swamp Eels in South Florida," news release (3 March 2000), on-line at http://biology.usgs.gov/pr/newsrelease/2000/3-3.html.
42. James T. Carlton, "Four Species of Marine Crabs Invade North America," *Aliens* (IUCN newsletter) 2 (1995): 5–6.
43. James D. Bland and Stanley A. Temple, "The Himalayan Snowcock: North America's Newest Exotic Bird," in *Biological Pollution: The Control and Impact of Invasive Exotic Species,* ed. Bill N. McKnight (Indianapolis: Indiana Academy of Science, 1993), 149–155.
44. Gary Turbak, "The Great State Animal Swap," *National Wildlife* (October–November 1992): 42–45.
45. Hugh A. Mulligan, "Turkeys Rebound from Near Extinction," Associated Press, 25 November 1999.
46. Leo G. Nico and Pam L. Fuller, "Spatial and Temporal Patterns of Nonindigenous Fish Introductions in the United States," *Fisheries* 24, no. 1 (1999): 16–27; Dan Ferber, "Will Black Carp Be the Next Zebra Mussel?" *Science* 292 (2001): 203.
47. John D. Varley and Paul Schullery, *The Yellowstone Lake Crisis: Confronting a Lake Trout Invasion* (Yellowstone National Park, Wyo.: U.S. National Park Service, 1995).
48. Francis M. Harty, "How Illinois Kicked the Exotic Habit," in McKnight, *Biological Pollution,* 195–209.
49. Nico and Fuller, "Spatial and Temporal Patterns."
50. Donald R. Strong and Robert W. Pemberton, "Food Webs, Risks of Alien Enemies, and Reform of Biological Control," in *Evaluating Indirect Ecological Effects*

of Biological Control, ed. E. Wajnberg, John K. Scott, and Paul C. Quimby (Wallingford, Oxon, England: CAB International, 2001), 57–79.

51. Nobuo Ishii, "Amami Mongoose," *Aliens* (IUCN newsletter) 7 (1998): 14.
52. Walter R. Courtenay Jr., "Biological Pollution through Fish Introductions," in McKnight, *Biological Pollution,* 35–61; Angela H. Arthington and Lance N. Lloyd, "Introduced Poeciliids in Australia and New Zealand," in *Ecology and Evolution of Livebearing Fishes (Poeciliidae),* ed. Gary K. Meffe and Franklin F. Snelson Jr. (Englewood Cliffs, N.J.: Prentice-Hall, 1989), 333–348.
53. U.S. Department of Agriculture, National Agricultural Statistics Service (USDA NASS), *Census of Horticultural Specialties* (Washington, D.C.: USDA NASS, 1998), on-line at http://www.usda.gov/nass/.
54. R. T. Isaacson, ed., *The Andersen Horticultural Library's Source List of Plants and Seeds: A Completely Revised Listing of 1993–1996 Catalogues,* 4th ed. (Chanhassen, Minn.: Andersen Horticultural Library, 1996).
55. Llewellyn C. Foxcroft, "A Case Study of Human Dimensions in Invasion and Control of Alien Plants in the Personnel Villages of Kruger National Park," in McNeely, *The Great Reshuffling,* 127–134.
56. David Sutasurya, "Alien Plant Invasion in Gunung Gede Pangrano National Park, Indonesia," *Aliens* (IUCN newsletter) 6 (1997): 8–9.
57. Peter Wyse Jackson, "Botanic Gardens: A Revolution in Progress," *World Conservation* (IUCN journal) 29, no. 2 (February 1998): 14–15.
58. Sarah Hayden Reichard and Peter White, "Horticulture as a Pathway of Invasive Plant Introductions in the United States," *BioScience* 51, no. 2 (2001): 103–113.
59. Harty, "How Illinois Kicked the Exotic Habit," 198.
60. Don C. Schmitz and Daniel Simberloff, "Biological Invasions: A Growing Threat," *Issues in Science and Technology* (summer 1997): 33–40.
61. China Council for International Cooperation on Environment and Development (CCICED) Biodiversity Working Group, *Working Group Report to the China Council for International Cooperation on Environment and Development* (Dujiangyan, Sichuan: CCICED, 2000), on-line at http://www.harbour.sfu.ca/dlam/WorkingGroups/biodiversity/00report.html; XIE Yan et al., "Invasive Species in China: An Overview," *Biodiversity and Conservation* 10, no. 8 (2001): 1317–1341.
62. Andrew Tolfts, "*Cordia alliodora:* The Best Laid Plans," *Aliens* (IUCN newsletter) 6 (1997): 12–13.
63. G. D. P. S. Augustus, M. Jayabalan, and Gerald J. Seiler, *Cryptostegia grandiflora: A Potential Multi-Use Crop,* interpretive summary (TEKTRAN, U.S. Department of Agriculture, Agricultural Research Service, 1999), on-line at http://www.nal.usda.gov/ttic/tektran/data/000010/83/0000108377.html.

64. George W. Cox, *Alien Species in North America and Hawaii: Impacts on Natural Ecosystems* (Washington, D.C.: Island Press, 1999).
65. National Research Council, *Lost Crops of the Incas: Little-Known Plants of the Andes with Promise for Worldwide Cultivation* (Washington, D.C.: National Academy Press, 1989); National Research Council, *Microlivestock: Little-Known Small Animals with a Promising Economic Future* (Washington, D.C.: National Academy Press, 1991).
66. National Research Council, *Little-Known Asian Animals with a Promising Economic Future* (Washington, D.C.: National Academy Press, 1983).

Chapter 6. The Making of a Pest

1. Rob Hengeveld, "Problems of Biological Invasions: An Overview," in *Norway/UN Conference on Alien Species,* ed. Odd Terje Sandlund, Peter Johan Schei, and Aslaug Viken (Trondheim, Norway: Directorate for Nature Management and Norwegian Institute for Nature Research, 1996), 18–29.
2. D. M. Richardson et al., "Plant and Animal Invasions," in *The Ecology of Fynbos,* ed. R. M. Cowling (Cape Town, South Africa: Oxford University Press, 1992), 271–308.
3. Daniel Simberloff, foreword to *The Ecology of Invasions by Animals and Plants,* by Charles S. Elton (1958; reprint, Chicago: University of Chicago Press, 2000), vii–xiv.
4. Elton, *Ecology of Invasions,* 31.
5. Ibid., 109.
6. Mark Williamson, *Biological Invasions* (London: Chapman and Hall, 1996).
7. Richard N. Mack et al., "Biotic Invasions: Causes, Epidemiology, Global Consequences, and Control," *Ecological Applications* 10, no. 3 (2000): 689–710.
8. Alfred W. Crosby, *Ecological Imperialism: The Biological Expansion of Europe, 900–1900* (Cambridge, England: Cambridge University Press, 1986); Daniel Simberloff and Betsy Von Holle, "Positive Interactions of Nonindigenous Species: Invasional Meltdown?" *Biological Invasions* 1 (1999): 21–32.
9. Vernon H. Heywood, "Patterns, Extents, and Modes of Invasions by Terrestrial Plants," in *Biological Invasions: A Global Perspective,* ed. J. A. Drake et al., SCOPE 37 (Chichester, West Sussex, England: John Wiley and Sons, 1989), 31–55.
10. Marcel Rejmánek, "Invasive Plant Species and Invasible Ecosystems," in Sandlund, Schei, and Viken, *Norway/UN Conference on Alien Species,* 60–68; Richard N. Mack, "Biotic Barriers to Plant Naturalization," in *Proceedings of the Ninth International Symposium on Biological Control of Weeds,* ed. V. C. Moran and J. H. Hoffmann (Stellenbosch, South Africa: University of Cape Town, 1996), 39–46.
11. Williamson, *Biological Invasions.*

12. Mack et al., "Biotic Invasions."
13. David M. Richardson and Steven I. Higgins, "Pines as Invaders in the Southern Hemisphere," in *Ecology and Biogeography of Pinus,* ed. David M. Richardson (Cambridge, England: Cambridge University Press, 1998), 450–473.
14. M. P. Moulton and S. L. Pimm, "Species Introductions to Hawaii," in *Ecology of Biological Invasions of North America and Hawaii,* ed. H. A. Mooney and J. A. Drake, Ecological Studies 58 (New York: Springer-Verlag, 1986), 231–249.
15. Richard N. Mack, "Predicting the Identity and Fate of Plant Invaders: Emergent and Emerging Approaches," *Biological Conservation* 78 (1996): 107–121; quote, 109–110.
16. Williamson, *Biological Invasions,* 61.
17. David M. Richardson et al., "Naturalization and Invasion of Alien Plants: Concepts and Definitions," *Diversity and Distributions* 6 (2000): 93–107.
18. Marcel Rejmánek and David M. Richardson, "What Attributes Make Some Plant Species More Invasive?" *Ecology* 77, no. 6 (1996): 1655–1661.
19. Sarah H. Reichard and Clement W. Hamilton, "Predicting Invasions of Woody Plants Introduced into North America," *Conservation Biology* 11, no. 1 (1997): 193–203.
20. IUCN Invasive Species Specialist Group, Global Invasive Species Database, online at http://www.issg.org/database/welcome/.
21. Reichard and Hamilton, "Predicting Invasions of Woody Plants."
22. Anthony Ricciardi and Hugh J. MacIsaac, "Recent Mass Invasion of the North American Great Lakes by Ponto-Caspian Species," *Trends in Ecology and Evolution* 15, no. 2 (2000): 62–65.
23. David Lodge, "Invasive Species Pathways, Science, and Prevention: US-Canadian Great Lakes Case Study" (paper presented at annual meeting of the American Association for the Advancement of Science, San Francisco, February 2001).
24. Mark Williamson, "Invasions," *Ecography* 22, no. 1 (1999): 5–12; quote, 5.
25. Cynthia S. Kolar and David M. Lodge, "Progress in Invasion Biology: Predicting Invaders," *Trends in Ecology and Evolution* 16, no. 4 (2001): 199–204.
26. Marcel Rejmánek, "Invasive Plants: Approaches and Predictions," *Austral Ecology* 25 (2000): 497–506.
27. W. Mark Lonsdale and C. S. Smith, "Evaluating Pest Screening Systems: Insights from Epidemiology and Ecology," in *Weed Risk Assessment,* ed. R. H. Groves, F. D. Panetta, and J. G. Virtue (Melbourne, Australia: CSIRO Publishing, 2001), 52–60.
28. Daniel S. Simberloff, "The Ecology and Evolution of Invasive Non-indigenous Species" (paper presented at conference, "Global Invasive Species Programme: Workshop on Management and Early Warning Systems," Kuala Lumpur, Malaysia, March 1999).

29. Richard N. Mack, "Motivations and Consequences of the Human Dispersal of Plants," in *The Great Reshuffling: Human Dimensions of Invasive Alien Species,* ed. Jeffrey A. McNeely (Gland, Switzerland, and Cambridge, England: IUCN, 2001).
30. Pierre Binggeli, "The Human Dimensions of Invasive Woody Plants," in McNeely, *The Great Reshuffling,* 145–160.
31. Elton, *Ecology of Invasions,* 116–117.
32. Simberloff, foreword to Elton, *Ecology of Invasions.*
33. David Tilman, "Community Invasibility, Recruitment Limitation, and Grassland Biodiversity," *Ecology* 78 (1997): 81–92.
34. Thomas J. Stohlgren et al., "Exotic Plant Species Invade Hot Spots of Native Plant Diversity," *Ecological Monographs* 69, no. 1 (1999): 25–46.
35. Ted J. Case, "Global Patterns in the Establishment and Distribution of Exotic Birds," *Biological Conservation* 78 (1996): 69–96.
36. Peter B. Moyle and Theo Light, "Fish Invasions in California: Do Abiotic Factors Determine Success?" *Ecology* 77, no. 6 (1996): 1666–1670.
37. Jonathan M. Levine and Carla M. D'Antonio, "Elton Revisited: A Review of Evidence Linking Diversity and Invasibility," *Oikos* 87 (1999): 15–26.
38. Richard N. Mack, "Predicting Communities Vulnerable to Plant Invasions" (paper presented at annual meeting of the Ecological Society of America, Madison, Wisc., August 2001).
39. Williamson, *Biological Invasions,* 56.
40. Levine and D'Antonio, "Elton Revisited"; Carla M. D'Antonio and T. L. Dudley, "Biological Invasions as Agents of Change on Islands versus Mainlands," in *Islands: Biological Diversity and Ecosystem Function,* ed. P. M. Vitousek, L. L. Loope, and H. Andersen, Ecological Studies (Berlin: Springer-Verlag, 1995), 103–121.
41. Jeffrey A. McNeely et al., eds., *A Global Strategy on Invasive Alien Species* (Gland, Switzerland, and Cambridge, England: IUCN and Global Invasive Species Programme, 2001).
42. Michael B. Usher et al., "The Ecology of Biological Invasions into Nature Reserves: An Introduction," *Biological Conservation* 44 (1988): 1–8; Ian A. W. Macdonald et al., "Wildlife Conservation and the Invasion of Nature Reserves by Introduced Species: A Global Perspective," in *Biological Invasions: A Global Perspective,* ed. J. A. Drake et al., SCOPE 37 (Chichester, West Sussex, England: John Wiley and Sons, 1989), 215–255.
43. W. M. Lonsdale, "Concepts and Synthesis: Global Patterns of Plant Invasions and the Concept of Invasibility," *Ecology* 80, no. 5 (1999): 1522–1536.
44. Jeff Crooks and Michael E. Soulé, "Lag Times in Population Explosions of Invasive Species: Causes and Implications," in Sandlund, Schei, and Viken, *Norway/UN Conference on Alien Species,* 39–46.

45. Mary Kay Solecki, "Cut-Leaved and Common Teasel (*Dipsacus Iaciniatus* L. and *D. sylvestris* Huds.): Profile of Two Invasive Aliens," in *Biological Pollution: The Control and Impact of Invasive Exotic Species,* ed. Bill N. McKnight (Indianapolis: Indiana Academy of Science, 1993), 85–92.
46. Ann Rodman, "Cooperative Weed Mapping in the Greater Yellowstone Area" (paper presented at conference, "Exotic Organisms in Greater Yellowstone: Native Biodiversity under Siege," Yellowstone National Park, Wyo., October 1999).
47. Martin Enserink, "Biological Invaders Sweep In," *Science* 285 (1999): 1834–1836.
48. Crooks and Soulé, "Lag Times in Population Explosions."
49. K. C. Kendall, K. Peterson, and D. Schirokauer, "White Pine Blister Rust: An Exotic Fungus among Us" (paper presented at conference, "Exotic Organisms in Greater Yellowstone: Native Biodiversity under Siege," Yellowstone National Park, Wyo., October 1999).
50. Jeffrey S. Dukes and Harold A. Mooney, "Does Global Change Increase the Success of Biological Invaders?" *Trends in Ecology and Evolution* 14, no. 4 (1999): 135–139.
51. David M. Richardson et al., "Plant Invasions: The Role of Mutualisms," *Biological Reviews* 75 (2000): 65–93.
52. Simberloff and Von Holle, "Positive Interactions of Nonindigenous Species," 29.
53. Richardson et al., "Plant Invasions."
54. Richard N. Mack, "Temperate Grasslands Vulnerable to Plant Invasions: Characteristics and Consequences," in Drake et al., *Biological Invasions,* 155–179.
55. Simberloff and Von Holle, "Positive Interactions of Nonindigenous Species."
56. Richardson et al., "Plant Invasions," 85.

Chapter 7. Taking Risks with Strangers

1. Agriculture Western Australia, *Weed Risk Assessment and Climate Analysis,* online at http://www.agric.wa.gov.au/progserv/plants/weeds/weedsci2.htm.
2. Clare Shine, Nattley Williams, and Lothar Gundling, *A Guide to Designing Legal and Institutional Frameworks on Alien Invasive Species* (Gland, Switzerland; Cambridge, England; and Bonn, Germany: IUCN Environmental Law Centre, 2000), 34.
3. Agriculture and Resource Management Council of Australia and New Zealand, Australian and New Zealand Environment and Conservation Council, and Forestry Ministers, *The National Weeds Strategy: A Strategic Approach to Weed Problems of National Significance* (Canberra: Commonwealth of Australia, 1999).
4. Malcolm E. Nairn et al., *Australian Quarantine: A Shared Responsibility* (Canberra, Australia: Department of Primary Industries and Energy, 1996).

5. Paul C. Pheloung, "Weed Risk Assessment of Plant Introductions to Australia," in *Weed Risk Assessment*, ed. R. H. Groves, F. D. Panetta, and J. G. Virtue (Melbourne, Australia: CSIRO Publishing, 2001), 83–92; C. S. Walton, "Implementation of a Permitted List Approach to Plant Introductions in Australia," in Groves, Panetta, and Virtue, *Weed Risk Assessment*, 93–99.
6. Agriculture and Resource Management Council of Australia and New Zealand, Australian and New Zealand Environment and Conservation Council, and Forestry Ministers, *National Weeds Strategy*.
7. Tim Low, *Feral Future: The Untold Story of Australia's Exotic Invaders* (Victoria, Australia: Viking, 1999), 302.
8. J. Morgan Williams, *New Zealand under Siege: A Review of the Management of Biosecurity Risks to the Environment* (Wellington, New Zealand: Parliamentary Commissioner for the Environment, 2000), on-line at http://www.pce.govt.nz/reports/allreports/0_908804_93_8.shtml.
9. Environmental Risk Management Authority (ERMA) New Zealand, *The Environmental Risk Management Authority*, on-line at http://www.ermanz.govt.nz.
10. Peter A. Williams, Euan Nicol, and Melanie Newfield, *Assessing the Risk to Indigenous New Zealand Biota from New Exotic Plant Taxa and Genetic Material* (Wellington, New Zealand: Department of Conservation, 2000).
11. Bas Walker, "New Organism Law Keeps Out Unwanted Aliens" (chief executive, Environmental Risk Management Authority [ERMA] New Zealand, news release [26 July 1999]), on-line at http://www.ermanz.govt.nz/NewsAndEvents/files/Features/fe19990726.htm.
12. Williams, *New Zealand under Siege*, 86.
13. National Plant Board, *Safeguarding American Plant Resources: A Stakeholder Review of the APHIS-PPQ Safeguarding System* (1999), on-line at http://www.aphis.usda.gov/ppq/safeguarding.
14. National Invasive Species Council, *Management Plan: Meeting the Invasive Species Challenge* (Washington, D.C.: National Invasive Species Council, 2001), 3, on-line at http://www.invasivespecies.gov.
15. U.S. Department of Agriculture, Animal and Plant Health Inspection Service (USDA APHIS), *APHIS Weed Policy, 2000–2001* (Washington, D.C.: USDA APHIS, 2000), on-line at http://www.aphis.usda.gov/ppq/weeds/nwpolicy2001.html.
16. U.S. Congress, Office of Technology Assessment, *Harmful Non-indigenous Species in the United States*, OTA-F-565 (Washington, D.C.: U.S. Government Printing Office, 1993).
17. Shine, Williams, and Gundling, *Guide to Designing Legal and Institutional Frameworks;* Geoffrey S. Becker, *Report for Congress, Agricultural Exports: Technical Barriers to Trade* (Washington, D.C.: Congressional Research Service, 1997).
18. Shine, Williams, and Gundling, *Guide to Designing Legal and Institutional Frameworks.*

19. James T. Carlton and Gregory M. Ruiz, "The Vectors of Invasions by Alien Species," in *Invasive Alien Species: Searching for Solutions,* ed. H. A. Mooney et al. (Washington, D.C.: Island Press, 2002); Faith T. Campbell, "International Plant Protection Convention: Send It Back to the Drawing Board," *Aliens* (IUCN newsletter) 6 (1997): 22.
20. Australian Quarantine and Inspection Service (AQIS), *The AQIS Import Risk Analysis Process Handbook* (Canberra: AQIS, 1998), 8.
21. U.S. Congress, Office of Technology Assessment, *Harmful Non-indigenous Species,* 114, 117.
22. U.S. Department of Agriculture, Forest Service, *Pest Risk Assessment of the Importation of Larch from Siberia and the Soviet Far East* (Washington, D.C.: U.S. Department of Agriculture, Forest Service, 1991), S-1.
23. U.S. Congress, Office of Technology Assessment, *Harmful Non-indigenous Species.*
24. Shine, Williams, and Gundling, *Guide to Designing Legal and Institutional Frameworks.*
25. U.S. Department of Agriculture, Forest Service, *Pest Risk Assessment,* 7-1.
26. Michael Wingfield, "Worldwide Movement of Forest Fungi, Especially in the Tropics and Southern Hemisphere" (paper presented at International Botanical Congress, St. Louis, Mo., August 1999); Michael Wingfield et al., "Worldwide Movement of Forest Fungi, Especially in the Tropics and Southern Hemisphere," *BioScience* 51, no. 2 (2001): 134–140.
27. Faith Thompson Campbell, "The Science of Risk Assessment for Phytosanitary Regulation and the Impact of Changing Trade Regulations," *BioScience* 51, no. 2 (2001): 148–153.
28. Mary Palm, "Invasion Pathways of Terrestrial Plant-Inhabiting Fungi" (paper presented at conference, "GISP Invasion Pathways: Analysis of Invasion Patterns and Pathway Management," Smithsonian Environmental Research Center, Edgewater, Md., November 1999).
29. Campbell, "Science of Risk Assessment."
30. National Invasive Species Council, *Management Plan,* 3.
31. U.S. Congress, Office of Technology Assessment, *Harmful Non-indigenous Species,* 117.
32. William E. Wallner, "Assessing Exotic Threats to International Forest Resources," in *National Stakeholder Timber Pest Conference* (Canberra: Australian Quarantine and Inspection Service, Timber Pest Coordination Unit, 1999).
33. U.S. Department of Agriculture, Animal and Plant Health Inspection Service (USDA APHIS), *2001 Vessel Inspection Guidelines: Asian Gypsy Moth (AGM),* on-line at http://www.aphis.usda.gov/oa/agm/agmguide.html.
34. Michael Hicks, *Changing Regulations on Packing Material: Will You Be Affected?* on-line at http://www.fas.usda.gov/info/agexporter/2001/jan/PackingRegulations.htm.

35. Ibid.
36. Ibid.
37. WTO News, *SPS Body Mulls Pests in Wood Crates, Studies "Equivalence"* (13 November 2000), on-line at http://www.wto.org/english/news_e/news_e.htm.
38. National Plant Board, *Exotic Longhorned Beetles on Imported Bamboo Stakes,* on-line at http://www.aphis.usda.gov/npb/issues.html.
39. Robin Salvage, "Timber Importation from an AQIS Operational Perspective," in *National Stakeholder Timber Pest Conference* (Canberra: Australian Quarantine and Inspection Service, Timber Pest Coordination Unit, 1999).
40. Williams, *New Zealand under Siege,* 70.
41. IUCN (World Conservation Union), *Biological Diversity Convention Draft* (Bonn, Germany: IUCN Environmental Law Centre, 1989).
42. IUCN Species Survival Commission, Invasive Species Specialist Group, *IUCN Guidelines for the Prevention of Biodiversity Loss Caused by Alien Invasive Species* (Gland, Switzerland: IUCN, 2000), 7, on-line at http://iucn.org/themes/ssc/pubs/policy/invasivesEng.htm.
43. Jeffrey A. McNeely et al., eds., *A Global Strategy on Invasive Alien Species* (Gland, Switzerland, and Cambridge, England: IUCN and Global Invasive Species Programme, 2001), 33.
44. Peter T. Jenkins, "Who Should Pay? Economic Dimensions of Preventing Harmful Invasions through International Trade and Travel," in *The Great Reshuffling: Human Dimensions of Invasive Alien Species,* ed. Jeffrey A. McNeely (Gland, Switzerland, and Cambridge, England: IUCN, 2001), 79–85.
45. Ibid., 81–83.
46. Charles Perrings, Mark Williamson, and Silvana Dalmazzone, "The Economics of Biological Invasions," in Mooney et al., *Invasive Alien Species.*

Chapter 8. Stemming the Tide

1. New Zealand Ministry of Agriculture and Forestry, "Hawkes Bay Outbreak of Exotic Mosquitoes Contained," *Quarantine Connection,* April–May 1999, 3.
2. J. Morgan Williams, *New Zealand under Siege: A Review of the Management of Biosecurity Risks to the Environment* (Wellington, New Zealand: Parliamentary Commissioner for the Environment, 2000), on-line at http://www.pce.govt.nz/reports/allreports/0_908804_93_8.shtml.
3. U.S. Department of Agriculture, Animal and Plant Health Inspection Service (USDA APHIS), *APHIS FY2001 Reference Book: Agricultural Quarantine Inspection* (Washington, D.C.: USDA APHIS, 2001), on-line at http://www.aphis.usda.gov/bad/refbook2001/.
4. Williams, *New Zealand under Siege.*
5. Ibid.
6. Ibid.

7. Ibid.
8. Ibid.
9. Ibid.
10. Gordon Hosking, "White-Spotted Tussock Moth: An Aggressive Eradication Strategy," *Aliens* (IUCN newsletter) 7 (1998): 4–5.
11. Williams, *New Zealand under Siege.*
12. New Zealand Ministry of Agriculture and Forestry, *Varroa Management Programme: Questions and Answers,* on-line at http://www.maf.govt.nz/MAFnet/press/archive/Varroa.html.
13. Williams, *New Zealand under Siege.*
14. Carolyn Whyte, *Effect of X-Ray Machines on the Risk of Fruit Fly Outbreaks in New Zealand* (Auckland: New Zealand Plant Protection Center, 1996).
15. Environmental Risk Management Authority (ERMA) New Zealand, "Instant Biosecurity Border Fines Approved," news release (26 December 2000), on-line at http://www.ermanz.govt.nz/NewsAndEvents/files/pressReleases/2001/pr20010110A.htm.
16. National Plant Board, *Safeguarding American Plant Resources: A Stakeholder Review of the APHIS-PPQ Safeguarding System* (1999), on-line at http://www.aphis.usda.gov/ppq/safeguarding.
17. Ibid.
18. Ibid.
19. USDA APHIS, *APHIS FY2001 Reference Book.*
20. Associated Press, "Noxious Weed Introduced in U.S. as a Pond Plant," 19 August 1999.
21. Malcolm E. Nairn et al., *Australian Quarantine: A Shared Responsibility* (Canberra, Australia: Department of Primary Industries and Energy, 1996).
22. National Research Council, *Stemming the Tide: Controlling Introductions of Nonindigenous Species by Ships' Ballast Water* (Washington, D.C.: National Academy Press, 1996).
23. James T. Carlton, "Pattern, Process, and Prediction in Marine Invasion Ecology," *Biological Conservation* 78 (1996): 97–106.
24. Williams, *New Zealand under Siege.*
25. Jose Matheickal, "Ballast Water Treatment Technology R&D in Singapore," *Aliens* (IUCN newsletter) 11 (2000): v–vi; Constance Holden, "From Ballast to Bouillabaisse," *Science* 289 (2000): 241.
26. National Research Council, *Stemming the Tide.*
27. Department of Agriculture, Fisheries and Forestry–Australia (AFFA), "New Ballast Water Arrangements to Protect Australia's Marine Environment," news release (3 July 2001), on-line at http://www.affa.gov.au/ministers/truss/releases/01/01179wt.html.

28. Keith Hayes, "Stemming the Tide: An Australian Perspective on Marine Biosecurity and the Role of Risk Assessment" (paper presented at annual meeting of the American Association for the Advancement of Science, San Francisco, February 2001).

Chapter 9. Beachheads and Sleepers

1. John Singe, *The Torres Strait: People and History* (St. Lucia, Queensland, Australia: University of Queensland Press, 1989).
2. E. Wright, K. Dunn, and A. Brown, *Report to the Minister for Agriculture, Fisheries, and Forestry on a Review of the Northern Australia Quarantine Strategy* (Canberra, Australia: Quarantine and Exports Advisory Council, 1998).
3. Peter Gadgil, *Responding to Incursions: A Generic Incursion Management Plan for the Australian Forest Sector* (Canberra: Australian Forest Health Committee, 2000).
4. Rudiger Wittenberg and Matthew J. W. Cock, eds. *Invasive Alien Species: A Toolkit of Best Prevention and Management Practices* (Wallingford, Oxon, England: CAB International, 2001).
5. J. Morgan Williams, *New Zealand under Siege: A Review of the Management of Biosecurity Risks to the Environment* (Wellington, New Zealand: Parliamentary Commissioner for the Environment, 2000), on-line at http://www.pce.govt.nz/Reports/allreports/0_908804_93_8.shtml.
6. Sandra Townsend, "Quarantine-Aware Public Helps Keep World's Worst Termites Out of WA," *AQIS Bulletin* (Australian Quarantine and Inspection Service) (November 2000), on-line at http://www.affa.gov.au/.
7. Australian Quarantine and Inspection Service, "Hit Parade," *AQIS Bulletin* (Australian Quarantine and Inspection Service) (May 2000), on-line at http://www.affa.gov.au/.
8. Wittenberg and Cock, *Invasive Alien Species.*
9. IUCN Invasive Species Specialist Group, Global Invasive Species Database, on-line at http://www.issg.org/database/welcome/.
10. Peter T. Jenkins, "Who Should Pay? Economic Dimensions of Preventing Harmful Invasions through International Trade and Travel," in *The Great Reshuffling: Human Dimensions of Invasive Alien Species,* ed. Jeffrey A. McNeely (Gland, Switzerland, and Cambridge, England: IUCN, 2001), 79–85.
11. National Plant Board, *Safeguarding American Plant Resources: A Stakeholder Review of the APHIS-PPQ Safeguarding System* (1999), on-line at http://www.aphis.usda.gov/ppq/safeguarding.
12. National Invasive Species Council, *Management Plan: Meeting the Invasive Species Challenge* (Washington, D.C.: National Invasive Species Council, 2001), on-line at http://www.invasivespecies.gov.
13. Jeff Crooks and Michael E. Soulé, "Lag Times in Population Explosions of

Invasive Species: Causes and Implications," in *Norway/UN Conference on Alien Species,* ed. Odd Terje Sandlund, Peter Johan Schei, and Aslaug Viken (Trondheim, Norway: Directorate for Nature Management and Norwegian Institute for Nature Research, 1996), 39–46.

14. Williams, *New Zealand under Siege.*
15. Peter A. Williams, Euan Nicol, and Melanie Newfield, *Assessing the Risk to Indigenous New Zealand Biota from New Exotic Plant Taxa and Genetic Material* (Wellington, New Zealand: Department of Conservation, 2000).
16. R. H. Groves et al., *Recent Incursions of Weeds to Australia, 1971–1995* (Adelaide: Cooperative Research Centre for Australian Weed Management, 1998).
17. Agriculture and Resource Management Council of Australia and New Zealand, Australian and New Zealand Environment and Conservation Council, and Forestry Ministers, *The National Weeds Strategy: A Strategic Approach to Weed Problems of National Significance* (Canberra: Commonwealth of Australia, 1999).
18. National Weeds Strategy, *Weeds of National Significance (Australia)* (2000), on-line at http://www.weeds.org.au.
19. Ian Atkinson, "Invasive Plants Not Wanted in Public or Private Gardens Identified," *The Nursery Papers* (Nursery and Garden Industry Australia), no. 12 (2000), on-line at http://www.ngia.com.au/np/index.html.
20. Richard N. Mack, "Assessing Biotic Invasions in Time and Space: A Second Imperative," in *Invasive Alien Species: Searching for Solutions,* ed. H. A. Mooney et al. (Washington, D.C.: Island Press, 2002).
21. Pacific Island Ecosystems at Risk (PIER), on-line at http://www.hear.org/pier/.
22. I. A. W. Macdonald and W. P. D. Gerbenbach, "A List of Alien Plants in the Kruger National Park," *Koedoe* (Research Journal for National Parks in the Republic of South Africa) 31 (1988): 137–150.

Chapter 10. Taking Control

1. B. W. van Wilgen and E. van Wyk, "Invading Alien Plants in South Africa: Impacts and Solutions," in *People and Rangelands, Building the Future: Proceedings of the Sixth International Rangeland Congress,* ed. David Eldridge and David Freudenberger (Townsville, Queensland, Australia, 1999), 566–571.
2. Brian J. Huntley, "South Africa's Experience regarding Alien Species: Impacts and Controls," in *Norway/UN Conference on Alien Species,* ed. Odd Terje Sandlund, Peter Johan Schei, and Aslaug Viken (Trondheim, Norway: Directorate for Nature Management and Norwegian Institute for Nature Research, 1996), 182–188.
3. van Wilgen and van Wyk, "Invading Alien Plants in South Africa."

4. Huntley, "South Africa's Experience," 185.
5. Yvonne Baskin, *The Work of Nature: How the Diversity of Life Sustains Us* (Washington, D.C.: Island Press, 1997).
6. van Wilgen and van Wyk, "Invading Alien Plants in South Africa."
7. B. W. van Wilgen et al., "The Sustainable Development of Water Resources: History, Financial Costs, and Benefits of Alien Plant Control Programmes," *South African Journal of Sciences* 93 (1997): 404–411.
8. van Wilgen and van Wyk, "Invading Alien Plants in South Africa."
9. D. C. Le Maitre, D. B. Versfeld, and R. A. Chapman, "The Impact of Invading Alien Plants on Surface Water Resources in South Africa: A Preliminary Assessment," *Water South Africa* 26, no. 3 (2000): 397–408, on-line at http://www.wrc.org.za.
10. van Wilgen et al., "Sustainable Development of Water Resources," 404.
11. D. M. Richardson et al., "Current and Future Threats to Plant Biodiversity on the Cape Peninsula, South Africa," *Biodiversity and Conservation* 5 (1996): 607–647.
12. D. C. Le Maitre et al., "Invasive Plants and Water Resources in the Western Cape Province, South Africa: Modelling the Consequences of a Lack of Management," *Journal of Applied Ecology* 33 (1996): 161–172.
13. B. W. van Wilgen, R. M. Cowling, and C. B. Burgers, "Valuation of Ecosystem Services: A Case Study from South African Fynbos Ecosystems," *BioScience* 46, no. 3 (1996): 184–189.
14. Dumisani Magadlela, "Social Challenges in the Working for Water Programme: Findings from a Study of Selected Projects," in *Best Management Practices for Preventing and Controlling Invasive Alien Species*, ed. Guy Preston, Gordon Brown, and Ernita van Wyk (Cape Town, South Africa: Working for Water Programme, 2000), 198–203.
15. Christo Marais, Jerry Eckert, and Carol Green, "Utilization of Invaders for Secondary Industries: A Preliminary Assessment," in Preston, Brown, and van Wyk, *Best Management Practices*, 141–155.
16. F. W. Prinsloo and David F. Scott, "Streamflow Responses to the Clearing of Alien Invasive Trees from Riparian Zones at Three Sites in the Western Cape Province," *Southern African Forestry Journal*, no. 185 (1999): 1–7; P. J. Dye and A. G. Poulter, "A Field Demonstration of the Effect on Streamflow of Clearing Invasive Pine and Wattle Trees from a Riparian Zone," *South African Forestry Journal*, no. 173 (1995): 27–30.
17. David F. Scott, "Managing Riparian Zone Vegetation to Sustain Streamflow: Results of Paired Catchment Experiments in South Africa," *Canadian Journal of Forestry Research* 29 (1999): 1149–1157.
18. D. M. Richardson, R. M. Cowling, and D. C. LeMaitre, "Assessing the Risk of Invasive Success in *Pinus* and *Banksia* in South African Mountain Fynbos," *Journal of Vegetation Science* 1 (1990): 629–642.

19. B. W. van Wilgen et al., "Big Returns from Small Organisms: Developing a Strategy for the Biological Control of Invasive Alien Plants in South Africa," *South African Journal of Science* 96 (March 2000): 148–152.
20. Ibid.
21. Rudiger Wittenberg and Matthew J. W. Cock, eds., *Invasive Alien Species: A Toolkit of Best Prevention and Management Practices* (Wallingford, Oxon, England: CAB International, 2001); Jeffrey A. McNeely et al., eds., *A Global Strategy on Invasive Alien Species* (Gland, Switzerland, and Cambridge, England: IUCN and Global Invasive Species Programme, 2001).
22. Daniel Simberloff, "Eradication of Island Invasives: Practical Actions and Results Achieved," *Trends in Ecology and Evolution* 16, no. 6 (2001): 273–274.
23. McNeely et al., *Global Strategy on Invasive Alien Species.*
24. Judith H. Myers et al., "Eradication Revisited: Dealing with Exotic Species," *Trends in Ecology and Evolution* 15, no. 8 (2000): 316–320.
25. Richard N. Mack et al., "Biotic Invasions: Causes, Epidemiology, Global Consequences, and Control," *Ecological Applications* 10, no. 3 (2000): 689–710.
26. Wittenberg and Cock, *Invasive Alien Species.*
27. Marcel Rejmánek et al., "Ecology of Invasive Plants: State of the Art," in *Invasive Alien Species: Searching for Solutions,* ed. H. A. Mooney et al. (Washington, D.C.: Island Press, 2002).
28. Myers et al., "Eradication Revisited."
29. Mack et al., "Biotic Invasions."
30. Subsidiary Body on Scientific, Technical, and Technological Advice (SBSTTA), *Comprehensive Review of Activities for the Prevention, Early Detection, Eradication, and Control of Invasive Alien Species* (Montreal: Convention on Biological Diversity, SBSTTA 6, February 2001).
31. Jeffrey K. Waage, "Biodiversity as a Resource for Biological Control," in *The Biodiversity of Microorganisms and Invertebrates: Its Role in Sustainable Agriculture,* ed. D. L. Hawksworth (Wallingford, Oxon, England: CAB International, 1991), 149–163.
32. P. DeBach, *Biological Control by Natural Enemies* (London: Cambridge University Press, 1974).
33. G. Hill and D. Greathead, "Economic Evaluation in Classical Biological Control," in *The Economics of Biological Invasions,* ed. C. Perrings, M. Williamson, and S. Dalmazzone (Cheltenham, England: Edward Elgar Publishing, 2000).
34. P. B. McEvoy and E. M. Coombs, "Why Things Bite Back: Unintended Consequences of Biological Weed Control," in *Nontarget Effects of Biological Control,* ed. P. A. Follett and J. J. Duan (Boston: Kluwer Academic Publishers, 2000), 167–194.
35. Wittenberg and Cock, *Invasive Alien Species;* Mark Williamson, *Biological Invasions* (London: Chapman and Hall, 1996).
36. Hans R. Herren and Peter Neuenschwander, "Biological Control of Cassava Pests in Africa," *Annual Review of Entomology* 36 (1991): 257–283; Peter Neuen-

schwander and Richard Markham, "Biological Control in Africa and Its Possible Effects on Biodiversity," in *Evaluating Indirect Ecological Effects of Biological Control,* ed. E. Wajnberg, John K. Scott, and Paul C. Quimby (Wallingford, Oxon, England: CAB International, 2001), 127–146.

37. Neuenschwander and Markham, "Biological Control in Africa," 129.
38. Jeffrey K. Waage, "Controlling Invasives: Biology Is Best," *World Conservation* (IUCN journal) 28, no. 4, and 29, no. 1 (April 1997–January 1998): 23–25.
39. Richard A. Malecki et al., "Biological Control of Purple Loosestrife," *BioScience* 43, no. 10 (1993): 680–685.
40. Wittenberg and Cock, *Invasive Alien Species.*
41. Kevin D. Lafferty and Armand M. Kuris, "Biological Control of Marine Pests," *Ecology* 77, no. 7 (1996): 1989–2000.
42. Anonymous, "Australia Declares Biological War on the Cane Toad," Environment News Service, 16 March 2001.
43. Dan Ferber, "Will Black Carp Be the Next Zebra Mussel?" *Science* 292 (2001): 203.
44. S. M. Louda et al., "Ecological Effects of an Insect Introduced for the Biological Control of Weeds," *Science* 277 (1997): 1088–1090.
45. Mari N. Jensen, "Silk Moth Deaths Show Perils of Biocontrol," *Science* 290 (2000): 2230–2231.
46. Francis G. Howarth, "Environmental Impacts of Classical Biological Control," *Annual Reviews of Entomology* 36 (1991): 485–509.
47. Donald R. Strong and Robert W. Pemberton, "Biological Control of Invading Species: Risk and Reform," *Science* 288 (2000): 1969–1970.
48. M. Laurie Henneman and Jane Memmott, "Infiltration of a Hawaiian Community by Introduced Biological Control Agents," *Science* 293 (2001): 1314–1316.
49. Jeff K. Waage, "Indirect Ecological Effects in Biological Control: The Challenge and the Opportunity," in Wajnberg, Scott, and Quimby, *Evaluating Indirect Ecological Effects,* 1–12.
50. Strong and Pemberton, "Biological Control of Invading Species"; Peter B. McEvoy, "Host Specificity and Biological Pest Control," *BioScience* 46, no. 6 (1996): 401–405.
51. Andre Gassmann and Svata M. Louda, "*Rhinocyllus conicus:* Initial Evaluation and Subsequent Ecological Impacts in North America," in Wajnberg, Scott, and Quimby, *Evaluating Indirect Ecological Effects,* 147–183.
52. Waage, "Indirect Ecological Effects in Biological Control."
53. Donald R. Strong and Robert W. Pemberton, "Food Webs, Risks of Alien Enemies, and Reform of Biological Control," in Wajnberg, Scott, and Quimby, *Evaluating Indirect Ecological Effects,* 57–79; quote, 58.
54. McEvoy and Coombs, "Why Things Bite Back," 177.

55. Daniel Simberloff and Peter Stiling, "How Risky Is Biological Control?" *Ecology* 77, no. 7 (1996): 1965–1974; McEvoy and Coombs, "Why Things Bite Back"; Strong and Pemberton, "Biological Control of Invading Species."
56. Neuenschwander and Markham, "Biological Control in Africa."
57. Strong and Pemberton, "Biological Control of Invading Species," 58.
58. Waage, "Controlling Invasives," 25.
59. Erika S. Zavaleta, Richard J. Hobbs, and Harold A. Mooney, "Viewing Invasive Species Removal in a Whole-Ecosystem Context," *Trends in Ecology and Evolution* 16, no. 8 (2001): 454–459.
60. McEvoy and Coombs, "Why Things Bite Back."
61. Myers et al., "Eradication Revisited."
62. Jason Van Driesche and Roy Van Driesche, *Nature Out of Place: Biological Invasions in the Global Age* (Washington, D.C.: Island Press, 2000).
63. Keith R. Edwards, "A Critique of the General Approach to Invasive Plant Species," in *Plant Invasions: Ecological Mechanisms and Human Responses*, ed. U. Starfinger et al. (Leiden, Netherlands: Backhuys Publishers, 1998), 85–94.
64. U.S. Congress, Office of Technology Assessment, *Harmful Non-indigenous Species in the United States*, OTA-F-565 (Washington, D.C.: U.S. Government Printing Office, 1993).

Chapter 11. Islands No Longer

1. Herman Melville, *The Piazza Tales* (New York: Dix, Edwards & Company, 1856).
2. William Beebe, *Galapagos, World's End* (New York: G. P. Putnam's Sons, 1924), 340.
3. Ibid.
4. Linda J. Cayot and Ed Lewis, "Recent Increase in Killing of Giant Tortoises on Isabela Island," *Noticias de Galápagos*, no. 54 (1994): 2–7.
5. Anonymous, "Giant Tortoise Breeding Program Reaches Milestone," *Galápagos News* (summer 2000): 5.
6. Linda J. Cayot and Peter Jenkins, "Galápagos Workshop: Eradication of Ungulates from Isabela Island," *Aliens* (IUCN newsletter) 7 (1998): 12–13.
7. Ibid.
8. Lloyd L. Loope, Ole Hamann, and Charles P. Stone, "Comparative Conservation Biology of Oceanic Archipelagoes," *BioScience* 38, no. 4 (1988): 272–282.
9. Gillian Key and Edgar Munoz Heredia, "Distribution and Current Status of Rodents in the Galápagos," *Noticias de Galápagos*, no. 53 (1994): 21–25.
10. Howard Snell and Solanda Rea, "The 1997–98 El Niño in Galápagos: Can 34 Years of Data Estimate 120 Years of Pattern?" *Noticias de Galápagos*, no. 60 (1999): 11–20.

11. Charlotte E. Causton, Carlos E. Zapata, and Lazaro Roque-Albelo, "Alien Arthropod Species Deterred from Establishing in the Galápagos, but How Many Are Entering Undetected?" *Noticias de Galápagos,* no. 61 (2000): 10–13.
12. Alan Tye, Monica Soria, and Mark R. Gardener, "A Strategy for Galápagos Weeds," in *Turning the Tide: The Eradication of Invasive Species,* ed. C. R. Veitch and M. N. Clout (Auckland, New Zealand: IUCN Invasive Species Specialist Group, forthcoming).
13. Alan Tye, "Invasive Plant Problems and Requirements for Weed Risk Assessment in the Galápagos Islands," in *Weed Risk Assessment,* ed. R. H. Groves, F. D. Panetta, and J. G. Virtue (Melbourne, Australia: CSIRO Publishing, 2001), 153–175.
14. Ibid, 29.
15. Tye, Soria, and Gardener, "Strategy for Galápagos Weeds"; Monica Soria, Mark R. Gardener, and Alan Tye, "Eradication of Potentially Invasive Plants with Limited Distributions in the Galápagos Islands," in Veitch and Clout, *Turning the Tide.*
16. Lazaro Roque-Albelo and Charlotte Causton, "El Niño and Introduced Insects in the Galápagos Islands: Different Dispersal Strategies, Similar Effects," *Noticias de Galápagos,* no. 60 (1999): 30–36; quote, 30.
17. Ibid.
18. Causton, Zapata, and Roque-Albelo, "Alien Arthropod Species Deterred."
19. Ibid.
20. Hernan Vargas and Heidi M. Snell, "The Arrival of Marek's Disease to Galápagos," *Noticias de Galápagos,* no. 58 (1997).
21. Gayle Davis-Merlen, "New Introductions and a Special Law for Galápagos," *Aliens* (IUCN newsletter) 7 (1998): 10–11.
22. Anonymous, "Canine Distemper Decimates Dogs," *Galápagos News* (summer 2001): 7.
23. Andre Mauchamp and Maria Luisa Munoz, "A Kudzu Alert in Galápagos: The Urgent Need for Quarantine," *Noticias de Galápagos,* no. 57 (1996): 22–23.
24. Soria, Gardener, and Tye, "Eradication of Potentially Invasive Plants."
25. Pablo Ospina, "Eradication and Quarantine: Two Ways to Save the Islands," *World Conservation* (IUCN journal) 28, no. 4, and 29, no. 1 (April 1997–January 1998): 43.

Chapter 12. Can We Preserve Integrity of Place?

1. Supporters of Tiritiri Matangi, *Tiritiri Matangi Island,* on-line at http://www.123.co.nz/tiri/default.htm.
2. Jeffrey A. McNeely et al., eds., *A Global Strategy on Invasive Alien Species* (Gland, Switzerland, and Cambridge, England: IUCN and Global Invasive Species Programme, 2001).

3. Subsidiary Body on Scientific, Technical, and Technological Advice (SBSTTA), *Recommendation VI/4: Alien Species That Threaten Ecosystems, Habitats, or Species* (Montreal: Convention on Biological Diversity, SBSTTA 6, February 2001).
4. Ibid., Guiding Principle 2.
5. Ibid., Guiding Principle 10.
6. National Invasive Species Council, *Management Plan: Meeting the Invasive Species Challenge* (Washington, D.C.: National Invasive Species Council, 2001), on-line at http://www.invasivespecies.gov.
7. National Research Council, *The Quarantine and Certification of Martian Samples* (Washington, D.C.: National Academy Press, 2001).
8. SBSTTA, *Recommendation VI/4: Alien Species,* Guiding Principle 3.
9. Clare Shine, Nattley Williams, and Lothar Gundling, *A Guide to Designing Legal and Institutional Frameworks on Alien Invasive Species* (Gland, Switzerland; Cambridge, England; and Bonn, Germany: IUCN Environmental Law Centre, 2000).
10. Greater Los Angeles County Vector Control District, *Exotic Species of Mosquito Discovered in Southern California* (Los Angeles: Greater Los Angeles County Vector Control District, 2001).
11. Sarah Hayden Reichard and Peter White, "Horticulture as a Pathway of Invasive Plant Introductions in the United States," *BioScience* 51, no. 2 (2001): 103–113; quote, 112.
12. Mark Sagoff, "Why Exotic Species Are Not as Bad as We Fear," *Chronicle of Higher Education* (23 June 2000); Michael Pollan, "Against Nativism," *New York Times Magazine* (15 May 1994): 52–55.
13. Reichard and White, "Horticulture as a Pathway."
14. John J. Ewel et al., "Deliberate Introductions of Species: Research Needs," *BioScience* 49, no. 8 (1999): 619–630.
15. James H. Brown, "Patterns, Modes and Extents of Invasions by Vertebrates," in *Biological Invasions: A Global Perspective,* ed. J. A. Drake et al., SCOPE 37 (Chichester, West Sussex, England: John Wiley and Sons, 1989), 85–109; quote, 105.
16. Peter M. Vitousek, "Diversity and Biological Invasions of Oceanic Islands," in *Biodiversity,* ed. Edward O. Wilson (Washington, D.C.: National Academy Press, 1988), 181–189; quote, 184.
17. Edward O. Wilson, *The Diversity of Life* (New York and London: W. W. Norton and Company, 1992), 341.
18. Michael L. Rosenzweig, "The Four Questions: What Does the Introduction of Exotic Species Do to Diversity?" *Evolutionary Ecology Research* 3 (2001): 361–367.

ACKNOWLEDGMENTS

This book was conceived in Trondheim, Norway, in the summer of 1996, at an international meeting known as the Norway/United Nations Conference on Alien Species. Representatives from eighty nations gathered there to hear dozens of scientists and technical experts detail the growing economic and ecological burdens posed by invasive alien species. I attended as a science journalist at the invitation of SCOPE, the Scientific Committee on Problems of the Environment, which had sponsored pioneering synthesis efforts on the phenomenon of biological invasions during the 1980s. More specifically, I was there at the urging of Harold A. (Hal) Mooney, who had led those earlier SCOPE efforts on bioinvasions and who now advocated the launch of a new program that would use the best science to develop practical strategies for coping with the problems of invasive species. I had just emerged from writing a book about insights gathered during a previous Hal-led SCOPE investigation (it was published in early 1997 by Island Press as *The Work of Nature: How the Diversity of Life Sustains Us*). As Hal's colleagues worldwide will recognize, he lured me to Trondheim with the hope of ensnaring me in another HRT–Hal-related task. This book is the happy result.

The new program Hal and others advocated at Trondheim was born the following year as the Global Invasive Species Programme (GISP), an international partnership committed to devising a global strategy to cope with species invasions. GISP has been nurtured to maturity by a consortium that includes SCOPE, the World Conservation Union (IUCN), and CAB International (CABI), in partnership with the United Nations Environment Programme (UNEP). GISP's scientific and policy development programs have been supported by voluntary contributions from its operating partners, as well as from individual participants, and by grants from the Global Environment Facility and a range of private and public sector institutions.

With support from SCOPE and the David and Lucile Packard Foundation, I shadowed the work of GISP as a journalist, attending workshops, interviewing participants, and traveling on my own during the following two years to see firsthand the invasive species problems that were being described in those meeting rooms and—more important—to meet the people on the front lines who are working on solutions. First and foremost, then, I am fundamentally indebted to the Packard Foundation and to hundreds of biologists, natural resource managers, economists, lawyers, and policy makers around the world who took part in the first two years of GISP deliberations.

I was especially fortunate to have the guidance and advice of the GISP Executive Committee: Hal, Jeffrey A. McNeely of IUCN, Peter Johan Schei of the Norwegian Directorate for Nature Management, and Jeff Waage, formerly of CABI and now of Imperial College at Wye. All four read and critiqued the first draft of the manuscript and provided invaluable criticisms and new insights that helped me to improve the final version. Throughout the project, they also allowed me the freedom to select, write, revise, and cast the material as I chose. That means that any errors or deficiencies that remain, as well as the opinions and interpretations expressed herein, are the author's responsibility alone.

I also owe special thanks to the other thirteen members of the GISP Scientific Steering Committee, who, along with the Executive Committee members, opened their deliberations to me, helped support my participation in workshops and conferences, and in some cases reviewed parts of the manuscript. They include, alphabetically, David A. Andow, James T. Carlton, Mick Clout, Richard Hobbs, Alan Holt, W. Mark Lonsdale, Richard N. Mack, Charles A. Perrings, Marcel Rejmánek, Dave Richardson, Gregory Ruiz, Nattley Williams, and Mark Williamson.

I'm also deeply indebted to many other experts, as well as people on the front lines of the fight against biological invasions, who provided professional, logistic, and often personal help, along with interviews, discussions, reviews, reactions, data, photographs, or tours in the course of my research and travels. I offer heartfelt thanks to the following: In the Galápagos Islands, Robert Bensted-Smith, Charlotte Causton, Desiré Cruz, Paola Diaz, Mark Gardener, Scott Henderson, Godfrey Merlen, Marc Patry, Brand Phillips, Poly Robayo, Heidi Snell, Howard Snell, and Alan Tye. In New Zealand, Peter Barnes, Michael Browne, Mick Clout, Tui DeRoy, Sarah Lowe, Maj de Poorter, and Dick Veitch. In Australia, Kylie Anderson, Evan Collis, Brian Cooke, Richard Davis, Richard Groves, Keith Hayes, Sandy Lloyd, Mark Lonsdale, Mike Nicholas, Paul Pheloung, Rod Randall, Peter Solness, Pedro Stephen, Barbara Waterhouse, and the community councils of Yorke and Warraber Islands. In South Africa, Jacqui Coetzee, Sandra Fowkes, Karoline Hanks, Brian Huntley, David Le Maitre, Derek Malan, Christo Marais, Guy Preston, Dave Richardson, and Brian van Wilgen. In the United Kingdom, Pierre Binggeli, Silvana Dalmazzone, Sean Murphy, Charles Perrings, Jeff Waage, and Mark Williamson. In Switzerland, Matthew Cock and Ruediger Wittenberg. In France, Alexandre Meinesz. In Malaysia, Soetikno Sastroutomo and Loke Wai Hong. In Canada, W. Stephen Price. In the Hawaiian Islands, Rob Cowie, Julie Denslow, Jack Ewel, Robert Hobdy, Paul Holthus, Jack Jeffrey, Fred Kraus, Lloyd Loope, Kim Martz, Jack Peterson and the Maui Invasive Species Committee, Don Reeser, Forest Starr, Philip Thomas, and Tim Tunison. And across the continental United States, Gordon Brown, Jim Brown, Faith Campbell, Jim Carlton, Peter Daszak, Ernest Delfosse, Peter Jenkins, Lynn Kinter, David Lodge, Linda McCann, Richard Mack, Mark Minton, Chester (Chet) Moore, Ann Murray, Robert Pemberton, Harriett Peuler, David Pimentel, Victor Ramey, John Randall, Jamie Reaser, Sarah Reichard, Greg Ruiz, Roger Sheley, Dan Simberloff, Jim Space, Thomas Stohlgren, Randy Westbrooks, Peter White, and Peter Yates.

There were, of course, dozens of others who shared their time and their insights with me. Some of them are named in the text; others are not. In many cases, space limitations kept me from using the material they helped me gather or the good examples they provided. Despite the necessary winnowing of an overwhelming amount of material on invaders and invaded places, this book reflects the experiences and insights of everyone who took time to talk to me and show me about in the course of my research and travels. To all those I've failed to name, my thanks for your contributions.

I want to express deep appreciation also to SCOPE for making this book possible, and particularly to SCOPE's executive director, Veronique Plocq-Fichelet, and editor in chief, John W. B. Stewart. GISP's coordinator, Laurie Neville, provided a great deal of logistic and practical advice, even while she was immersed in the task of shepherding GISP's many technical and policy undertakings into publication. Finally, Jonathan Cobb, executive editor at Shearwater Books, provided much-needed faith and enthusiasm during the writing of this book and very thoughtful and helpful editing of the rough manuscript.

Once again, I'm overwhelmingly grateful for the daily support of my husband, Mike Gilpin, who graciously quit reminding me as I muttered "Never again" that I had often declared the same intent during the writing of the previous book.

INDEX